Lecture Notes in Computer Science 6670

Marina L. Gavrilova C.J. Kenneth Tan
Alexei Sourin Olga Sourina (Eds.)

Transactions on Computational Science XII

Special Issue on Cyberworlds

Springer

Editors-in-Chief

Marina L. Gavrilova
University of Calgary, Department of Computer Science
2500 University Drive N.W., Calgary, AB, T2N 1N4, Canada
E-mail: marina@ucalgary.ca

C.J. Kenneth Tan
Exascala Ltd.
Unit 9, 97 Rickman Drive, Birmingham B15 2AL, UK
E-mail: cjtan@exascala.com

Guest Editors

Alexei Sourin
Nanyang Technological University, School of Computer Engineering
Division of Computer Science, Singapore
E-mail: assourin@ntu.edu.sg

Olga Sourina
Nanyang Technological University, School of Computer Engineering
Division of Information Engineering, Singapore
E-mail: eosourina@ntu.edu.sg

ISSN 0302-9743 (LNCS)
e-ISSN 1611-3349 (LNCS)
ISSN 0302-9743 (TCOMPSCIE)
e-ISSN 1611-3349 (TCOMPSCIE)
ISBN 978-3-642-22335-8
ISBN 978-3-642-22336-5 (eBook)
DOI 10.1007/978-3-642-22336-5
Springer Heidelberg Dordrecht London New York

Library of Congress Control Number: Applied for

CR Subject Classification (1998): H.5, I.2, H.3-4, C.2, I.3, I.4

Typesetting: Camera-ready by author, data conversion by Scientific Publishing Services, Chennai, India

Printed on acid-free paper

Springer is part of Springer Science+Business Media (www.springer.com)

LNCS Transactions on Computational Science

Computational science, an emerging and increasingly vital field, is now widely recognized as an integral part of scientific and technical investigations, affecting researchers and practitioners in areas ranging from aerospace and automotive research to biochemistry, electronics, geosciences, mathematics, and physics. Computer systems research and the exploitation of applied research naturally complement each other. The increased complexity of many challenges in computational science demands the use of supercomputing, parallel processing, sophisticated algorithms, and advanced system software and architecture. It is therefore invaluable to have input by systems research experts in applied computational science research.

Transactions on Computational Science focuses on original high-quality research in the realm of computational science in parallel and distributed environments, also encompassing the underlying theoretical foundations and the applications of large-scale computation. The journal offers practitioners and researchers the opportunity to share computational techniques and solutions in this area, to identify new issues, and to shape future directions for research, and it enables industrial users to apply leading-edge, large-scale, high-performance computational methods.

In addition to addressing various research and application issues, the journal aims to present material that is validated – crucial to the application and advancement of the research conducted in academic and industrial settings. In this spirit, the journal focuses on publications that present results and computational techniques that are verifiable.

Scope

The scope of the journal includes, but is not limited to, the following computational methods and applications:

- Aeronautics and Aerospace
- Astrophysics
- Bioinformatics
- Climate and Weather Modeling
- Communication and Data Networks
- Compilers and Operating Systems
- Computer Graphics
- Computational Biology
- Computational Chemistry
- Computational Finance and Econometrics
- Computational Fluid Dynamics

- Computational Geometry
- Computational Number Theory
- Computational Physics
- Data Storage and Information Retrieval
- Data Mining and Data Warehousing
- Grid Computing
- Hardware/Software Co-design
- High-Energy Physics
- High-Performance Computing
- Numerical and Scientific Computing
- Parallel and Distributed Computing
- Reconfigurable Hardware
- Scientific Visualization
- Supercomputing
- System-on-Chip Design and Engineering

Editorial

The Transactions on Computational Science journal is part of the Springer series *Lecture Notes in Computer Science*, and is devoted to the gamut of computational science issues, from theoretical aspects to application-dependent studies and the validation of emerging technologies.

The journal focuses on original high-quality research in the realm of computational science in parallel and distributed environments, encompassing the facilitating theoretical foundations and the applications of large-scale computations and massive data processing. Practitioners and researchers share computational techniques and solutions in the area, identify new issues, and shape future directions for research, as well as enable industrial users to apply the techniques presented.

The current volume is devoted to the topic of cyberworlds – information worlds and communities created on cyberspace – and is edited by Alexei Sourin and Olga Sourina. It is comprised of 13 best papers selected from the International Conference on Cyberworlds 2010, Singapore, a leading international event devoted to cyberworlds and their applications to e-business, e-commerce, e-manufacturing, e-learning, e-security, and cultural heritage.

We would like to extend our sincere appreciation to special issue guest editors, Alexei Sourin and Olga Sourina, for his dedication and insights in preparing this high-quality special issue. We would also like to thank all of the authors for submitting their papers to the special issue and the associate editors and referees for their valuable work. We would like to express our gratitude to the LNCS editorial staff of Springer, in particular Alfred Hofmann, Ursula Barth, and Anna Kramer, who supported us at every stage of the project.

It is our hope that the fine collection of papers presented in this special issue will be a valuable resource for Transactions on Computational Science readers and will stimulate further research into the vibrant area of computational science applications.

March 2011

Marina L. Gavrilova
C.J. Kenneth Tan

Guest Editors' Preface
Special Issue on Cyberworlds

Cyberworlds are information worlds or communities created on cyberspace by collaborating participants either intentionally or spontaneously. As information worlds, they accumulate information regardless of whether or not anyone is involved, and they can be with or without 2D or 3D visual graphics. This area of research is essential at our time of globalization of economy and competition for resources. Cyberspace provides access to virtual lands and unlimited opportunities for entrepreneurs, researchers, engineers, students, and in fact any network users.

The 10th in the series, the 2010 International Conference on Cyberworlds, addressed a wide range of research and development topics. CW 2010 accepted 49 full and 14 short papers out of 130 submitted. The 13 articles appearing in this special issue are revised and extended versions of a selection of papers presented at CW 2010. The papers have been selected based on their reviewers' comments, on the quality of the oral presentations, and the conference delegates' feedback.

The first paper, "*Bridging Digital and Physical Worlds Using Tangible Drag-and-Drop Interfaces*", presents a study of tangible drag-and-drop remote control interfaces. This interaction technique aims at reducing the gap between the digital and physical worlds, enabling the transfer of digital data from one device to another.

Virtual puppetry is the subject of the paper "*Puppet Playing: An Interactive Character Animation System with Hand Motion Control*". To implement a puppet playing scenario in virtual environments using a new input device SmartGlove, the authors designed an interactive animation system based on both the user's hand motion and the constraints of puppet and environment.

The paper "*Reconstructing Multiresolution Mesh for Web Visualization Based on PDE Resampling*" addresses the topic of using Partial Differential Equations (PDEs) for approximating surfaces of geometric shapes. It proposes an algorithm for automatically deriving PDE-boundary conditions from the surface of the original polygon mesh.

The topic of using PDEs in geometric modeling is continued in the paper "*On the Development of a Talking Head System Based on the Use of PDE-Based Parametric Surfaces*", which proposes a talking head system based on animating facial expressions using a template face generated from PDEs.

Current active 3D range sensors, such as time-of-flight cameras, enable us to acquire range maps at video frame rate. The paper "*Real-Time Spatial and Depth Upsampling for Range Data*" presents a pipeline to enhance the quality as well as improve the spatial and depth resolution of range data in real time.

Virtual molecular docking with a haptic device is addressed in the paper "*Six-Degree-of-Freedom Haptic Rendering for Biomolecular Docking*", which proposes a haptic rendering algorithm for biomolecular docking with force-torque feedback. It lets the user experience six-degree-of-freedom haptic manipulation in molecular docking process.

A multiuser virtual trade fair developed using a 3D game engine is presented in the paper "*Design of a Multiuser Virtual Trade Fair Using a Game Engine*". Users represented by avatars can interact with each other while they are visiting the virtual fair. The stands at the fair include a number of interactive objects providing information about the exhibitors.

Just like it is necessary to be able to accurately authenticate the identity of human beings, it is becoming essential to be able to determine the identify of non-biological entities. The paper "*Applying Biometric Principles to Avatar Recognition*" presents the current state of the art in virtual reality security, focusing specifically on emerging methodologies for avatar authentication.

The paper "*Range-Based Cybernavigation in Natural Known Environments*" considers navigation of a physical robot in real natural environments which have been previously scanned in considerable detail so as to permit virtual exploration by cybernavigation prior to mission replication in the real world.

A computational model of situation awareness for the bots in Military Operations on Urban Terrain (MOUT) simulations is proposed in the paper "*Generating Situation Awareness for Time-Critical Decision Making*". The model forms up situation awareness quickly with key cues.

Generating human-like behaviors for virtual agents has become increasingly important in many applications. One of the challenging issues in behavior modeling is how virtual agents make decisions given some time-critical and uncertain situations. This is discussed in the paper "*HumDPM: A Decision Process Model for Modeling Human-Like Behaviors in Time-Critical and Uncertain Situations*".

The paper "*Group-Agreement as a Reliability Measure for Witness Recommendations in Reputation-Based Trust Protocols*" presents an approach allowing agent-based trust frameworks to leverage information from both trusted and untrusted witnesses that would otherwise be neglected. An effective and robust voting scheme based on an agreement metric is presented and its benefit is shown through simulations.

Finally, automatic emotion recognition is considered in the paper "*Real-Time EEG-Based Emotion Recognition and Its Applications*", which describes a real-time fractal dimension-based algorithm of quantification of basic emotions using the Arousal-Valence emotion mode.

We express our thanks and appreciation to the authors, the reviewers, and the staff working on the Transactions of Computational Science.

March 2011

Alexei Sourin
Olga Sourina

LNCS Transactions on Computational Science – Editorial Board

Table of Contents

Bridging Digital and Physical Worlds Using Tangible Drag-and-Drop Interfaces

Mathieu Hopmann, Mario Gutierrez, Daniel Thalmann, and Frederic Vexo

VRLab, EPFL, 1015 Lausanne, Switzerland
{mathieu.hopmann,daniel.thalmann}@epfl.ch,
{mgutierrez,frederic.vexo}@logitech.com
http://vrlab.epfl.ch

Abstract. The last ten years have seen an explosion in the diversity of digital-life devices, e.g. music and video players. However, the interaction paradigm to use these devices has remained mostly unchanged. Remote controls are still the most common way to manage a digital-life device. Moreover, the interaction between devices themselves is still very limited and rarely addressed by a remote control interface. We present in this paper a study of *tangible drag-and-drop*, a remote control interface based on the well-known paradigm coming from the graphical user interface. This interaction technique aims at reducing the gap between the digital and physical worlds, enabling the transfer of digital data from one device to another. To validate such a concept, we present two prototypes, along with user studies and a general discussion about the tangible drag-and-drop technique.

Keywords: Interaction techniques, Input devices, Tangible interaction, Remote control, Drag-and-drop, Multimedia content.

1 Introduction

At the end of the 90's, digital content at home was limited and mainly stored in home computers. But with the explosion of digital formats and the Internet, our "digital life" has hugely gained in importance. Pictures, songs, movies, almost all of our multimedia content could be dematerialized. The diversity of devices to manage this digital life has also drastically increased: laptops, digital cameras, MP3 players, digital photo frames, Smartphones ...

A problem resulting from this device multiplication is the content dispersal: having many devices often results on having different content on each device, and synchronizing everything is a baffling problem. For this reason and with the multiplication of devices connected to the Internet, the Web becomes more and more the aggregator of our digital data. We store our emails on a webmail provider, share our pictures on an online photo gallery or on a social network, and listen to music via webradios or music recommender systems. However, the main device to access and manage our digital data locally and on the web remains the same: a computer, a mouse and a keyboard.

M.L. Gavrilova et al. (Eds.): Trans. on Comput. Sci. XII, LNCS 6670, pp. 1–18, 2011.

Mouse devices, which have accompanied computers for more than 20 years, are still the main device to select, move and interact with digital content. The reason is probably because it combines efficiency, easiness, and low physical effort. Standard interaction tasks offered by the mouse are selection, pointing, drag-and-drop, copy/cut and paste... These tasks are now massively adopted by computer users, and we use them everyday in a natural way. A question we asked ourselves was: among these interaction tasks, which are the ones more closely related to everyday life gestures.

Pointing is one of them. Every child points his/her finger to pick up something, and everyday we are pointing devices with remote controls to interact with them. It is natural, but also limited: pointing allows to select, but generally not to perform an action. On the other hand, copy and paste allows to transfer text, files or objects from a source to a destination, but is hardly comparable to an everyday life gesture: we are still not able to clone things right away ! On the contrary, drag-and-drop is an action performed continually in everyday life: when we take an object to put it in a different location, it is kind of a drag-and-drop action.

In human-computer interaction, drag-and-drop is the action of clicking on a virtual object, and dragging it to a different location. This is a task broadly used in graphical user interfaces: it is natural, efficient, and with a low learning curve. That is why we chose to adapt this in a tangible way, and use it to bridge the gap between digital data and physical devices. In this paper, we present an interaction technique, called tangible drag-and-drop, and two prototypes that have been developed in order to evaluate our concept. The first one is intended to ease the interaction when sharing music with friends, whereas the second one aims at managing a multi-room audio system. We describe for both of them the implementation and the feedback we got from users. The paper ends with a discuss about the tangible drag-and-drop experience.

2 Related Work

Despite the fact that our digital life has hugely gained in importance over the last years, the mechanisms to interact with it have poorly evolved. It is still difficult to share digital content between devices and to interact with it. In the smart home[2], which could be a vision of our future home, our environment is surrounded of interconnected devices, which respond smartly to our different actions. This interconnection gives us the possibility to create new kinds of interaction between the different devices. The goal of these new interaction concepts could be to make our digital life more tangible, but also to add efficiency.

The concept of using a tangible user interface to facilitate the manipulation of digital content has been introduced in the middle of the 90's by Fitzmaurice et al. with Bricks[4]. With Bricks, they presented a "Graspable User Interface" to control virtual objects through physical handles. Ishii and al. continued in this way with Tangible Bits[9], where they suggested to bridge the gap between bits and atoms using graspable objects.

Several works put forward the idea of taking advantage of a tangible user interface in order to manage digital multimedia content. About music, Alonso et al. created the MusicCube[1]. Users interact with the MusicCube using gestures to shuffle music and a rotary dial for song navigation and volume control. With iCandy[5], Graham et al. designed a tangible interface to restore the benefits of physical albums for digital music. About pictures, Nunes et al.[13] implemented a system that allow people to link digital photo with physical souvenirs. In [8], Hsu et al. designed a tangible device which supported gestures to browse digital pictures.

Another aspect of our multimedia life which suffers of the lack of tangibility is the communication between the devices. Many remote controls have been commercialized in order to interact with several devices[1][2], and the idea of the universal remote control has more than 20 years[15]. However, the goal of these current universal remote controllers remains basic: to replace several remotes by a single one. There is no possibility of interaction between the different devices, the goal is only to limit the number of remote control or the number of steps you need to perform an action (using an "Activity" based remote).

The concept of enhancing our digital life in the “smart home” has been addressed in [17][16] with CRISTAL, a system to simplify “the control of our digital devices in and around the living room”. In these papers, they described a system based on an interactive multi-touch surface and a camera. Users can interact with the different devices in the living room by manipulating their representation on the screen. It is a interesting solution for mixing digital files and physical devices on a single surface, but cumbersome and invasive, having a live camera in the living room may be unacceptable for some users.

Different works suggest to use tangibility to facilitate the transfer of digital data between devices. In [14], Rekimoto et al. pointed out two main problems of using multiple computers: the input device, which is often tethered to a single computer, and user interfaces, which are not designed to manage a multiple-computer environment. With pick-and-drop, they tackled this problem by using a pen to pick a virtual object on a screen and drop it on another display. A close idea is described in [18] with Slurp, an eyedropper with haptic and visual feedback which extracted digital media from physical objects in order to inject it into other devices. In [11][10], the authors suggested to use hand gestures in order to connect devices and transfer files from a device to another. In [7], we presented a first tangible drag-and-drop prototype focused on digital pictures and the interaction with a digital photo frame. This first prototype gave encouraging results in terms of interaction between multiple devices, and for this reason we decided to further further develop this concept. This paper presents our latest results.

[1] http://www.logitech.com/index.cfm/remotes/universal_remotes

[2] http://www.remotecontrol.philips.com/

3 Tangible Drag-and-Drop

Drag-and-drop in computer graphical user interfaces (GUI) is intended to drag virtual objects from a location to another in a natural way. However, drag-and-drop is not limited to the action of moving files to a new location. In GUI, you can drag a window border to increase its size and drop it to validate your choice. You can drag a file on a program icon to execute an action on this file. You can drag a picture from a web browser to the desktop in order to create a file which contains this picture. Besides being a fast interaction task to adopt, drag-and-drop is flexible and allows for diverse functionality.

Our idea is to take advantage of this efficient and easy-to-learn interaction task and adapt it to the physical world. Using a 'drag-and-drop remote control', we want to be able to select digital content, drag it physically with the remote and drop it on another physical device.

In order to illustrate our idea, we designed two prototypes. The first one is dedicated to share music with friends, the second is intended to manage a multi-room audio system.

3.1 Prototype #1: Sharing Music by Drag-and-Dropping between Devices and a Virtual Shelf

Despite the fact that digital media has already invaded our homes, interaction models have not changed: it is still hard to interact with our music collection without a computer. Our main goal is to make this interaction more simple and efficient, even for people with few computer skills. We focus this first prototype on three points:

- Browsing our music collection in a familiar way
- Sharing music between friends
- Sharing music between devices

In order to browse our music collection, we choose to display it in a familiar context: as we had CDs at home stored in a shelf, our application displays albums on a screen inside a virtual shelf. The user doesn't need to enter a login or a password. Indeed, the application detects the user's cell phone, scans the music database and displays her collection inside the virtual shelf on the TV screen. With the remote control, she can choose the album he wants to play, and drag-and-drop it on the sound player close to her and listen to it. If a friend enjoys the music, she can use the application too: a second virtual shelf appears on the TV with her music collection, she just has to select the album in the other shelf and drop it in her collection.

Our application adapts the drag-and-drop interaction task to transfer music files from the digital to the physical world (from the user's music collection to the selected sound player) and within digital worlds, allowing music sharing between friends.

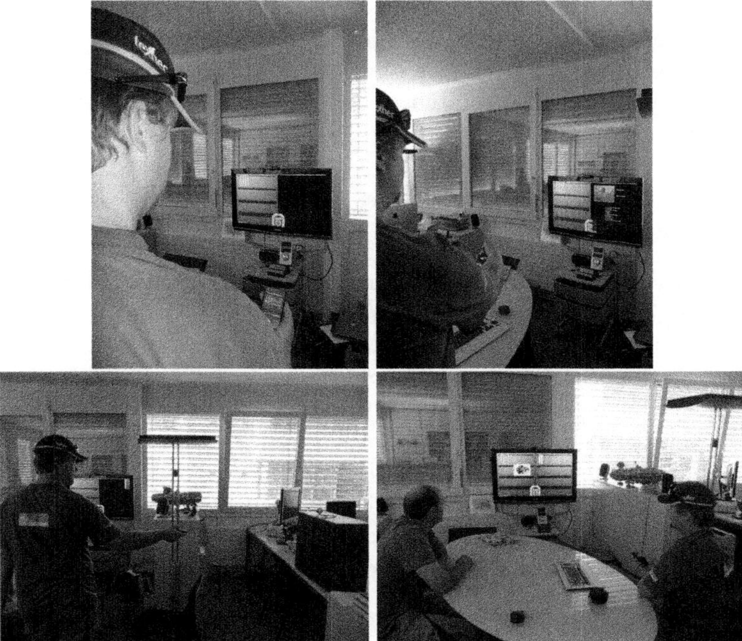

Fig. 1. From top left to bottom right: (I) User's cell phone is detected, a shelf appears with his music collection. (II) The user is browsing his music collection using the MX Air mouse. (III) The user drops an album on the sound player after selecting it in the virtual shelf. (IV) Two users are sharing albums from their music collection.)

Music centralization. The storage of digital music is a real challenge. Keeping our music collection ordered in a single location requires too much time. With this in mind, we decided to use a single place to store user's music collection. For our prototype, we have stored all the music files on a web server, to make it accessible from everywhere.

We have created two MySQL databases to manage the music collections: the first one stores all the basic information about songs (band, album, cover, song name, song number) and their location (a hyperlink to the song path on the server); the second one stores information about users (first name, last name, bluetooth MAC address of the cellphone) and the songs they have in their music collection (a reference to the first database). This is a technical choice above all: sharing music with a friend just consists in adding an entry in the database. In the discussion section, we get back to that point based on feedback from the testers.

Hardware description. The user interacts with our application using a remote control, which is a custom Logitech MX Air^{TM}. This mouse is a cordless air mouse which maps hand motions to cursor control: the mouse follows the user's hand, using a gyroscopic sensor and an accelerometer. This functionality is interesting to control the mouse pointer on a TV for example, from the sofa.

We have stuck an infrared LED under the mouse in order to detect where the user is pointing the mouse. A webcam connected to a computer near the sound player is necessary to detect the mouse. Technically, the sound player is connected to this additional computer, but this is invisible to the user. The user must have the feeling of pointing the shelf and dropping the digital CD in the sound player directly.

At last, the virtual shelf computer has bluetooth capability in order to detect user's bluetooth cell phone.

User identification. To access to her music collection, the user doesn't need to enter a login or password: she is identified with her bluetooth cell phone. As almost everybody has a cell phone, it seems to be the best way to identify transparently the user. To allow this possibility, a Java software is running on the virtual shelf computer. It permanently calls a linux application (hcitools) in order to scan the bluetooth devices in the room, and if a new one is detected, the Java application sends the bluetooth MAC address to the flash-based virtual shelf application using a TCP connection. The virtual shelf application checks if the user is present or not inside the centralized music database. If yes, a new shelf appears with the user's music collection. We limited the application to two shelves (thus, two users): firstly for a size reason, and secondly because it seems useless to allow more users on a single screen (sharing music is often between two persons).

Browsing music. We have designed the virtual shelf to be simple and familiar, with albums stored inside the shelf. Browsing a musical collection is really easy, even for people with limited computer skills. At this point, we have mainly worked on facilitating the browsing and we have taken advantage of the MX Air: pointing a CD with the mouse cursor displays information about it (album and artist name, songs in this album, CD cover), whereas clicking on the CD animates it as if you remove a CD from a real shelf. Then you can share your CD with your friends, listen to it, or release it and an animation replaces the CD at its initial position.

As we explained before, the music information is stored on a distant database, and not directly on the computer. And because the virtual shelf has been developed in flash, it could be run on other computers with a web browser and flash capabilities (the only additional part to install is the bluetooth java module to detect the user's cell phone). Furthermore, new interactive TVs such as GoogleTV include a Web browser with the Adobe Flash plug-in, this enables virtual shelf to directly run on a TV without any additional computer.

Sharing music. Sharing our music with our friends is not an easy task nowadays. You need to parse the music files as any other file type, and sometimes it is even impossible (with ipod or iphone for example if you are not on the computer synchronized with the device). With our application, you can drag-and-drop music from your friend's music collection to your own music collection: you pick up a CD from her shelf and release it on your shelf, and the copy is done. Visually, it appears as a cloning: when you release the mouse button, a new CD appears under the mouse cursor, and goes next to your other CDs, whereas the original CD goes to its initial position. Technically, the virtual shelf application adds new entries in the user's music database, which are the same entries already present inside the other user's music database. This way, the copy is done instantaneously: the music files are not duplicated, only the database entries are.

Playing music. To play the CD selected on the shelf, the user has two possibilities:

- Dropping the CD on the jukebox displayed between the shelves.
- Spatially drag-and-dropping the CD from the flash application to the physical sound player

For the first point, the virtual shelf application just plays the CD on the TV speakers. There is no technical challenge with this solution: Flash has sound capabilities and could play distant MP3s without any problem. So when a user drops a CD in the jukebox, we play the album by sending the MP3 hyperlink of the first track to the Flash application. The mouse wheel allows to change the album tracks, and the mouse click on the jukebox will turn off the current music.

The second point was more challenging to implement. In a previous work [7], we already developed an application to transfer files from TV to a digital photo frame. We have adapted our software to the virtual shelf application. An additional application dedicated to the webcam image analysis is running on the sound player computer, and informs the virtual shelf application about it. If the user selects a CD and releases the button when the mouse is pointing to the sound player device, then the sound player computer plays the corresponding album. When the sound player is on, the user can stop the music by pointing the device with the mouse, clicking on the left button and releasing the button outside of the webcam area: this way, the user drag-and-drops the sound outside of the music player. Technically, in order to perform the mouse LED detection, we have used cvBlobsLib, a blob detection API based on the OpenCV API. Coupled with an infrared LED and a webcam that filters visible light, detecting if the user is pointing or not the mouse in front of the sound player is functional.

3.2 Prototype #2: Tangible Drag-and-Drop to Manage a Multi-Room Audio System

The dematerialization of the music content brings advantages: you can carry all your music collection on a single device, which could be your mp3 player or your cell phone for example. Digital music allows you to have access to your entire music collection everywhere. Paradoxically, playing music at home has not changed that much. People who want to play their digital music are often using computers to do that, and so they are limited to one room.

Different commercial audio systems are now widely available to enjoy a digital music collection in different rooms. For example, Philips with the Streamium

TMwireless music center[3] and Logitech with the SqueezeboxTMaudio system[4]. Technically, these solutions are fast and easy to deploy, they can play a lot of different digital formats. On the other hand, navigating in these devices is not so comfortable. You have to navigate through a classical linear menu to select artists, albums or playlists, using the buttons on the device or a basic remote control.

Our idea is to use the tangible drag-and-drop technique in order to control a multi-room audio system. Our system is based on the Logitech SqueezeboxTM server for the multi-room audio system, and on the Nintendo WiiTMRemote (also nicknamed Wiimote) for the drag-and-drop remote control. We want to fully control the audio player next to us, but also the players situated in all the other rooms of the house, while sitting in our living room. Thus, we need to be able to:

Browse and select content. The music album collection has to be accessible and browsable. We need to be able to select an album, a web radio or a playlist with our remote control, and to create playlists with it.

Play selected content. After selecting a music content, we could choose to play it on the device in the same room, but we have also the possibility to play it on a further device.

Control the devices. Our remote control has to support all the basic features of a sound system, such as play, pause, next or previous song, etc.

Transfer content. The devices have to communicate with each other. For example, we want to be able to transfer the playlist currently played on a device to another device.

A possible use case could be the following: You are in your living room, sitting in your sofa, and wish to listen to some music. Your entire music collection is displayed on your TV. You select the album you want to play, and drag-and-dropping it from the television to the music player. After some tracks, you want to go to the bathroom to take a shower. Thus, you target your music player, and drag-and-dropp its content to the representation of the bathroom displayed on the screen. The album (or the playlist if you choose several albums) are now playing in the bathroom.

Implementation. This second tangible drag-and-drop prototype is not based on the first prototype. For this experiment, we have selected products which are already available on the market, and we have designed the software to make them communicate with each other.

This approach has the following advantages: Firstly, we did not have to develop a complete multi-room audio system. As explained before, commercial solutions already exist to perform that and they work well. Secondly, users are already familiar with these devices, they wont be facing an unknown environment.

---

[3] www.streamium.com

[4] http://www.logitechsqueezebox.com/

Fig. 2. Graphical user interface of the multi-room audio system

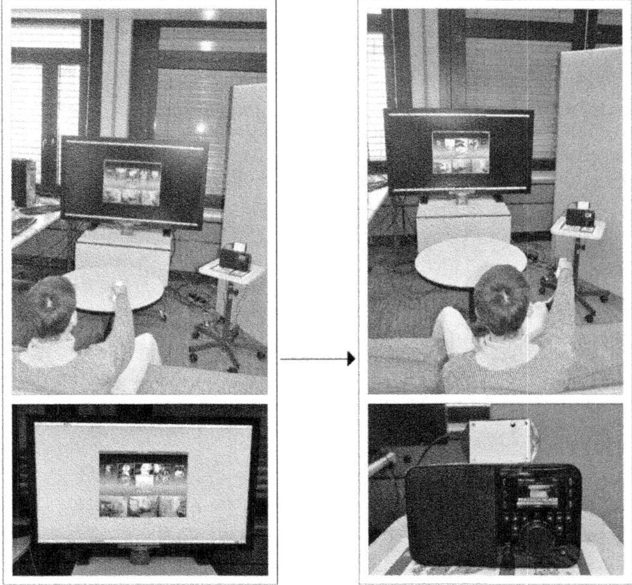

Fig. 3. Concept and hardware of the second prototype

On the other hand, we need the devices to be fully customizable. We choose the Squeezebox™ of Logitech to be the multi-room audio system, and the Nintendo's Wiimote™[5] to be our tangible drag-and-drop remote control. Both devices can be controlled and customized with open source software libraries.

[5] http://www.nintendo.com/wii/what/controllers

Audio system. Our audio installation is composed of a computer running the Squeezebox^{TM}Server, and three Squeezebox^{TM} Radios.

The server provides the access to the music library, and can control the entire system. From the server, we can create a playlist, choose an album/radio/playlist and on which player we want to listen to it, or control the players (play, pause, stop, increase volume, etc.). This is exactly what we want. To communicate with the server, the Squeezebox Server provides a command-line interface via TCP/IP. Commands and queries could be sent using TCP/IP in order to fully control the server. This way, we are able to ask programmatically the album list, the covers, and interact with all the players. We choose to use javaSlimServer[6], a Java API used to control the Squeezebox Server via the command-line interface, to easily communicate with the server.

Remote control. As tangible drag-and-drop remote control, we choose this time the Nintendo Wii^{TM}Remote. The reasons are the following: Firstly, the Wiimote integrates a 1024 x 768 100Hz infrared camera. This is a good solution to detect which device the user is currently pointing at. It favorably replaces the multiple webcams of our first prototype. Secondly, the Wiimote has some output possibilities: it can rumble for a few milliseconds or seconds, there is a speaker so it can play sounds, and there is four LEDs. We take advantage of these Wiimote capacities to give a feedback when the pointed device has changed. Now, when the user moves the remote control in order to point another device, the Wiimote rumbles 100 milliseconds to indicate that the new device has been detected. Finally, another advantage of the Wiimote is that several API already exist to access to the Wiimote abilities. As we are using Java to control the Squeezebox Server, we choose a Java library to communicate with the Wiimote: motej[7].

The main challenge of this second prototype was to detect where the user is pointing the remote control. Our first idea was to put a different number of infrared LED on the devices. The Wiimote camera can detect four different LEDs at the same time, so we wanted to take advantage of this ability. For example, putting one LED on the television, two on the first Squeezebox Radio, etc., and then identifying which device is targeted depending on the number of LEDs detected. But this solution did not work. At a distance of two meters and more, the Wiimote mixes the different LEDs into a single one. In order to avoid this problem, the different LEDs have to be separated by at least 20 cm. (which is actually the size of the Wii^{TM}Sensor Bar). It is acceptable for one or two LEDs, but too invasive for three or four.

Our second idea was to put a single infrared LED on each device, and make each LED blink at a different rate. This was the way we implemented the system. In order to make our LED blink, we use Arduino Duemilanove boards[8]. Arduino is a platform intended to easily create electronic prototypes. We just have to plug our infrared LED on the board and program the blink rate. We equipped each

[6] http://code.google.com/p/javaslimserver/

[7] http://motej.sourceforge.net/

[8] http://arduino.cc/

Squeezebox Radio with an Arduino board and a LED blinking at a particular rate.

The television does not have a blinking infrared LED. Indeed, the user needs to know where she is pointing on the graphical user interface, and we do not want to have a blinking pointer on the screen. For this reason, the screen is equipped with a standard Wii Sensor Bar. This way, there is no device identification problem (the sound systems have LEDs with different blink rates while the screen has a standard infrared LED) and our application can precisely detect where the user is pointing the remote on the screen.

Graphical user interface. The graphical user interface is displayed on the television. This is the central part of our application, as it allows to control all the devices.

Concerning the visualization of the media, we choose to display the albums using the cover flow interface. Cover flow is an interface created by Andrew Coulter Enright[3], but is mainly known for its integration in iTunes 7.0. We select this interface because it is adapted to our prototype environment (television + Wiimote), and also because it is near to the physical representation of a music album. The view is more limited in comparison to the virtual shelf (only 5 albums are visible at the same time), but this interface is less demanding in terms of precision. Indeed, you don't need to point precisely at the album, the selected album is always the one in the center. In order to go to the next or previous album, the user just has to target the television and press the left and right button of the Wiimote.

The first version of the main screen of the graphical user interface was limited to the cover flow. The idea was to display the sound devices only when the drag-and-drop button was pressed. So, the user had to point at the television, select the album through the cover flow interface, press the drag-and-drop button on the Wiimote (at this time, the sound devices appear on the visual interface) and release it when she points at the desire device on the screen. But after some early tests, we noticed that this interface had a major drawback: it did not allow to directly interact with the distant devices (for example, to stop a device that you forget to turn off).

The second version tackles this drawback. We add the devices directly on the main screen of the graphical interface, under the cover flow. In order to represent the sound devices on the screen, we choose to display a representation of the room where they are located. Indeed, it is easier to identify the destination of the selection (bathroom, living room, bedroom) rather than three identical pictures of the Squeezebox Radio with different names. So, when the user wants to interact with the device of a distant room, he/she just has to target the room on the T.V. and press the corresponding Wiimote button.

Controlling the tangible drag-and-drop audio system. To give an idea of how to control the system, the following list partially presents a possible action sequence of the application.

Browse the album collection. Point at the screen and press Wiimote's left or right button

Play an album. Use the drag-and-drop button (Wiimote's trigger) to transfer the album from the collection to the physical device or its representation on the screen

Change volume. Point at a playing device (physical or on the screen) and press the plus or minus button of the Wiimote

Transfer content from a player to another. Point at the playing device, press the drag-and-drop button and release it on the destination

Stop playing/clear playlist. Point at the playing device, press the drag-and-drop button and release it on the album collection

4 Evaluation and Discussion

In this section we present the tests we performed to assess the usability and user acceptance of the previously described drag-and-drop applications.

4.1 Evaluation of Prototype 1: Music Sharing

Twelve participants tested our application, and gave their feedback about it. Users enjoyed the simplicity of our application, and the efficiency of drag-and-dropping albums from a collection to another and from the collection to the sound device. During a post-test interview, different remarks and concerns were made about some points of our application.

Browsing on the virtual shelf has been described as really easy and natural. Of course, with this first prototype of virtual shelf, the possibilities are limited, and not adapted for browsing big music collections. Albums are now classified by names, and other criteria should be added. For example, one shelf level could be dedicated to one genre. Creating and managing playlists is another option to add, without reducing the ease of use.

Sharing music between two collections has also been described as really easy and natural, but raised some legal concerns. Our research does not address the legal aspect of content sharing.

Furthermore, our centralized system for music collection could be adapted to deal with these problems: for example, instead of copying the album from one shelf to another, the system could ask to the user if he wants to buy this album on a digital music store. Another example : on a music store with a flat monthly fee (such as the Zune Pass on the Zune Marketplace[9]), subscribers can download unlimited music. Thus, copying a track from one shelf to another could also inform this music store about the transfer, and then adapt the remuneration of each artist depending on the number of downloads.

Finally, our testers have really enjoyed the music transfer from the shelf to the music player using the drag-and-drop paradigm. They described it as natural, convenient and efficient. One of the main lesson we have learned from this first

[9] http://www.zune.net/en-US/software/zunepass

prototype study is that it is possible to mix the on-screen and spatial drag-and-drop without disturbing the user. Our participants used it naturally, going from shelf to shelf or from the shelf to the music device without any problem.

4.2 Evaluation of Prototype 2: Multiroom Audio Management System

In order to get a feedback and evaluate our second prototype, fifteen participants (5 women, 10 men) aged between 25 and 45 (mean 32.5) participated in our experiment. Two-thirds are working in the computer science field, and the others frequently use computers at work or at home (all of them were aware of the drag-and-drop concept in graphical user interfaces). Six of the participants do not use frequently the Wiimote, but they all had at least a little experience with it.

After explaining the tangible drag-and-drop concept and how to control the system, the user had some minutes to play with it and be familiar with the prototype. After this time, we asked them to perform a list of actions:

- Select an album in the GUI and play it on the player in this room (i.e. the living room)
- Change the volume of the device
- Transfer the content from the current player to the bedroom player
- Transfer the content from the bedroom to the living room

During the test, we observed where the user was pointing during the drag-and-drop operations (on the device or on the representation of the device on the screen), and after we asked them to fill up a form. The form was divided in three parts: one about the prototype, one about the tangible drag-and-drop concept in general, and one with questions. The first and second part was questions with answers from 1 to 5, 1 for "not at all" and 5 for "definitely yes".

About the prototype questions, it was to ensure that the quality of the experience was high enough to not interfere with the opinion about tangible drag-and-drop. Indeed, if a person did not like the experience, we want to know if the reason was the prototype or the concept. We asked if the remote control and the graphical user interface were easy to use, reactive and precise enough. All the answers are between 4 to 5, which is very good, except for the pointing precision of the Wiimote at the screen (some 3 and one 2, but an average of 4). Thus, the experience was good enough to fairly evaluate the tangible drag-and-drop concept.

The user evaluation of tangible drag-and-drop was encouraging as well. We asked five questions (check-boxes from 1 to 5) to users about the concept. The questions were, in this order:

1. Is this concept an interesting idea?
2. Are you convinced by this concept?
3. Is this concept natural?
4. Is this concept efficient?
5. Is this concept not demanding/tiring?

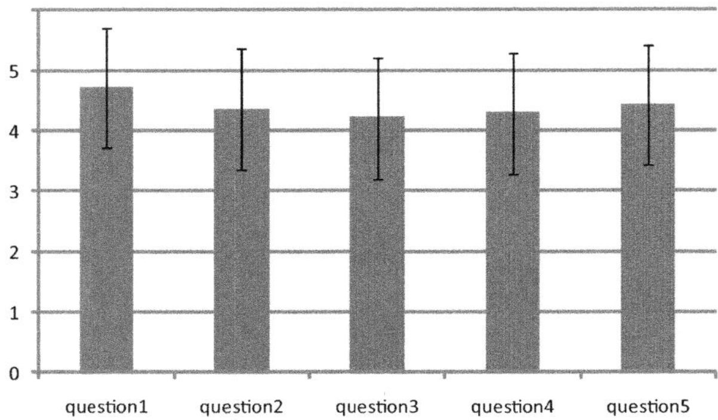

Fig. 4. User answers for the five questions about the concept. Columns represent the means of the 15 answers, the Y error bars are the standard deviations.

Results are presented in figure 4. As we can see, users were satisfied about all these points of the application.

Paradoxically, users pointed at the physical device only 57% of the time. This value moderates the natural aspect of the concept. After informally asking some questions to the users, several reasons can explain this result.

Firstly, it is important to consider the difficulty of fully breaking the border between digital and physical worlds. Indeed, all the participants were familiar with graphical user interface, and keept their focus on this interface in order to perform all possible actions.

Secondly, the feedback of the music device does not require to look at it. In [7], users looked at the photo frame to check if the picture transfer were done. In this audio prototype, users can "hear" if he successfully transfer the album, and for this reason can stay focused on the graphical interface.

Finally, the Fitt's law[12] index of performance is inferior for tangible drag-and-drop than for the graphical user interface. Indeed, this value directly depends on the distance from the starting point to the center of the target, and the distance from the album selection interface to the digital representation of the device was smaller.

Despite that, it is important to take into account that the graphical interface was really uncluttered, with only three rooms, one device per room, and few options. With five devices or more on the GUI, or more than one device per room, the graphical interface becomes more complicated and the tangible drag-and-drop gains in interest. Some preliminary tests with two squeezebox in the same room are going in favor of this assumption.

In order to better assess the usability and applicability of the drag-and-drop interaction technique we performed a third experiment focused on evaluating the two variations of the technique: on-screen and spatial. Next subsection presents our results.

4.3 Evaluation of the Tangible Drag-and-Drop Technique

With this final evaluation, we wanted to compare separately what we call *on-screen* drag-and-drop technique -inside a graphical interface, and the *spatial* tangible drag-and-drop technique.

We created two applications out of prototype 2. In both applications, the concept was to play music, while sitting in a couch in the living room in front of the TV. The first application implemented an *on-screen* drag-and-drop interface (GUI-based). We kept the coverflow in order to display the albums, but we displayed only a picture of a Squeezebox device instead of the pictures of the rooms. If the user wanted to play an album, we had to drag-and-drop it from the coverflow to the device icon, and all the interaction regarding to the device (volume up/down, previous/next track, etc) was performed by pointing at this icon. The second interface was based on the tangible *-spatial-* drag-and-drop technique. The graphical interface on the TV was limited to the coverflow, and nothing else. Compared to prototype 2, we removed all the pictures of the rooms. If the user wanted to play an album, we had to drag-and-drop it from the TV to the device using the remote control.

For this study, we asked 10 participants (4 women, 6 men) to test our 2 interfaces, and to evaluate it with an AttrakDiff[6] questionnaire. All the participants were different from the previous tests. After a short training session, we asked them to play at least 4 albums with each system. After the test, we asked them their opinion concerning the two interfaces, and then to fill in the questionnaire. Figure 5 presents the results of the questionnaire.

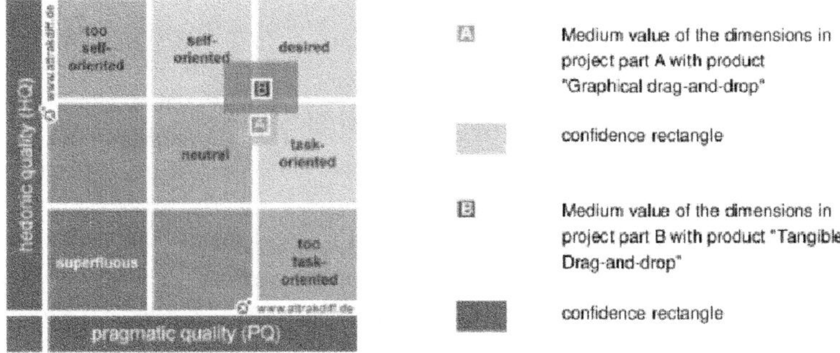

Fig. 5. Results from the AttrakDiff evaluation

From the AttrakDiff questionnaire, we can see that the tangible drag-and-drop technique performs better than the *on-screen* drag-and-drop technique. The difference in terms of pragmatic quality is not statistically significant, but the difference in terms of hedonic quality is important. The tangible drag-and-drop technique is perceived as much more human, innovative and attractive, but less predictable. The two interfaces are perceived as equally efficient, which explained the close result concerning the pragmatic quality.

During this evaluation, we learned that both interfaces are not mutually exclusive. Even if the interaction task - drag-and-drop - is the same, the two techniques are quite different.

Tangible drag-and-drop was appreciated because of its spatial aspect: you don't need to be precise, you don't even need to look at the device when you transfer an album on it. It is a natural way to interact when involved devices are within the field of view.

The graphical interface is more demanding in terms of precision: you have to be focused on the screen to interact with the system. The *on-screen* drag-and-drop was appreciated because of its iconic representation. You can interact with devices which are not in the field of view, like we did in the section 3.2. Another advantage is that you can create symbolic icons: for example, one user suggested implementing a Wikipedia icon that could be used to display the information of the album dropped on it. This kind of interaction is not possible using tangible drag-and-drop, as Wikipedia is not a physical interface.

To conclude this evaluation, we can say that the two interfaces are more complementary than in competition: one is using an iconic representation of the devices, the other one is more suitable for actions to be performed within a well-delimited area.

5 Conclusions and Future Work

Breaking the border between the digital and the physical world is difficult, especially for computer users who are used to stay focused on the graphical user interface in order to manipulate their digital content. Physical devices have to manifest their presence, to show that the tangible drag-and-drop action is possible when a particular content is selected. For example, a solution could be to illuminate the devices when the user is pressing the button. This way, it will attract her attention and the user will know that the drag-and-drop is possible on these devices. Another solution could be to add additional information directly on the remote control itself. Indeed, as users of the first prototype were complaining about the lack of feedback, we tried to add a picoprojector on the remote control. The idea is to display the selected content during the drag-and-drop action. There are two benefits of such an approach: besides providing visual feedback, it also helps to maintain the continuum between the virtual and the physical worlds. We conducted a small user experiment with this prototype, and these early tests are promising. Figure 6 is a photo of this prototype.

The action performed after a tangible drag-and-drop movement has to be clear and consistent. For example, if a user wants to visualize photos located on a digital camera, he/she will drag-and-drop the pictures from the camera to the television. But what will happen to the pictures located on the digital camera? Will they be erased or just transferred? Generally, it is better to divide the action as much as possible. If the user wants to erase the pictures, he will perform another action (for example, drag-and-drop the pictures from the digital camera to a trash can). Another solution to inform the user about what will happen after a tangible drag-and-drop action could be to use a remote control equipped with a screen.

Fig. 6. Drag-and-drop remote control with pico projector

Tangible -spatial- drag-and-drop is interesting when all the devices are in range. It is interesting for example in the living room where there is often a large number of devices (television, VCR, media center, etc.). To transfer content physically from one room to another, a different paradigm such as on-screen drag-and-drop could fit better. Finally, one interesting point we have learned is that the on-screen drag-and-drop and the spatial drag-and-drop are not mutually exclusive, and could be used together to create a good user experience.

References

1. Alonso, M.B., Keyson, D.V.: MusicCube: making digital music tangible. In: CHI 2005 Extended Abstracts on Human Factors in Computing Systems (CHI 2005), pp. 1176–1179. ACM, New York (2005)
2. Edwards, W., Grinter, R.: At Home with Ubiquitous Computing: Seven Challenges. In: Abowd, G.D., Brumitt, B., Shafer, S. (eds.) UbiComp 2001. LNCS, vol. 2201, pp. 256–272. Springer, Heidelberg (2001)
3. Enright, A.C.: Dissatisfaction Sows Innovation, `http://tinyurl.com/2zr7bw`
4. Fitzmaurice, G.W., Ishii, H., Buxton, W.A.S.: Bricks: laying the foundations for graspable user interfaces. In: Proceedings of the SIGCHI Conference on Human Factors in Computing Systems (CHI 1995), pp. 442–449. ACM Press, New York (1995)
5. Graham, J., Hull, J.J.: Icandy: a tangible user interface for itunes. In: CHI 2008 Extended Abstracts on Human Factors in Computing Systems (CHI 2008), pp. 2343–2348. ACM, New York (2008)

6. Hassenzahl, M., Burmester, M., Koller, F.: AttrakDiff: Ein Fragebogen zur Messung wahrgenommener hedonischer und pragmatischer Qualitt [AttracDiff: A questionnaire to measure perceived hedonic and pragmatic quality]. In: Ziegler, J., Szwillus, G. (eds.) Mensch & Computer 2003. Interaktion in Bewegung, pp. 187–196. B. G. Teubner, Stuttgart (2003)
7. Hopmann, M., Thalmann, D., Vexo, F.: Tangible Drag-and-Drop: Transferring Digital Content with a Remote Control. In: Chang, M., Kuo, R., Kinshuk, Chen, G.-D., Hirose, M. (eds.) Learning by Playing. LNCS, vol. 5670, pp. 306–315. Springer, Heidelberg (2009)
8. Hsu, S.H., Jumpertz, S., Cubaud, P.: A tangible interface for browsing digital photo collections. In: Proceedings of the 2nd International Conference on Tangible and Embedded Interaction (TEI 2008), pp. 31–32. ACM, New York (2008)
9. Ishii, H., Ullmer, B.: Tangible bits: towards seamless interfaces between people, bits and atoms. In: Proceedings of the SIGCHI Conference on Human Factors in Computing Systems (CHI 1997), pp. 234–241. ACM, New York (1997)
10. Lee, H., Jeong, H., Lee, J., Yeom, K.-W., Park, J.-H.: Gesture-Based Interface for Connection and Control of Multi-device in a Tabletop Display Environment. In: Proceedings of the 13th International Conference on Human-Computer Interaction. Part II: Novel Interaction Methods and Techniques, pp. 216–225. Springer, Heidelberg (2009)
11. Lee, H., Jeong, H., Lee, J., Yeom, K.-W., Shin, H.-J., Park, J.-H.: Select-and-point: a novel interface for multi-device connection and control based on simple hand gestures. In: CHI 2008 Extended Abstracts on Human Factors in Computing Systems (CHI 2008), pp. 3357–3362. ACM, New York (2008)
12. MacKenzie, I.S.: Fitts' law as a research and design tool in human-computer interaction. Human-Computer Interaction 7, 91–139 (1992)
13. Nunes, M., Greenberg, S., Neustaedter, C.: Sharing digital photographs in the home through physical mementos, souvenirs, and keepsakes. In: Proceedings of the 7th ACM Conference on Designing Interactive Systems (DIS 2008), pp. 250–260. ACM, New York (2008)
14. Rekimoto, J.: Pick-and-drop: a direct manipulation technique for multiple computer environments. In: Proceedings of the 10th Annual ACM Symposium on User Interface Software and Technology (UIST 1997), pp. 31–39. ACM, New York (1997)
15. Rumbolt, R.B., McIntyre Jr., W.R.: Universal remote control unit. US Patent 4,774,511 (1988)
16. Seifried, T., Haller, M., Scott, S.D., Perteneder, F., Rendl, C., Sakamoto, D., Inami, M.: CRISTAL: a collaborative home media and device controller based on a multi-touch display. In: Proceedings of the ACM International Conference on Interactive Tabletops and Surfaces (ITS 2009), pp. 33–40. ACM, New York (2009)
17. Seifried, T., Rendl, C., Perteneder, F., Leitner, J., Haller, M., Sakamoto, D., Kato, J., Inami, M., Scott, S.D.: CRISTAL, control of remotely interfaced systems using touch-based actions in living spaces. In: ACM SIGGRAPH 2009 Emerging Technologies (SIGGRAPH 2009), ACM, New York (2009) Article 6
18. Zigelbaum, J., Kumpf, A., Vazquez, A., Ishii, H.: Slurp: tangibility spatiality and an eyedropper. In: CHI 2008 Extended Abstracts on Human Factors in Computing Systems (CHI 2008), pp. 2565–2574. ACM, New York (2008)

Puppet Playing: An Interactive Character Animation System with Hand Motion Control

Zhiqiang Luo, Chih-Chung Lin, I-Ming Chen, Song Huat Yeo, and Tsai-Yen Li

School of Mechanical & Aerospace Engineering
Nanyang Technological University, Singapore
{zqluo,michen,myeosh}@ntu.edu.sg
Computer Science Department
National Chengchi University
Taipei, Taiwan
{s9421,li}@cs.nccu.edu.tw

Abstract. Puppetry is a popular art form involving the process of animating the inanimate performing puppets. Puppet playing not only is controlled by artist's hand motions but also follows physical laws and engineering principles. To implement a puppet playing scenario in virtual environments, an interactive animation system is designed by taking into account both the user's hand motion and the constraints of puppet and environment. Here the hand motion is real-time captured and recognized through a new input device, namely SmartGlove. The animation system adapts IMHAP testbed for procedural animation and generates puppet animation based on both the procedural animation technique and motion capture data from SmartGlove. Thus the physical hand motion can either activate the designed procedural animation through motion recognition or tune the parameters of the existing procedural animation to generate new puppet motions. This system allows a user to directly control puppet animation while preserving the high accuracy of motion control. The animation results show that the association of hand motion can smoothly plan and generate each procedure animation and seamlessly blend the gap between key frames. The potential application and improvement of the current animation system are discussed.

Keywords: Character Animation, SmartGlove, Motion Planning, Puppetry.

1 Introduction

Character (specifically humanoid) animation is an oldest but still challenging research topic in computer animation. Two of challenges are emphasized here. First, people (e.g. viewers of animation) are most familiar with the characteristics (satisfying physics and biological requirements) of human motion in daily life. Thus the generation of natural and accurate motions should take into account of various parameters affecting human motion. Second, the character animation is usually predefined and is difficult to be generated and controlled in real time in order to adapt new tasks and environments. Interactive control of character animation can extend the adaption of predefined animation procedure and significantly save the time and effort to build a new animation. To

M.L. Gavrilova et al. (Eds.): Trans. on Comput. Sci. XII, LNCS 6670, pp. 19–35, 2011.

respond to these two challenges, one interactive animation system is designed by employing both motion capture and procedure animation techniques. The interactive animation system can plan, generate, and control natural character animations in real time. One task scenario, puppet playing, is chosen to illustrate the background, procedure, and test results of our animation system.

Puppetry and Puppet Theater are popular art forms having a long and fascinating heritage in many cultures. The merit of the art form is the fascination with the inanimate object animated in a dramatic manner and humans' curiosity in producing an exact artificial copy of themselves. People are largely interested in the theatrical and artistic content of puppetry, though basic puppet fabrication and manipulation techniques follow physical laws and engineering principles. Most types of puppets in use today fall into four broad categories [1]. First, the glove puppet is used like a glove on a user's hand. Second, the rod puppet is held and moved by rods, usually from below but sometimes from above. Third, the marionette is a puppet on strings, suspended from a control mechanism held by the puppeteer. Last, the shadow puppets are usually flat cut-out figures held against a translucent, illuminated screen. In the present study, the first category, the glove puppet, will be animated in computer through the interaction with the user's hand.

Glove puppet animation is one type of character animation. Currently, the mainstream methods for generating character animations can be divided into three categories. The first method is to generate character animation by sampled data. Specifically, the sampled data can be obtained by the motion capture technology where the motions are performed by a real actor [6]. Animations made by this technology are more plausible. However, without knowing the meaning and structure of the motion, it is also difficult to modify a capture motion to fit the constraints of a new environment. The second method is to use the commercial 3D animation software to set the key frames of a character by animation professionals and interpolate between key frames to generate the final animations. However, even for good animators, producing an animation in this way is time consuming and the result is less realistic. The third method is to generate animations by simulation or planning procedures [2][11]. It is also called knowledge-based animation since it usually requires a dynamics model of the object under simulation or motion-specific knowledge to plan and generate the motions. Due to the complexity of the underlying computational model, the knowledge-based animation is usually generated in an off-line manner and displayed in real time. However, even along with a good knowledge or simulation system it is still difficult to generate the realistic animation by the computer procedural. It should be noted that the methods in above three categories are somewhat complementary to some degree and each of them may be more suitable for certain kinds of animations.

Being the intricate and prehensile parts of the human body, our hands are the primary agent for physical interaction with the external world. We use our hands in varies of aspects to perform our everyday activities. In order to capture the human hand's motion, researchers have developed many sensing methods in the past few decades. Besides the optical [4], acoustic and magnetic [10] sensing techniques, innovative techniques such as fiber-optic [3], strain gauge and hall-effect sensing are introduced. Nevertheless, these sensing methods still have rooms for improvement to achieve the stringent requirements from applications in various fields, such as the medicine, training, entertainment, and virtual reality. Some criteria for a glove to capture the hand motion include the sensing accuracy (stability of the sensor signals, repeatability & reliability of movements), ease of wear and removal, rapid calibration, adaptation for different hand sizes, no electromagnetic interference, no temperature variation, and low cost. Furthermore, to manipulate the glove puppet investigated in the present study, it is crucial to accurately capture the finger motion in order to control the specific component of a puppet. A wearable glove-based multi-finger motion capture device, called the SmartGlove, is developed in the present study. The user's hand motion, especially the finger's motion, is captured by SmartGlove in order to drive the glove puppet animation implemented by the computer.

The present study introduces an interactive animation system to generate, plan and control a glove puppet animation by using the SmartGlove. With the user's hand motion input from the SmartGlove, a user can directly control the puppet's two arms and head motions. Furthermore, the animation is created based on the method of the procedural animation. The procedural animation is used to design some motions on the puppet's feet and the body, which compensates the input from SmartGlove. The input from SmartGlove can in turn help to tune the parameters of the animation procedure. The seamless integration between the hand motion input and the procedural animation ensures the fluent motion of the puppet.

The rest of the paper will be organized as follows. The system architecture is first introduced to provide the big picture of the current glove puppet animation system. Then, the design of the SmartGlove and the puppet model will be described, respectively. After that, the technique for the motion recognition and the implementation of the procedural animation is presented, followed by the experimental results to test the animation. Finally, we will conclude the paper and list the future work in the last section.

2 System Architecture

The present puppet animation system, called the IMPuppet, is constructed based on an experimental testbed for procedural animation, IMHAP (Intelligent Media Lab's Humanoid Animation Platform, [7]). The system architecture of IMPuppet is shown in Fig. 1. The structure of the system is based on the model-view-control (MVC) design pattern. With the MVC design pattern, the system is divided into three main models: *model*, *view* and *controller*. Model is to encapsulate data; view is to interact with the user; and controller is the communication bridge between the former two. In Fig. 1, each square stands for a module, and the arrow between the modules means the data flow from one module to the other one. The modules of animation dispatcher and animation generator are the main components in IMPuppet, and play the role of "model" in MVC. The module of animation controller plays the role of "controller", receives events from the user, and controls how animation is played. The modules of user interface and 3D browser play the roles of "view" in MVC, which responds to interaction with the user and draws the scene on the screen. The function of each module is described in more details as follows.

The modules of animation dispatcher and animation generator are the core of the IMPuppet system. They are defined with abstract interfaces in order to perform experiments on animation procedures. The animation dispatcher functions as a global planner which directs one or more animation generators. Specifically, the animation dispatcher receives high-level commands, such as a path that the animated character should follow, from the graphical user interface of the system or sequence of motion data from the data glove. Then, the animation dispatcher maps the high-level commands into a sequence of locomotion according to the kinematics and geometric models of the puppet character as well as the environment. The animation dispatcher defines the target of a motion generation and dispatches relevant commands and data to the appropriate animation generators. An animation generator is both a local planner in charge of generating animation motions with the predefined procedure and a motion editor utilizing and modifying the motion capture data.

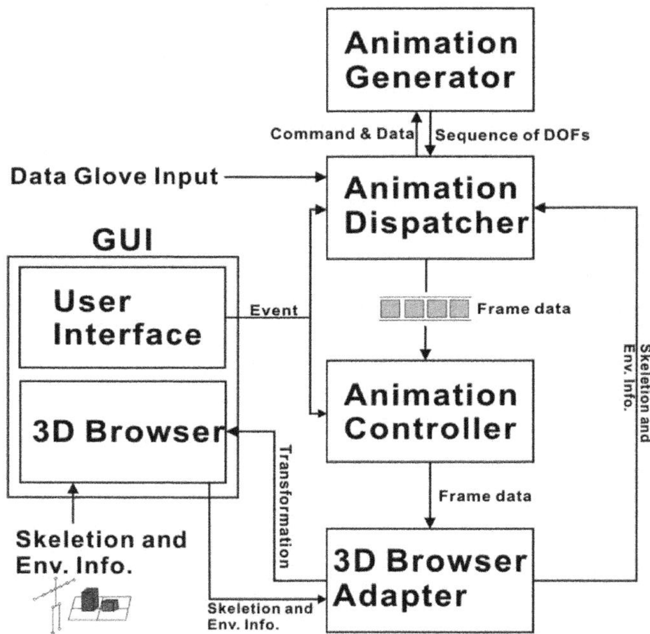

Fig. 1. IMPuppet system architecture

The detail structure of animation generator and animation dispatcher is shown in Fig. 2. The motion recognition system and planning system are put together in the animation dispatcher. Once a motion is recognized, the corresponding procedural motion can be sent to the frame data with the help of planning system. The animation generator module consists of all generators for an animation frame. The module of animation controller reads the queue of frames sent from the animation dispatcher and controls the playing of animation. The user controls the animation by either the SmartGlove (the main input) or the command (which is activated as an event) through the user interface.

The module of animation controller reads the queue of frames generated by the animation dispatcher and controls the playing of animation. The user controls the animation by accessing the given user interface and then the user interface calls the animation controller to invoke proper actions.

The 3D browser is treated as a plug-in in IMPuppet to read 3D models from files and provide External Authoring Interface (EAI) to other modules. EAI is a programming interface defined in most VRML browsers that allows a programmer to retrieve and set the attributes of an arbitrary object or skeleton joint, but not all 3D browsers support exactly this interface. To adapt to those browsers that support an interface with different naming or argument conventions, another module called "browser adapter" is designed to act as a wrapper of the 3D browser. It encapsulates the 3D browser and provides a uniform interface to other modules for the rest of the system. As a result, the 3D browser can be replaced easily without modifying the rest of the system as long as appropriate browser adapter has been implemented. In the present IMPuppet system, the open-source 3D browser JMonkey [5] is chosen as the 3D browser.

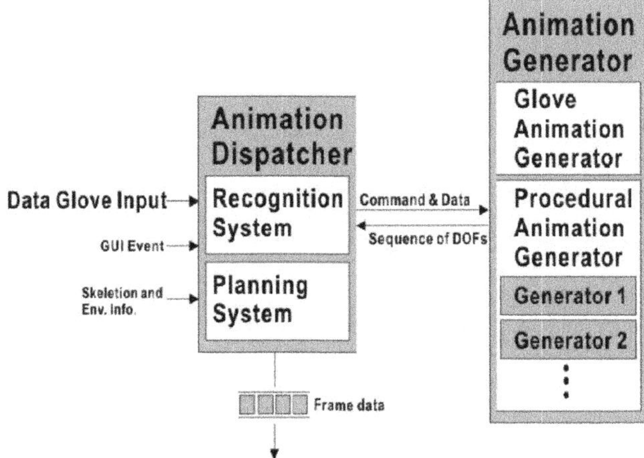

Fig. 2. Detail structure of animation generator (right) and animation dispatcher (left)

3 SmartGlove

The SmartGlove is aimed to achieve high performance hand/finger motion tracking and monitoring at an affordable cost for wide adoption. SmartGlove employs a new optical linear encoder (OLE) to capture the finger motion. The SmartGlove system consists of five multi-OLE strips (each includes two OLEs) and a microcontroller. The multi-OLE strips will capture and send the appropriate motion data of each finger joint to the microcontroller that synthesizes all the information sent to it from the multiple OLEs. Then, using a forward human hand kinematics model embedded into the gateway, the microcontroller will transmit the captured motion data to a remote robot, a virtual reality system or a computer for further analysis and application through wired or wireless communication. Compared to currently available hand capturing devices, the critical OLE sensing element is low-cost, compact, light-weighted, and immune to temperature or electromagnetic interferences.

3.1 Hand Kinematic Modeling

Human hand is modeled with a hierarchical tree structure that consists of rigid links and joints. This hierarchical structure is represented in Fig.3, and the position of each joint is described using the D-H transformation with reference to the heel of the hand (the world coordinate system ($x_0^{'}, y_0^{'}, z_0^{'}$)). [8] The posture of each finger ray (labeled as 1 to 5 from the thumb to the little finger as shown in Figure 3.) is represented under a local coordinate system. Now with the D-H transformation, the position of each joint can be transformed from the local coordinates to the world coordinates sequentially. There are 23 internal degrees of freedom (DOF's) located in the hand skeleton model [13][12].

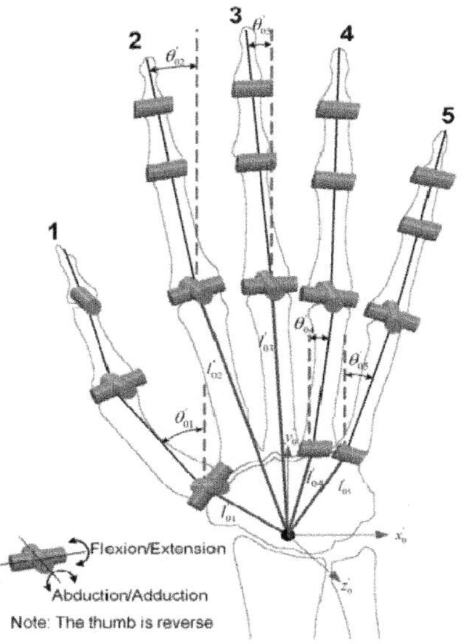

Fig. 3. The kinematic model of a user hand

As shown in Fig.5, the five finger rays can be divided into three different groups in terms of the kinematic structure. Each of the four fingers has four DOF's. The Distal Interphalangeal (DIP) joint and the Proximal Interphalangeal (PIP) joint both have one DOF and the remaining two DOF's are located at the Metacarpophalangeal (MCP) joint. Different from the four fingers, the thumb has five DOF's. There are two DOF's at the Trapeziometacarpal (TM) joint (also referred as Carpometacarpal (CMC) joint), and two DOF's at the Metacarpophalangeal (MCP) joint. The remaining one DOF of the thumb is located at the Interphalangeal (IP) joint. The basic flexion/extension (f/e) and abduction/adduction (a/a) motions of the thumb and fingers are performed by the articulation of the aforementioned 21 DOF's. The abduction/adduction motions only occur at each finger's MCP joint as well as the thumb's MCP and TM joints. Another two internal DOF's are located at the base of the 4th and 5th (ring and little finger's) metacarpals which performs the curve or fold actions of the palm. It should be noted that the present SmartGlove only captures fifteen DOF's of hand motion in the present project. The motions of the end joints (near the finger tips) of index, middle, ring and little fingers were not captured as the range of the joint motion is quite small. The motions of the first joints of the thumb, ring and little fingers will be captured in the future system.

3.2 Multiple OLEs

The single-point OLE sensor detects joint flexion displacement through the 1-DOF linear movement of the sensor outputs [9]. Two or three OLE sensors put together can detect joint motions of multiple DOF's. The basic working principle of SmartGlove uses a variation of OLE principle by placing multiple OLE's in series on different finger segments to capture the displacements of different detecting points on a single encoder strip. This encoder strip passes through all OLE's on the same finger. Thus, it is termed as Multi-point OLE. As shown in Figure 4, three disks (from left to right) represent three in-line joints with radius of R1, R2 and R3, respectively. Denote their bending angles as φ1, φ2 and φ3, respectively. Three OLE's are placed and fixed at positions (A), (B) and (C), as shown in Fig.4. Assume that the displacement readings obtained by these three OLEs are D_1, D_2 and D_3, respectively.

Due to the accumulated displacement at the distal joints, we use the following equations to calculate the angle of each finger joint.

$$D_1 = L_1 = \frac{2\pi R_1 j_1}{360} \tag{4}$$

$$D_2 = L_1 + L_2 = \frac{2\pi R_1 \varphi_1}{360} + \frac{2\pi R_2 \varphi_2}{360} \tag{5}$$

$$D_3 = L_1 + L_2 + L_3 = \frac{2\pi R_1 \varphi_1}{360} + \frac{2\pi R_2 \varphi_2}{360} + \frac{2\pi R_3 \varphi_3}{360} \tag{6}$$

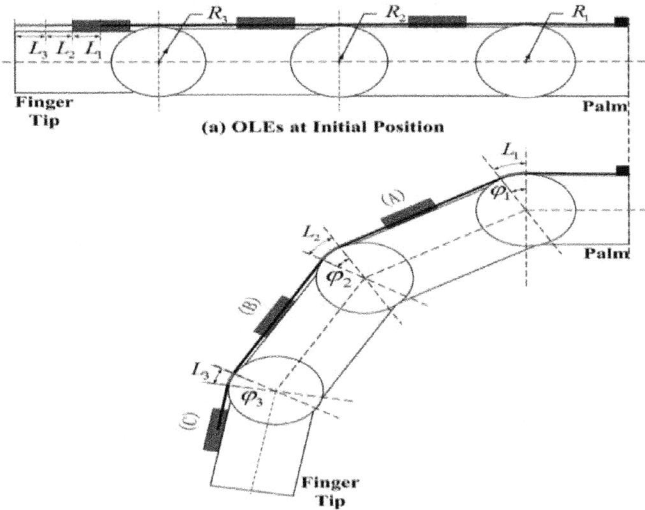

Fig. 4. Multi-point sensing principle for multiple OLEs

Because of the natural arches of a hand, the multi-point sensing can be adopted in finger motion capture. As introduced in the hand kinematics above, there is at least 14 joints' FE motions need to be captured in order to perform basic multi-finger sensing, and all these 14 joints are all within the five longitudinal arches. Hence, by introducing one strip for each longitudinal finger arch, we are able to use the multi-point sensing method to capture the finger's movement.

3.3 Prototype

The hardware of SmartGlove comprises two main components, the OLE module and the microcontroller. All these hardware components are mounted on a glove [14].

As shown in Fig. 5, the OLE module is the sensing module in the system which includes three basic units: *the sensing unit* (sensor and lens), *the interface unit* (the customized PCB board), and *the housing unit* (the customized base plate & strip). The sensing unit is fixed in the housing unit to obtain the displacement of strip and to communicate with the microcontroller through the interface unit.

The sensor used in OLE is Avago's optical mouse sensor product ADNS-3530, which is based on Optical Navigation Technology that measures changes in position by optically acquiring sequential surface images (frames) and mathematically determining the direction and magnitude of movement. In order to make the size of the OLE compact, the ADNS-3530 is designed for surface mounting on a PCB board.

The housing unit is the holder for the optical navigation sensor and the moving strip made of DelrinTM. According to the performance of ADNS-3530, the distance between the lens and the moving strip determines the resolution of the result. Based on the datasheet, in order to get high resolution of the sensor, the distance should be within the range of 0.77mm to 0.97mm. Furthermore, the surface material of the strip also affects the sensor's resolution. To make sure that the strip slides smoothly in the housing unit, there must be a gap between the strip and the base plate. Consequently, for the stable readout, white Formica is the ideal choice for surface material of the strip because the mean resolution is very stable within the pre-defined range.

SmartGlove uses the Arduino Diecimila/Bluetooth which is based on the Atmega168. It is an open-source physical computing platform based on a simple I/O board. The programming language for the Arduino microcontroller is an implementation of Wiring/Processing language. The microcontroller communicates with the OLE's via SPI protocol and sending out all the motion data to PC via USB/Bluetooth.

For the ease of the replacement and maintenance of the sensors, the OLE's are mounted onto the glove using Velcro, and the microcontroller connects OLE's by ribbon wires. Thus, the glove can be separated from the OLE's and all the hardware for cleaning. This feature takes a big leap toward using such data gloves in common daily living. Several photos of the SmartGlove prototype are shown in Fig.5.

Furthermore, in order to make the glove type OLE's sensitive, the glove should fit nicely on the human hand. On the other hand, the glove should not hinder free motion of the hand. Therefore, soft and stretchable fabric is used for the SmartGlove. In this project, two different fabrics are used: the semi-stretching fabric, which can be stretched only in a single direction, and the stretching fabric, which stretches in all directions. The glove uses stretching fabric for backside of the MCP joints and semi-stretching fabric for the palm side to avoid stretching along the finger direction. Thus, the glove has good elasticity to fit the hand of the users.

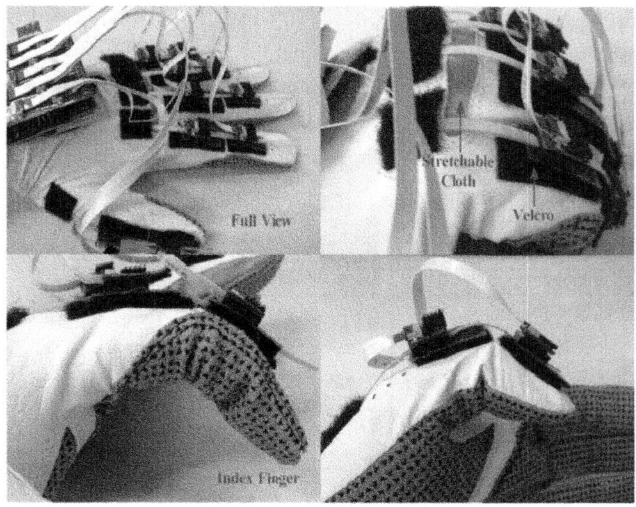

Fig. 5. SmartGlove prototype

3.4 Experimental Test

The Grip Test (uses a gripped hand position) and the Flat Test (uses a flat hand position) are carried out to analyze the repeatability and reliability. Data is collected from five healthy male students aged 22-27 years with comparable hand size and no hand movement disorders. Five sets of measurement are performed in each test on each subject and each set of measurement includes ten grip/release actions.

Repeatability is indicated by the range and standard deviation (SD). The average range and SD obtained from each OLE across Subjects 1 to 5 for each test are shown in the histogram in Fig.6.

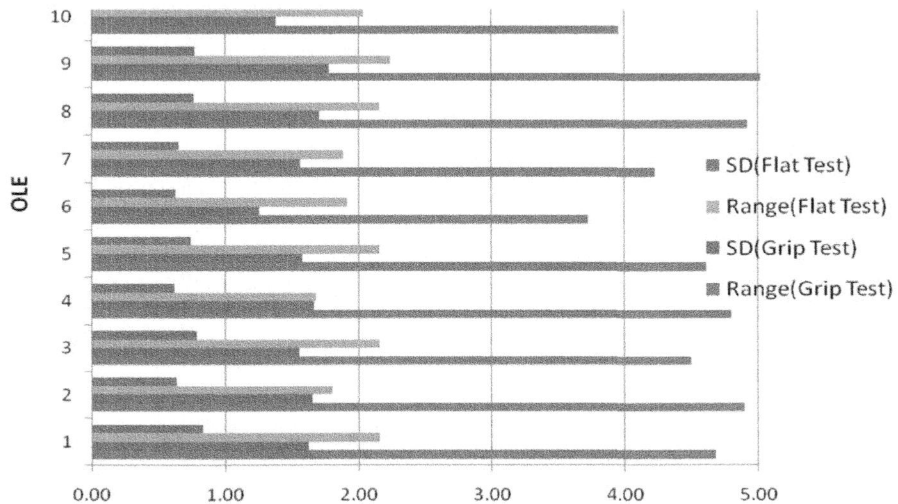

Fig. 6. Histogram of averaged rang and SD for each OLE

Reliability is indicated by the intraclass correlation coefficient (ICC). [15] The ICC analysis is performed for each test and for each OLE individually (ICC is calculated using Excel). The ICC values in Table 1 show consistency from one data block to another with no particular OLE showing significant lower reliability than the overall mean.

Table 1. ICC of reliability

	thumb		index		middle		ring		little		Aver-age
	MCP	IP	MCP	PIP	MCP	PIP	MCP	PIP	MCP	PIP	
Grip Test	0.937	0.954	0.882	0.963	0.913	0.987	0.948	0.957	0.969	0.964	0.947
Flat Test	0.955	0.968	0.893	0.966	0.908	0.976	0.955	0.968	0.958	0.979	0.953
Over-all											0.950

4 Puppet Model

The puppet model we use in our system is based on the traditional Chinese puppet. There are three reasons that we chose traditional Chinese puppet rather than other kinds of puppets. The first reason is that it fit our desired application for SmartGlove due to its manipulation method. The second reason is the rich performance that the traditional Chinese puppet has, compared to other kinds of puppetry. This performing art not only combines elements of Chinese opera, music, delicate costumes and props but also skillful manipulation of the puppet. This also makes the design of our system a great challenge. The third reason is that the motion of traditional Chinese puppet is suitable for procedural modeling. Because of the elements from the Chinese opera, most motions of traditional Chinese puppet follow strict rules about how it should be performed. These rules make it possible to design the motions in a procedural way.

4.1 Kinematics Model

The kinematics structure of a puppet is similar to a human figure with minor differences as show in Fig. 6. For example, the major active joints of a puppet are the neck, shoulder, left/right hips, and pelvis. The joints of elbows and knees are usually ignored. The motions of the legs are triggered either by another hand or by the gravity during swinging. The clothes are the passive part that moves with the legs as they swing. As show in Fig. 7, the additional links and joints are attached to the clothes at the front and back of the legs. Thus the clothes can moved when the legs are lifted.

4.2 Control of Puppet

SmartGlove can be used to directly control the puppet's hand and head movements. As shown in Fig. 8, the position of point O on the hand corresponds to the center of puppet. Currently SmartGlove cannot capture the abduction/adduction (a/a) motions between each finger. The angles of the two joints in Fig. 8, θ_1 andθ_2, represent the angle

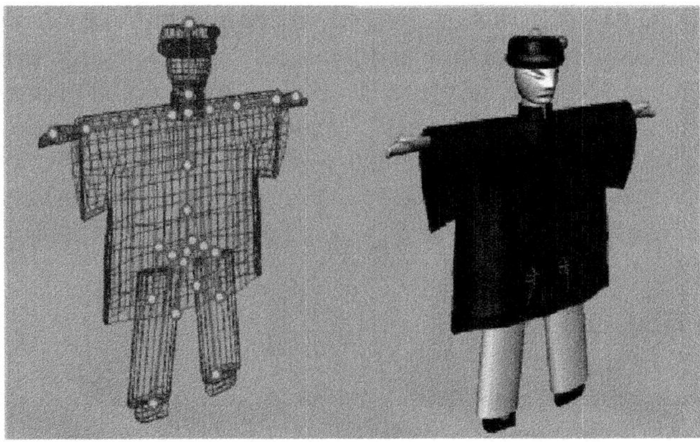

Fig. 7. Kinematics model of the hand puppet including the additional linkages with clothes

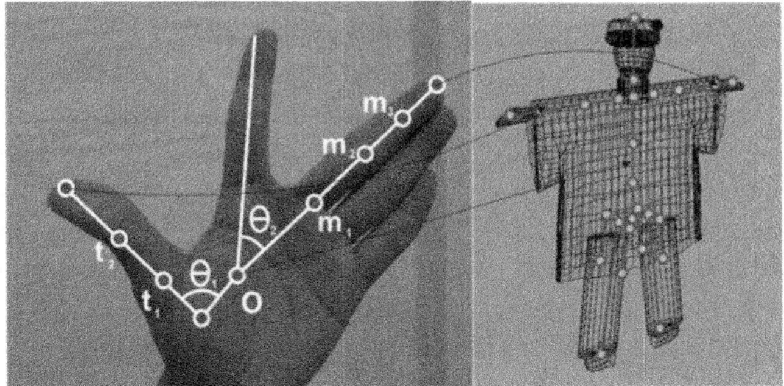

Fig. 8. Use thumb and middle finger to control the two arms of a puppet

of the joint between the thumb and the index finger and the angle of the joint between the index finger and the middle finger, respectively. Beside these two joints, SmartGlove can capture the bending angles of the other four joints, t_1, t_2, m_1 and m_2. According to our experiments, we found that it is reasonable to assume the bending angle of joint m_3 be equal to the angle of m_2. Then the end-point position of the two fingers can be computed by using forward kinematic based on the above joint angle and the length between the two joints. The end-point positions for the fingers of the puppeteer can then be used to update the hand positions of the puppet. Finally, by using inverse kinematic, the rotation matrix of the puppet's shoulder and elbow on the puppet's two hands can be determined. Similarly, SmartGlove can be used to detect the flexion/extension (f/e) of the index finger to control the head of the puppet.

With the method described above, we can move the puppet's hands and head in a way that reflects the physical motion of the user's fingers. And this is exactly the way that we control traditional Chinese puppet in most cases. However, it is not enough to only use this method to control and generate puppet motion if we want our virtual puppet to have full features of motions as the real one does. Hence, a computer procedure that can help on generating the modified motions is highly desirable. The idea behind the design of our computer-assisted procedure can be divided into two steps. The first step is motion recognition and the second step is motion generation. In the motion recognition step, the computer is used to recognize the motion that a user intends to perform. If the motion can be recognized and an appropriate animation procedure is also available, then the motion will be generated with the corresponding procedure. Otherwise, the motion mapped directly to the hand gesture will be generated. These two steps will be described in more details in the following subsections.

5 Motion Recognition

The motion recognition module here is used to recognize the motion that the user intends to perform. With this module, the system can ask for the help of the motion generation procedure whenever needed. For example, when the user performs a walking motion with SmartGlove, the computer can be used to automatically generate the motion of swinging legs that are not controlled by the user. Or when a walking motion with the glove is recognized, we can use the procedural animation generator to generate a walking puppet motion for a given set of parameters. However, recognizing a puppet's motion such as walking and jumping by using hand gestures detected by SmartGlove is difficult. Instead, we recognize a puppet's motion based on the sequence of common puppet poses. In the present study, we find that a meaningful motion of a puppet usually consists of a continuous motion connecting two distinct key poses. For example, in shown in Fig. 9, the puppet moves to the right, to the left and with both hands, facing left and right and nodding.

To recognize the feature of a key pose, the rotation value on each joint is inputted into a finite state machine. The rotation value on each joint may trigger the state change of the state machine. When a state machine reaches its final state, the feature is recognized and the values are recorded.

Two types of motions can be recognized by the present recognition system. The first type of motion can be recognized right after a motion is inputted. This type of motion, such as walking, cannot be recognized until a complete sequence of the motion has been perceived. The second type of motion can be recognized after only parts of the motion are inputted. For example, jumping is an example. Once a fast movement from nodding down to nodding up is detected, jumping procedural is followed.

A basic motion for a puppet can be divided into parts. Assume that each part is represented by an alphabet. For example, moving right hand can be divided into parts such as moving right hand in and moving right hand out. We can use 'a' to stand for the motion segment of the right hand in, and 'b' stands for the motion of moving right out. Through this method, each motion can be represented by a finite string (e.g. 'acbd' or 'abde'). Then we can define a similarity function to compute the similarity between the input motion and the motion in the database. The similarity function is computed by comparing the two strings which represents the input and the stored motions. If the similarity value is larger than the pre-defined threshold, the current input motion is substituted by the motion previously stored in the database.

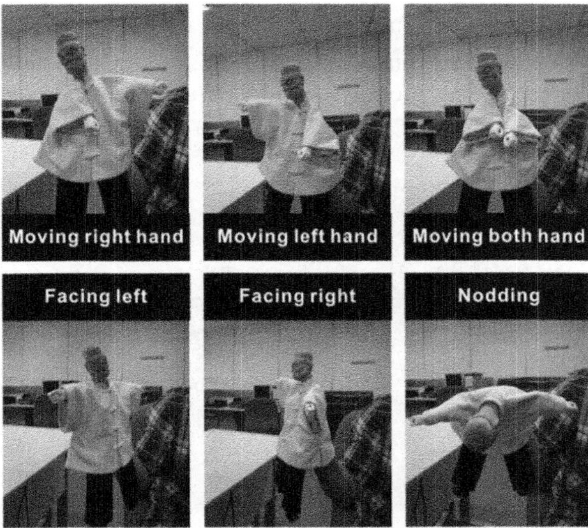

Fig. 9. Six base pose for motion recognition

6 Puppet Procedural Animation

Procedural animation is one type of the knowledge-based animation since it uses the knowledge of glove puppet manipulation to design animation procedures. A simple motion can be divided into multiple phases delimited by key frames with distinct spatial constraints in each phase. Then the motion between key frames can be computed by appropriate interpolation functions satisfying either the temporal or spatial constraints or both. Animations produced by this approach have the advantages of being flexible and computationally efficient for real-time environments. Nevertheless, it is also a great challenge to design a procedure with appropriate parameters that can produce natural and accurate motions.

In procedural animation, different motions require different animation generators. Here the walking motion is analyzed to illustrate how the motions are generated in a procedural manner. Nevertheless, walking can be performed in various ways. As advised by our puppet master, we have chosen the normal walking as an example, as shown in Fig. 10. In this walking motion, the adjustable high-level parameters for the animation procedure include step length, step height, and body swing angle. Four key frames are defined according to the given motion parameters.

The process of a walking animation procedure is carried out by three steps. First, the motions performed by a master puppeteer are recorded and the motion parameters are extracted to describe the characteristics of the motion. Second, the motion is decomposed into several phases by defining key frames to separate these phases according to the above motion parameters. For example, walking animation illustrated in Fig.9 is divided into four phases based on the three aforementioned motion parameters. Phase 1 is determined by the body swing angle; phase 2 is by the step height, phase 3 is by the step length, and phase 4 is by the body swing angle too. Lastly, the procedure for interpolation between key frames is defined. More specifically, two types of interpolations are used to produce plausible motions. The first one is to interpolate along a curve for trajectory following, and the second defines how a motion is timed. In the current system, linear interpolation is used for simple motions while Bezier curve are used for sophisticate motions.

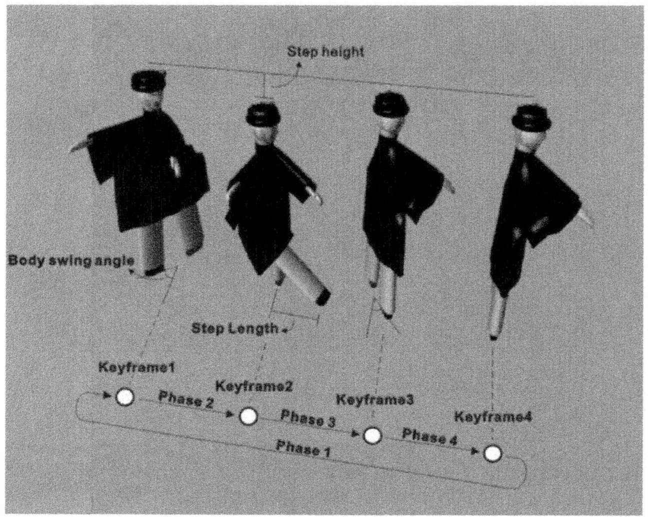

Fig. 10. The keyframes and the procedural parameters for the walking

7 Animation Results

As mentioned in Section 2, the system is constructed based on an IMHAP experimental testbed for procedural animation. The 3D browser is JMonkey. The programming language is Java. The geometry of the virtual characters was modeled with the Alias Maya package. In Fig. 11, the animation controlled by the motion of three fingers is illustrated. Specifically, the thumb controls the animation for the right arm of the character (see Fig. 11(a)); the index finger controls the head animation of the character (see Fig. 11(b)); and the middle fingers control the left arm animation of the character(see Fig. 11(c)). These three controls lead to three featured motions recognized by using the real puppet (see Fig. 9). Each finger motion is recognized by the system and the three corresponding motions are implemented by the procedural animation.

Fig. 12 shows the result of procedural animation by taking into account the input of the finger motions. The user attaches the thumb and the middle finger together and the system recognizes the activation of a specific motion: "moving puppet hand together." After that, the animation system continues to generate the walking motion by swinging the puppet's legs with appropriate motion parameters for each key frame. The user can then abducts the thumb in order to generate the specific animation: "waving right hand". As shown in Fig. 12, the animations for different parts are performed simultaneously to synthesize the final animation.

(a) Controlling the right arm animation through the thumb motion

(b) Controlling the head animation through the index finger motion

(c) Controlling the left arm animation through the middle finger motion

Fig. 11. Control of the puppet motion by three fingers, thumb (a), index finger (b) and the middle finger (c)

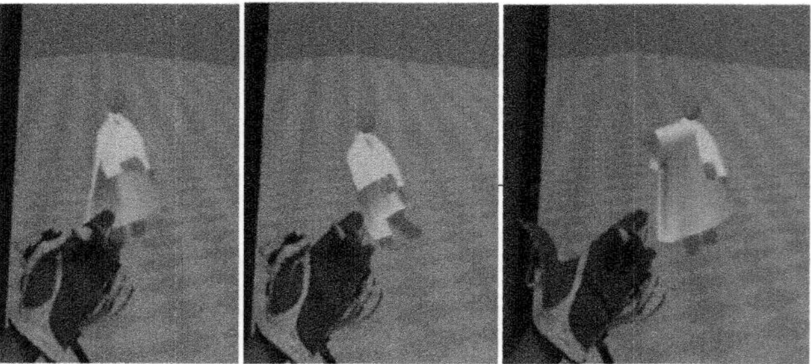

Fig. 12. Walking with hand motion input

8 Conclusion

The present study proposes an interactive character animation system to integrate the procedural animation technique and the input of motion data from the user's hand. The input of the motion data from SmartGlove can help user generate and control the puppet animation in real time. One efficient way is to generate the desired procedural animation is by recognizing the motion based on the stored motion database. When the motion is not recognized, the motion input can still be used to control the specific puppet joint to build the puppet animation or tune the parameters of the existing procedural animation. The gaps between key frames generated by different methods are blended by additional blending frames between motions.

The future work is two-fold. The immediate work is to build a glove puppet animation with personality. This can be done by automatically adjusting the parameters of the motion procedures, such as setting up the dynamic timing of animation to fit the motion feature data. Another work is to help a puppet learner practice during their training. The puppet learner can use SmartGlove to practice their manipulation skills virtually (without wearing the puppet), get the animation feedback, and then try the real puppet after the virtual practice.

Acknowledgement

This work was supported by Media Development Authority, Singapore under NRF IDM004-005 Grant.

References

1. Currell, D.: Puppets and Puppet Theatre. Crowood Press, Wiltshire (1999)
2. Chen, F., Li, T.Y.: Generating Humanoid Lower-Body Motions with Real-time Planning. In: Proc. of Computer Graphics Workshop, Taiwan (2002)
3. Fifth Dimension Technologies, `http://www.5dt.com`
4. Goebl, W., Palmer, C.: Anticipatory Motion in Piano Performance. J. of the Acoustical Society of America 120(5), 3004 (2006)
5. jMonkey Engine, `http://www.jmonkeyengine.com`

6. Kovar, L., Gleicher, M., Pighin, F.: Motion Graphs. In: Proc. of ACM SIGGRAPH (2002)
7. Liang, C.H., Tao, P.C., Li, T.Y.: IMHAP – An Experimental Platform for Humanoid Procedural Animation. In: Proc. of the Third International Conference on Intelligent Information Hiding and Multimedia Signal Processing, Tainan, Taiwan (2007)
8. Lin, J., Wu, Y., Huang, T.S.: Modeling the Constraints of Human Hand Motion. In: Proc. of the Workshop on Human Motion, pp. 121–126. IEEE, Los Alamitos (2000)
9. Lim, K.Y., Goh, Y.K., Dong, W., Nguyen, K.D., Chen, I.-M., Yeo, S.H., Duh, H.B.L., Kim, C.G.: A Wearable, Self-Calibrating, Wireless Sensor Network for Body Motion Processing. In: Proc. of IEEE ICRA, Pasadena, California, pp. 1017–1022 (2008)
10. Mitobe, K., Kaiga, T., Yukawa, T., Miura, T., Tamamoto, H., Rodger, A., Yoshimura, N.: Development of a Motion Capture System for a Hand Using a Magnetic Three-Dimensional Position Sensor. In: ACM SIGGRAPH Research Posters: Motion Capture and Editing, Boston, USA (2006)
11. Perlin, K.: Real Time Responsive Animation with Personality. IEEE Transactions on Visualization and Computer Graphics 1(1), 1–16 (1995)
12. Rhee, T., Neumann, U., Lewis, J.P.: Human Hand Modeling from Surface anatomy. In: Proc. of the 2006 Symposium on Interactive 3D Graphics and Games, Redwood City, California, USA, pp. 27–34 (2006)
13. Wu, Y., Huang, T.S.: Human Hand Modeling, Analysis and Animation in the Context of HCI. In: Proc. of the International Conference on Image Processing., vol. 3, pp. 6–10 (1999)
14. Li, K., Chen, I.-M., Yeo, S.H.: Design and Validation of a Multi-Finger Sensing Device based on Optical Linear Encoder. In: Proc. of ICRA, pp. 3629–3634 (2010)
15. Dipietro, L., Sabatini, A.M., Dario, P.: Evaluation of an Instrumented Glove for Hand Movement Acquisition. J. of Rehabilitation Research and Development 40(2), 179–190 (2003)

Reconstructing Multiresolution Mesh for Web Visualization Based on PDE Resampling

Ming-Yong Pang[1], Yun Sheng[2], Alexei Sourin[2], Gabriela González Castro[3], and Hassan Ugail[3]

[1] Department of Educational Technology, Nanjing Normal University, China
[2] School of Computer Engineering, Nanyang Technological University, Singapore
[3] Centre for Visual Computing, University of Bradford, UK
panion@netease.com, {shengyun,assourin}@ntu.edu.sg, {g.gonzalezcastro1,h.ugail}@bradford.ac.uk

Abstract. Various Partial Differential Equations (PDEs) have been used in computer graphics for approximating surfaces of geometric shapes by finding solutions to PDEs, subject to suitable boundary conditions. The PDE boundary conditions are defined as 3D curves on surfaces of the shapes. We propose how to automatically derive these curves from the surface of the original polygon mesh. Analytic solutions to the PDEs used throughout this work are fully determined by finding a set of coefficients associated with parametric functions according to the particular set of boundary conditions. When large polygon meshes are used, the PDE coefficients require an order of magnitude smaller space compared to the original polygon data and can be interactively rendered with different levels of detail. It allows for an efficient exchange of the PDE shapes in 3D Cyberworlds and their web visualization. In this paper we analyze and formulate the requirements for extracting suitable boundary conditions, describe the algorithm for the automatic deriving of the boundary curves, and present its implementation as a part of the function-based extension of VRML and X3D.

Keywords: partial differential equations, surface modeling, surface approximation, 3D reconstruction, web visualization.

1 Introduction

Complex geometric shapes are often represented as large polygon meshes which require a considerable amount of memory for their handling and also have a fixed Level of Detail (LOD). It is therefore problematic to use these shapes in 3D shared virtual worlds due to the Internet bandwidth limitation. To address this problem, we previously proposed to replace polygon meshes with models obtained from the use of Partial Differential Equations (PDEs) generally represented by relatively small sets of PDE coefficients [1]. Mathematically, given a 3D geometric shape s defined by vertices $P = \{p_i \in R3 \mid 1 \leq i \leq Mp\}$ on its surface, we derive a 3D geometric shape S

M.L. Gavrilova et al. (Eds.): Trans. on Comput. Sci. XII, LNCS 6670, pp. 36–55, 2011.

which is a close parametric approximation of the shape s. The shape S is defined by a group of PDEs with M_c coefficients such that $M_c << M_p$. However, in [1] the required boundary conditions were extracted from the corresponding original polygon mesh semi-automatically which delayed the reconstruction of the resulting PDE shape.

We have also previously proposed how to exchange and visualize PDE-defined shapes in 3D shared virtual worlds using the function-based extension of VRML and X3D [2, 3]. The proposed methods were based either on direct writing of the PDE solutions (parametric equations) straight in the VRML/ X3D source code or on more complex computational procedures which were plugged-into VRML/X3D browser as binary DLL libraries. However, it required additional design and development to be able to interactively render complex PDE shapes with branches and irregular and sharp geometric details.

The paper is organized as follows. Section 2 summarizes our previous work on patchwise approximation of large polygon meshes and formulates the requirements to the algorithm for automatically deriving PDE boundary curves. In Section 3, we propose how to automatically derive these curves as boundaries outlining curved patches on the surface of the original polygon mesh. In Section 4, we discuss some practical issues of exchanging and visualizing complex PDE shapes in 3D shared virtual worlds based on VRML and X3D. Section 5 concludes the work and outlines the future directions of our research.

2 A PDE for Patchwise Approximation of Large Polygon Meshes

Bloor *et al* [4] pioneered the research on using PDEs in computer graphics by producing a parametric surface $S(u, v)$ defined as a solution to an elliptic fourth order PDE:

$$\left(\frac{\partial^2}{\partial u^2}+a^2\frac{\partial^2}{\partial v^2}\right)^2 S(u,v)=0 \tag{1}$$

where $a \geq 1$ is a parameter that controls the relative rates of smoothing between the u and v parameter directions. The partial differential operator in Eqn. (1) represents a smoothing process in which the value of the function at any point on the surface can be understood as a weighted average of the surrounding values. Thus, a single PDE patch of a surface is obtained as a smooth transition between the boundary conditions.

In [1], we used a patchwise PDE method which differed from other PDE methods in the configuration of the PDE patches. Commonly, the PDE boundary curves are extracted in a circular though not necessarily co-planar way around the object so that the whole reconstructed object is a single entity in its parametric space, and that each PDE patch is adjacent up to two other patches (Fig.1a). In the patchwise PDE method, each shape is represented by patches with their own parametric coordinate systems (Fig. 1b). This configuration allows for representing shapes with branches and for preservation of irregular and sharp details on the object surface by matching the respective patches with various sizes and orientations to the surface.

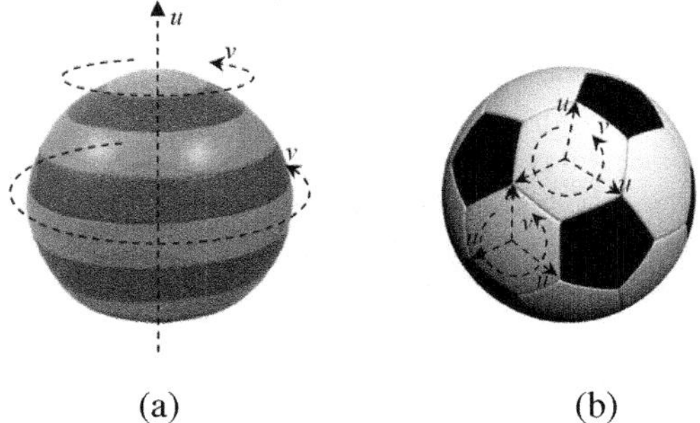

(a) (b)

Fig. 1. Illustration of PDE patch configurations: (a) Conventional and (b) the patchwise configurations

Each PDE patch $S(u, v)$ can be formulated by Eqn. (1). Assuming that the effective region in the uv space is restricted to $0 \le u \le 1$ and $0 \le v \le 2\pi$, and using the method of separation of variables, the analytic solution to Eqn. (1) is generally given by

$$S(u,v) = A_0(u) + \sum_{n=1}^{\infty}[A_n(u)\cos(nv) + B_n(u)\sin(nv)] \tag{2}$$

where

$$A_0(u) = \alpha_{00} + \alpha_{01}u + \alpha_{02}u^2 + \alpha_{03}u^3 \tag{3}$$

$$A_n(u) = \alpha_{n1}e^{anu} + \alpha_{n2}ue^{anu} + \alpha_{n3}e^{-anu} + \alpha_{n4}ue^{-anu} \tag{4}$$

$$B_n(u) = \beta_{n1}e^{anu} + \beta_{n2}ue^{anu} + \beta_{n3}e^{-anu} + \beta_{n4}ue^{-anu} \tag{5}$$

and vector-valued PDE coefficients α_{00}, α_{01}, …, α_{n3}, α_{n4} and β_{11}, β_{12},…, β_{n3}, β_{n4} are determined by the boundary conditions. In Eqn. (3), $A_0(u)$ takes the form of a cubic polynomial curve with respect to u. In Eqn. (2), the term $A_0(u)$ is regarded as the "spine" of the surface, while the remaining terms represent a summation of "radius" vectors that give the position of $S(u, v)$ relative to the "spine". Therefore, the PDE surface patch $S(u, v)$ may be pictured as a sum of the "spine" vector $A_0(u)$, a primary "radius" vector $A_1(u)\cos(v) + B_1(u)\sin(v)$, a secondary "radius" vector $A_2(u)\cos(2v) + B_2(u)\sin(2v)$ attached to the end of the primary "radius", and so on. The amplitude of the "radius" term decays as the frequency increases. It can be observed that the first few "radii" contain the most essential geometric information while the remaining ones can be neglected. Therefore, Eqn. (2) can be rewritten as:

$$S(u,v) = A_0(u) + \sum_{n=1}^{N}[A_n(u)\cos(nv) + B_n(u)\sin(nv)] \tag{6}$$

where only N "radii" are selected and the number of vector-valued PDE coefficients is M_c=4×(2N+1). Strictly speaking, this equation needs a remainder term to guarantee that the boundary conditions are satisfied. The boundary conditions imposed on Eqn. (6) take the following forms:

$$S(0,v) = C_0(v) \quad S(u_1,v) = C_1(v)$$
$$S(u_2,v) = C_2(v) \quad S(1,v) = C_3(v) \tag{7}$$

where C_0, C_1, C_2 and C_3 are isoparametric boundary curves on the surface patch at $u = 0$, u_1, u_2, and $u = 1$, respectively, and $0<u_1, u_2<1$. Fig.2 illustrates the layout of the PDE boundary curves for an individual PDE patch, where C_0 degenerates into one point.

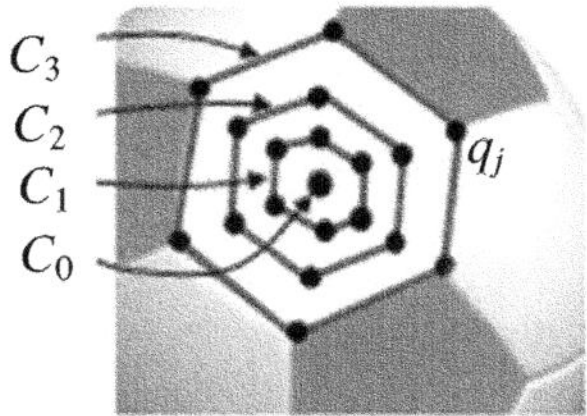

Fig. 2. Boundary curves for an individual PDE patch

Theoretically, the PDE patch can be of any shape provided:

1. It can be mapped from a planar *u,v* domain to the 3D surface,
2. The sampling points of its boundary curves can be derived from the surface of the object; and
3. The PDE patches can be blended together seamlessly.

In [1] we proposed how to blend and render the patches. Here we formulate the requirements to the automatic algorithm for deriving the boundary curves from the original polygon mesh to be approximated by PDEs. We aim to achieve a reasonable compromise between the precision of the approximation and the number of PDE coefficients which still has to be significantly smaller than the number of vertices in the original polygon mesh $M_c << M_p$.

If we use points $Q = \{q_j \in \mathbf{R}^3 \mid 1 \leq j \leq M_q\}$ to define the sampling points of these boundary curves, then we can write C_0, C_1, C_2 and $C_3 \in Q$. In order to derive the boundary curves following the original polygon mesh as precise as possible, the automatic boundary curve deriving algorithm should utilize as much as possible of the vertices of the original polygon mesh, $P = \{p_i \in \mathbf{R}^3 \mid 1 \leq i \leq M_p\}$. Additional points can be interpolated from points M_p. The number of q_j, i.e. M_q, defines the precision of the approximation.

The compression rate δ brought by the patchwise PDE method is defined as the ratio of the total size of vertices of the original shape against the total size of PDE coefficients used for the original shape, which is of the form

$$\delta = \frac{total\ size\ of\ vertices}{total\ size\ of\ coefficients} \tag{8}$$

Assume that each vertex consumes the same size as one PDE coefficient. Then, Eqn. (8) can be rewritten approximately as the ratio of the sum of the original vertices represented by each PDE patch over that of coefficients of each PDE patch,

$$\delta \cong \frac{\sum_{L} \textit{vertices per patch}}{\sum_{L} \textit{coefficients per patch}} \tag{9}$$

where L indicates the PDE patch number of a shape.

Now we will discuss to what extent we can achieve the compression rate.

Ideally, we expect $Q \subseteq P$. As discussed in [1], in such cases we can achieve a compression rate

$$\delta \geq \frac{3N_b + 1}{4 \times (2N + 1) + H} \tag{10}$$

where N_b, N and H symbolize the number of the sampling points in each boundary curve, the number of "radii" in Eqn. (6), and the number of what we call "position fixers" for each PDE patch, respectively. As Eqn (6) is a spectral approximation of Eqn (2), the position fixers, which are a group of spatial points automatically picked out of C_3, are introduced to prevent the PDE patch from distortion. The position fixers are particularly useful when blending PDE patches. During the PDE patch reconstruction the position fixers are used to help linearly interpolate the vertices on the outermost boundary curve of the PDE patch. To that end, in order to make the resulting PDE shape look more realistic, there must be a reasonable number of position fixers provided subject of the object surface smoothness.

A minimum value of δ is obtained when all the M_p mesh vertices are used as the boundary curve sampling points. Theoretically, it only occurs when the geometry of the original shape is too irregular so that we need a large number of PDE patches with all the vertices of the original mesh representation employed forming the boundary curves in order to preserve the original geometry, or when there is only one PDE patch required with all the vertices of its original mesh used as the boundary curves. The compression rate in such scenarios is not associated with the patch number as all the vertices are used up.

Besides the discussed case of $Q \subseteq P$, the boundary curve sampling points to be extracted often outnumber the existing vertices of the original mesh when handling a shape with complex features, i.e. $Q \not\subset P$. Thus, the compression rate in such cases can still be formulated using Eqn.(9), but it follows the constraint

$$\delta < \frac{3N_b + 1}{4 \times (2N + 1) + H} \tag{11}$$

which is not difficult to deduce.

Generally, the compression rate and precision which the patchwise PDE method offers are subject to several factors:

(a) Mesh partitioning,
(b) Deriving of boundary curves,
(c) The number of frequency modes, or "radii" as called before, N,
(d) The position fixers required, and
(e) The geometric complexity of the shape.

As we know that the PDE method is a boundary value approach, mesh partitioning, as well as deriving of the PDE boundary curves, plays a crucial role in the developed patchwise method. Mesh partitioning involves a process which determines how many PDE patches will be used to represent the original shape according to its geometric information on the surface, while boundary curve deriving refers to an optimal extraction of geometric information. In our previous work [1], a semi-automatic approach was utilized for partitioning the mesh and deriving the boundary curves. As a result, user interaction and off-line commercial software are demanded at certain processing stages. For example, for a shape with a flat or smooth surface the algorithm should use less PDE patches, whereas a sharp and detailed surface requires more PDE patches. However, to the best of our knowledge there is no such a partitioning algorithm satisfying the above demand. Moreover, an ideal boundary curve deriving process should sample boundary curve points straight from the vertices of mesh grid of the original shape, i.e. $Q \subseteq P$. Nevertheless, in most cases the sampling points outnumber the existing vertices of the original mesh when handling a shape with complex features. In [1], we simply linearly interpolated the vertices extracted out of the original shape. However, this may cause a loss of general geometric information of the original shape. Therefore, a more sophisticated solution has to be sought, which should sample as many boundary curve points as possible from the original shape surface in order to minimize the loss of geometric information.

It is worth mentioning that Eqn. (6) expresses a spectral approximation to the PDE solution (Eqn. (2)), where the first N low frequency modes or "radii" as called previously, contain the most essential energy or geometric information for reconstruction, while the following high frequency modes are selectively discarded. This spectral property of the PDE method is the one we explore here, since a proper selection of N not only affects the accuracy of the total information preserved, but also determines the storage cost of the patchwise PDE method. These two factors form the cornerstones and are of primary concern when we study and test the effectiveness of our algorithm in exchanging data in 3D shared virtual worlds.

As we studied, for each PDE patch we need $4\times(2N+1)$ vector-valued coefficients to approximate its solution in the patchwise PDE method. Accuracy, as well as the storage cost of each patch approximation, is affected by the number of N low frequency modes. Generally speaking, the amplitude of the nth mode decays by a factor e^{-1} over a length scale $O[(1/an)]$, while the approximation with n frequency modes demands 8 more vector-valued coefficients than the approximation with $(n-1)$ frequency modes to the PDE solution.

As the presented patchwise PDE method adopts a spectral approximation to its PDE solution, it describes 3D geometric information of original shapes in a lossy compression sense. It is worth noting that the developed method is only concerned with the preservation of geometry data, and that connectivity data are not retained. Thus, it is understandable that the PDE method may end up with a topological change of the mesh.

Moreover, our scheme can also be implemented in a progressive manner in order to handle massive geometry data transmission in 3D shared virtual worlds. For instance, assuming that the number of PDE patches remains the same, a rough approximation of the shape may be represented using a few low frequency modes with a small number of PDE coefficients. It can be progressively refined by increasing gradually the number of frequency modes.

3 Automatic Deriving of the Boundary Curves

In this section, we present a mesh decimation-based method that can automatically derive PDE boundary curves from a triangular mesh with a specified decimation level. Our method first decimates the initial mesh to a simplified mesh at a certain level with a lesser number of vertices, and then, partition the initial mesh based on the simplified mesh. The method re-samples and calculates the boundary curves for each patch, with technical maneuvers, such as geodesics computation, mesh subdivision and re-sampling , *etc*.

In the past two decades, there were many algorithms proposed to simplify large triangular meshes. Schroeder et al. [5] proposed a mesh decimation method based on removing vertices from the dense mesh. The method first evaluates the importance of each vertex in the current mesh model and then selects and removes the vertex with the least importance. The hole left by the removing operation is subsequently triangulated. Hoppe et al. [6] presented a mesh simplification algorithm based on edge collapse. The algorithm first evaluates the costs of collapsing for all edges in the initial mesh, and then uses the energy optimization technique to compute a new position for the collapsed edge in each decimation step. Rossignac et al. [7] presented a vertex clustering method to simplify a given mesh model. Their algorithm first obtains the bounding box of mesh models to be simplified, and then adaptively subdivides the box into a set of cells. Collecting the vertices in a single cell and defining a new substitute vertex for them, the initial mesh model can be simplified. With advances of digital geometry processing, some new technologies, such as re-meshing and mesh parameterization, have also been used in constructing simple models. Turk [8] presented a physically based method to optimize shape representation by retiling the surface model with a given number of vertices, according to the local curvature information on the model. However, this method can only deal correctly with models with sharp features. Lounsbery [9] introduced the wavelet based signal processing method to the fields of geometry processing. This method decomposes a given geometric shape to the combination of a series of signals with various frequencies, where the low-frequency components define the base shape of the model and the high-frequency components represent the details of the shape. Such a model can be simplified just by removing the high-frequency components from the model. However, this method can only handle the mesh models with subdivision connectivity.

3.1 Building Base Mesh Using Decimation

First, we consider, as a foundation of our method, the work introduced by Garland et al. in [10]. Their surface simplification algorithm can rapidly and efficiently produce high quality approximations of complex polygonal models by using iterative contractions of vertex pairs in models while maintaining surface error approximations using quadric matrices. By contracting arbitrary vertex pairs (not just edges), the algorithm is able to join unconnected regions of models. However, Garland's method has two weak points for our subdivision connectivity mesh construction. One is that it introduces many new vertices in the simplified mesh. The other is that it cannot keep the surface topology of the simplified mesh consistent with that of its original mesh.

To solve the first problem, our algorithm uses a special edge collapse as an atomic operator of mesh decimation rather than the edge collapse employed in the original algorithm. As shown in Fig.3, when the edge v_0v_1 is collapsed, there is no new vertex created and the edge collapses to one of its two ends, v_0 or v_1, which leads to a smaller simplified error.

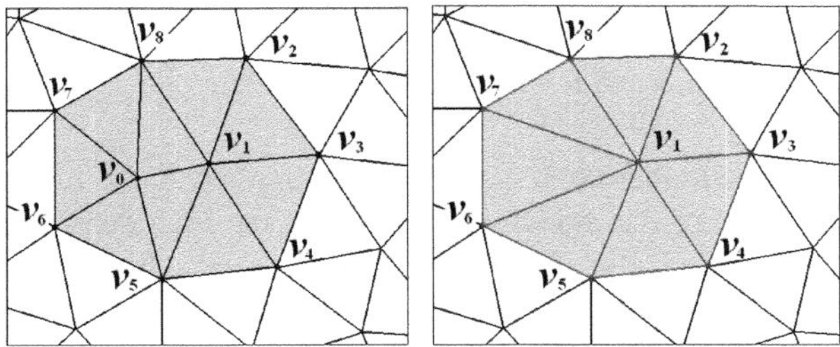

Fig. 3. Half-edge collapse operation

To solve the second problem, the algorithm imposes a rigorous restriction on the edge collapse scheme to keep the simplified surface topology consistent with the initial surface when the decimation operation is performed. In Fig. 3, once the edge v_0v_1 is collapsed to vertex v_1, all the highlighted edges of the facets neighboring to v_1 are "frozen", i.e., these edges cannot be collapsed anymore until all the edges in the current model are checked. Once all the edges are checked, we say that the algorithm finishes a level of decimation. At this time, the algorithm unfreezes all the edges in the current mesh and starts its decimation work towards the next level. This decimation scheme can keep the surface topology preserved and can distribute the triangles of the initial complex mesh uniformly.

It is important to decide which edge to be collapsed in each level of decimation. In [10], Garland et al. selected the edge with a smallest decimation error to perform the atomic operator. They used the quadric error metrics to evaluate the error for each edge in the current mesh. In our work, we use a light technique presented in [10] to evaluate collapse errors for the edges. The basic idea behind the technique is to use fewer triangles in flat areas on mesh model while more triangles in the areas with a higher curvature. So, our algorithm evaluates the curvature of an edge as:

$$C = \max_{T_i \in T(u)} \{ \min_{T_j \in T(u,v)} \{(1 - T_i.norm \cdot T_j.norm)/2\} \} \tag{12}$$

where, $T(\boldsymbol{u})$ is the set of triangles neighboring the vertex $\boldsymbol{u}$, $T(\boldsymbol{u},\boldsymbol{v})$ is the set of triangle neighboring both the vertices $\boldsymbol{u}$ and $\boldsymbol{v}$, and $T_i.norm$ is the unit normal of facet T_i.

Because $T_i.norm \cdot T_j.norm$ is equal to $\| T_i.norm\| \cdot \| T_j.norm \| \cdot \cos\alpha$, where α is the angle between $T_i.norm$ and $T_j.norm$, we have

$$C = \max_{T_i \in T(u)} \{ \min_{T_j \in T(u,v)} \{(1-\cos\alpha)/2\}\} \tag{13}$$

for $\| T_i.norm \| \cdot \| T_j.norm \| = 1$, and we further have

$$C = \max_{T_i \in T(u)} \{ \min_{T_j \in T(u,v)} \sin^2 \frac{\alpha}{2}\} \tag{14}$$

The length of each edge should also be considered as an important factor of the edge in mesh, so the error or weight to collapse the edge can be defined as

$$weight(\boldsymbol{u},\boldsymbol{v}) = \| \boldsymbol{u}-\boldsymbol{v} \| \times \max_{T_i \in T(u)} \{ \min_{T_j \in T(u,v)} \sin^2 \frac{\alpha}{2}\} \tag{15}$$

On the other hand, the higher degree a vertex has in a mesh, the more power it has to control the local shape of the model, i.e., removing the high degree vertex with edge collapse can cause big change to the model appearance. So our algorithm should prevent the vertices with higher degrees from being deleted during simplification. We can therefore rewrite the weight of edge collapse as

$$weight(\boldsymbol{u},\boldsymbol{v}) = \| \boldsymbol{u}-\boldsymbol{v} \| \times \max_{T_i \in T(u)} \{ \min_{T_j \in T(u,v)} \sin^2 \frac{\alpha}{2}\} \times D^2 \tag{16}$$

where, D is the degree of vertex $\boldsymbol{u}$ or $\boldsymbol{v}$ to be deleted.

A heap data structure is used in our method to manage the errors of edges for rapidly accessing the edge with the smallest error. Once an edge is collapsed, the errors of edges related to the collapsed edge are updated immediately and the heap is also adjusted at the same time.

Since the half-edge collapse operator always contracts an edge to one of its ends, it can easily create triangles with a bad aspect ratio. These thin triangles can result in a low quality mesh. Thus, we should find a method to restrain the thin triangles. For this purpose, we introduce a shape measure of the triangles:

$$Q = \frac{4 \cdot a \cdot \sqrt{3}}{l_0^2 + l_1^2 + l_2^2} \tag{17}$$

where, a and $l_i (i = 0,1,2)$ are the area and edge lengths of the triangle, respectively. Q is 1 for an equilateral triangle while 0 when the triangle is degenerated. Hence, we can define a threshold degree for the triangles. If Q is less than this threshold the half-edge collapse operation has to be canceled.

As mentioned above, most of the existing mesh decimation algorithms are unable to preserve mesh topology. For example, Garland's mesh decimation algorithm can simplify an initial manifold surface to a non-manifold one [10]. It may create problems in sampling PDE boundary curves. In our method, we impose a criterion to prevent topological change. Before an edge is collapsed, we check if it satisfies the condition, otherwise the edge collapse will be refused.

The criterion is as follows: Without loss of generality, suppose the edge to be collapsed is v_0v_1, and $N(v_0)$ and $N(v_1)$ are the respective sets of neighboring vertices.

If $\| N(v_0) \cup N(v_1) \| < (\deg(v_0) + \deg(v_1) - 2)$, where $\deg(v_i)$ $(i = 0,1)$ are the degrees of v_0 and v_1, and $\| S \|$ is the number of the components in Set S, we consider the edge collapse may change the topology of the mesh surface. In Fig. 4, two simplified polygon meshes built with our algorithm are illustrated.

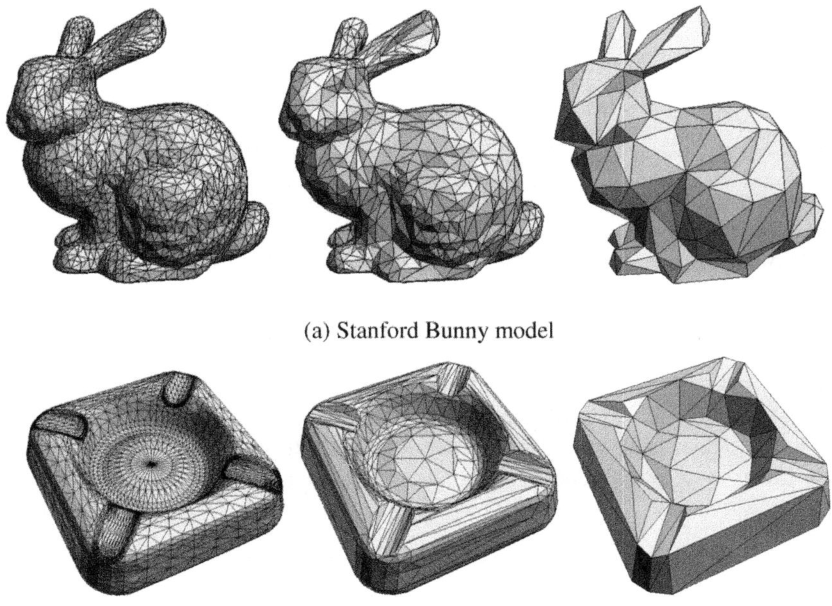

(a) Stanford Bunny model

(b) Ashtray model (http://www.oyonale.com).

Fig. 4. Examples of the simplified models created by our decimation algorithm

3.2 Parameterizing Patches

Given two arbitrary surfaces with similar topology, it is possible to find a one-to-one mapping between them [11]. If one of these surfaces is represented by a triangular mesh, the problem of computing such a mapping is referred to as mesh parameterization [12]. The surface which the mesh is mapped to is typically referred to as the parameter domain. The input surfaces which are homeomorphic to a disk can be directly parameterized on a planar domain. Many planar paramerterization methods have been proposed ever since Tutte's theorem [14]. Although all these methods can be used for this paper, we preferred Floater's mean-value theorem [13] because of its angle preserving and robustness.

Let $P = (V, E, F)$ be a local mesh patch with the vertex set $V = \{\boldsymbol{v}_i = (x_i, y_i, z_i), 1 \le i \le N\}$, where $\boldsymbol{v}_i (i = 1,...,n)$, $(n < N)$ are the internal vertices and $\boldsymbol{v}_{n+1}, \boldsymbol{v}_{n+2}, ..., \boldsymbol{v}_N$ are the boundary vertices in any counter-clockwise sequence, E and F are edge and facet set of the patch, respectively. We now choose some points, $\boldsymbol{p}_{n+1}, \boldsymbol{p}_{n+2}, ..., \boldsymbol{p}_N$, to be the vertices of any planar triangle $D \subset \boldsymbol{R}^2$ in an counter-clockwise sequence. For each $\boldsymbol{v}_i$ $(i = 1,..., n)$, we choose a set of real numbers $\lambda_{i,j}$ for $\boldsymbol{v}_i (i = 1,..., N)$, such that

$$\lambda_{i,j} = \begin{cases} 0 & edge(\boldsymbol{v}_i \boldsymbol{v}_j) \in E \\ > 0 & edge(\boldsymbol{v}_i \boldsymbol{v}_j) \notin E \end{cases} \tag{18}$$

and

$$\sum_{j=1}^{N} \lambda_{i,j} = 1 \tag{19}$$

Further, we define $\boldsymbol{p}_1, \boldsymbol{p}_2, ..., \boldsymbol{p}_n$ ψto be the solutions of the linear system of equations,

$$\boldsymbol{p}_i = \sum_{j=1}^{N} \lambda_{i,j} \boldsymbol{p}_j, \ i = 1,..., n \tag{20}$$

In order to establish a one-to-one mapping from a local mesh patch on the original mesh to the planar parameter domain, i.e., the planar triangle, we need to solve a large sparse linear system with coefficients, i.e. an equation of the form, derived from Eqn. (20):

$$\boldsymbol{AX} = \boldsymbol{b} \tag{21}$$

where, $\boldsymbol{A}$ is a $n \times n$ non-singular square matrix, $\boldsymbol{b}$ is a vector of dimension n, and $\boldsymbol{X}$ is the unknown vector of dimension n we need. Many solver tools can be used for doing it, such as Successive *Over-Relaxation*, *Bi-Conjugate Gradient* [18], and Matlab API functions. Now we proceed to fix the boundary vertices of the original local mesh patch to the boundary of the planar triangle. Then, all "inner vertices" of the patch and all "inner points" of the triangle are a one-to-one mapping through solving a sparse linear system, Eqn.(21).

3.3 PDE Patch Partitioning

Once a simplified version of the initial dense mesh is derived, we can partition the initial mesh into patches with assistance of the simplified mesh, and subsequently, parameterize all the patches onto a planar triangle in a plane. Based on the patch parameterization method mentioned above, partitioning PDE patches becomes straightforward.

Because there are no new vertices introduced in the simple mesh during decimation, it is straightforward to find correspondence of the vertices between the initial and simplified meshes. This gives us an easy way to partition the initial mesh according to the simplified mesh. We take the initial mesh as a graph embedded in a 3D space. For each edge of the triangular facets in the simplified mesh, we can use the Dijkstra algorithm to find the consecutive edges of the initial complex mesh to connect the ends of the edge [15]. However, the consecutive edges do not form the shortest path between the two vertices of the initial mesh model. The shortest path should be a geodesic path, which typically cuts across the facets in the mesh and therefore cannot be found by the traditional graph-based Dijkstra algorithm for the shortest paths. In order to extract the boundary contour of each patch of the initial mesh accurately, we employ a method of geodesics calculation to find an optimized curve for the patch boundary contour.

The exact geodesic algorithm was described by Chen et al. in [16] and partially implemented by Kaneva et al. [17]. The algorithm has a complexity of $O(n^2)$. Another algorithm with $O(n \log^2 n)$ time for the "single source, single destination" geometric path between two given mesh vertices was described by Kapoor [18]. However, this method is too complicated to use. There are also some other approximating algorithms [19] with pre-given error bounds where some extra edges are added and the traditional Dijkstra algorithm is employed to obtain the approximating shortest paths. Martinez et al. [20] presented an iterative optimization algorithm, but its results significantly depend on the initial approximation path. A detailed survey of approximation algorithms for graph and geodesic search can be found in [21].

To calculate geodesics on the mesh model, we use Surazhsky's implementation of the continuous Dijkstra-like algorithm presented in [22]. In the algorithm, the shortest paths can be visualized as rays cast from the source vertex in all tangent directions. Within a triangle, the shortest path is a straight line. When crossing an edge, the shortest path must correspond to a straight line if the two adjacent facets are unfolded into a common plane. The basic idea of the algorithm is to track together groups of the shortest paths that can be parameterized atomically and this is achieved by partitioning each mesh edge into a set of intervals called windows. The windows are then propagated across the facets of the mesh in Dijkstra-like sweep. In order to optimize the performance of the algorithm, all propagating windows are organized as a wave-front by ordering them in the queue according to minimal distances from the source vertex.

Once all the edges are covered by the windows representing geodesic distance, it is very easy to trace the shortest path from any surface point to the source by back-tracing. For each triangular facet of the simplified mesh, it is easy to compute the PDE patch boundary contour based on the initial mesh by calculating the geodesics between vertices of the facet. The examples of the curve patches created by calculating geodesics are given in Fig.5.

Fig. 5. PDE patches created by calculating geodesics from the simplified Stanford Bunny model with 200 triangles

3.4 Sampling PDE Boundary Curves

Now, since we have partitioned the PDE patches based on the initial mesh, we need to sample the PDE boundary curves for each patch. Our method first defines a set of isometric polylines as parametric curves in the triangular facets of the simplified mesh, as illustrated in Fig. 6.

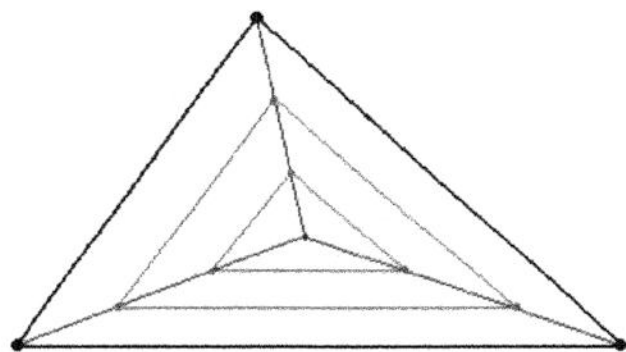

Fig. 6. Boundary curves in triangular domain

Next, the algorithm samples the polylines and maps the sampling points onto the surface of the initial mesh via the patch parameterization. For example, let $\boldsymbol{p}$ be an arbitrary sampling point on the parametric curves and $\boldsymbol{p}_i$, $\boldsymbol{p}_j$, $\boldsymbol{p}_k$ are its three nearest points in the planar parameterized mesh. It is not difficult to evaluate the barycenter coordinates, c_1, c_2, c_3, of $\boldsymbol{p}$ with respect to $\mathbf{p}_i, \mathbf{p}_j, \mathbf{p}_k$, that is,

$$\mathbf{p} = c_i\mathbf{p}_i + c_j\mathbf{p}_j + c_k\mathbf{p}_k \tag{22}$$

Obviously, $\boldsymbol{p}_i$, $\boldsymbol{p}_j$, $\boldsymbol{p}_k$ are the parametric points of the mesh vertices $\boldsymbol{v}_i, \boldsymbol{v}_j, \boldsymbol{v}_k$. Thus, it is easy to calculate the 3D position $\boldsymbol{v}$ on the original mesh for the parametric point $\boldsymbol{p}$, that is,

$$\boldsymbol{v} = c_i\boldsymbol{v}_i + c_j\boldsymbol{v}_j + c_k\boldsymbol{v}_k \tag{23}$$

All the boundary curve data including the PDE patch boundary contours extracted early, are stored in a file which is then processed by the PDE engine for PDE coefficient calculation and rendering.

In Fig.7, we illustrate two examples of the boundary condition curves created by our algorithm.

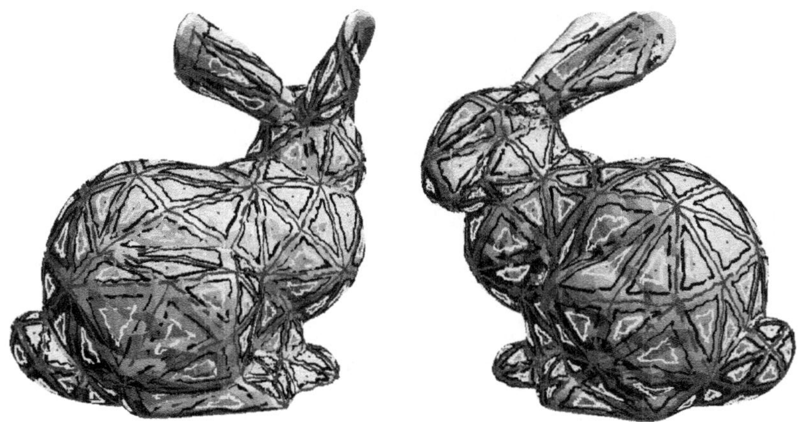

Fig. 7. PDE boundary curves of the Stanford Bunny model

3.5 Subdivision Connectivity Polygonization

With the boundary curves sampled, our algorithm automatically solves Fourier coefficients of the PDE patches according to Eqn. (3)-(5). Once the coefficients are derived, the PDE solution in Eqn. (6) naturally gives a parameterization of each patch. It also provides a convenient way for polygonization to visualize the patches.

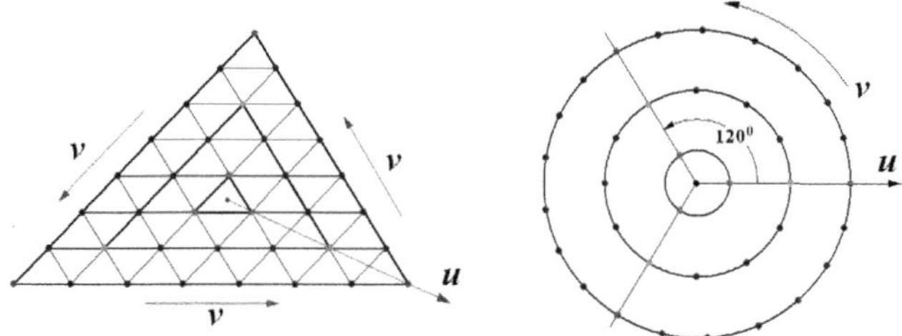

Fig. 8. Defining parametric (u,v)-mask for vertices of polygonization mesh of a single patch: Polygonization mask (left) and vertex parametric points on a polar parametric plane (right)

For all the patches, our algorithm defines a uniform mask of subdivision connectivity polygonization with a user-defined resolution. The mask allows for keeping the same mesh topology for patches. In Fig. 8, we illustrate how the mask is defined. In the mask, some edges consist of loop curves around the center of the base triangle of the patch. The loop curves are defined as a set of concentric circles on the polar parameter plane. For each edge of the base triangle, the vertices are equally-sampled on a one-third arc of the parametric circles. Once the (u, v) parametric coordinates are derived, their 3D Cartesian geometric coordinates can easily be computed according to Eqn. (6). Examples of such isoparametric curves on PDE patches are given in Fig. 9.

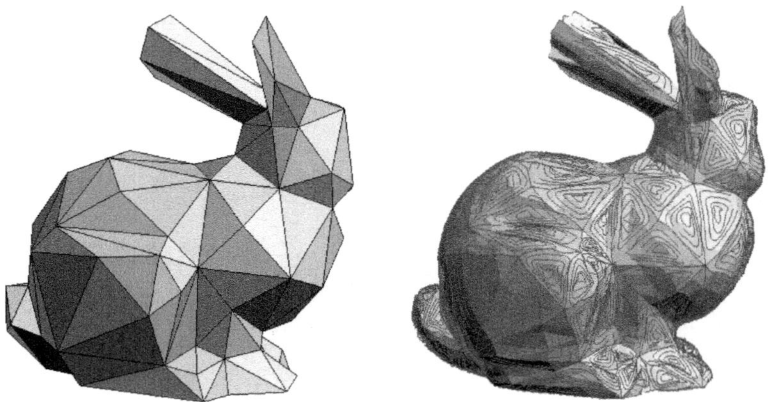

Fig. 9. Iso-parametric curves on PDE

4 Implementation in 3D Shared Worlds

We have implemented the algorithm using C++ and OpenGL library on a PC. In Fig. 4, Stanford Bunny and the ashtray models are simplified for constructing the base meshes. Fig. 5 illustrates the boundary curves of patches created from the simplified Bunny base mesh by computing geodesics. A sampling example of boundary curves is show in Fig.7. Once the PDE patches are constructed from the original models, we can calculate isoparametric curves on the patches for polygonizing the surface with subdivision connectivity. In Fig. 9, a base mesh and the iso-parameter curves are illustrated. The iso-parametric curves are computed from the PDE patches by Eqn. (6). In Fig. 10, three subdivision connectivity models with different resolutions are finally constructed. The experiments illustrate that our method can create connectivity meshes not only from dense mesh models, but also from very simple models, e.g., a simple cube mesh in Fig. 10. We also compare the PDE reconstruction of a sphere using the developed automatic decimation algorithm with a result using a perfect manually polygonized base mesh. As shown in Fig. 11, the upper row illustrates the original shape of the sphere, the base mesh after decimation and the PDE reconstruction with flat shading, while the lower row shows the elaborately created base mesh with a regular grid and its PDE reconstruction. It can be seen that resolutions of the two base meshes are manually set to be close enough that the size of the triangles within both the meshes are roughly equal. The same procedure including patch parameterization (Section 3.2), PDE patch partitioning (Section 3.3) and boundary curve sampling (Section 3.4), and polygonization (Section 3.5), is carried out with both the base meshes. The figure shows the developed decimation algorithm can generate a satisfactory precision as using the perfectly polygonized base mesh.

We have also implemented the proposed algorithms as a part of the function-based extension plug-in to VRML and X3D [23]. The automatic boundary curve deriving software is a standalone program which has to be run first. It reads polygon mesh data files (e.g. in the commonly used .OBJ data format) and creates data files with the coordinates of the boundary curves. The user can only define the degree of representation which will result in various numbers of the boundary curves derived. These curve files serve as an input data for the PDE shape defined in VRML/X3D and they can be located either on a client computer or on a web server. When the VRML/X3D code is run, it will first detect whether the PDE plug-in is installed. If the case, it will process the PDE shape definition given by the PDE curves file name and a few rendering parameters. The PDE plug-in software is implemented as two binary DLL library functions. The first one is be called by the VRML/X3D viewer only once to read the PDE curve data files and to calculate the PDE coefficients. The calculation time depends on the number of the PDE curves read and in many cases is performed at interactive rates. After the PDE coefficients are calculated, they are stored in a file on the hard drive which will be read by another DLL function that will perform rendering of the PDE shape. This function calculates the polygons interpolating the PDE shape and sends them into the VRML/X3D rendering pipeline making the PDE shape a legitimate part of the shared virtual scene that can be further treated as any other standard VRML/X3D shapes.

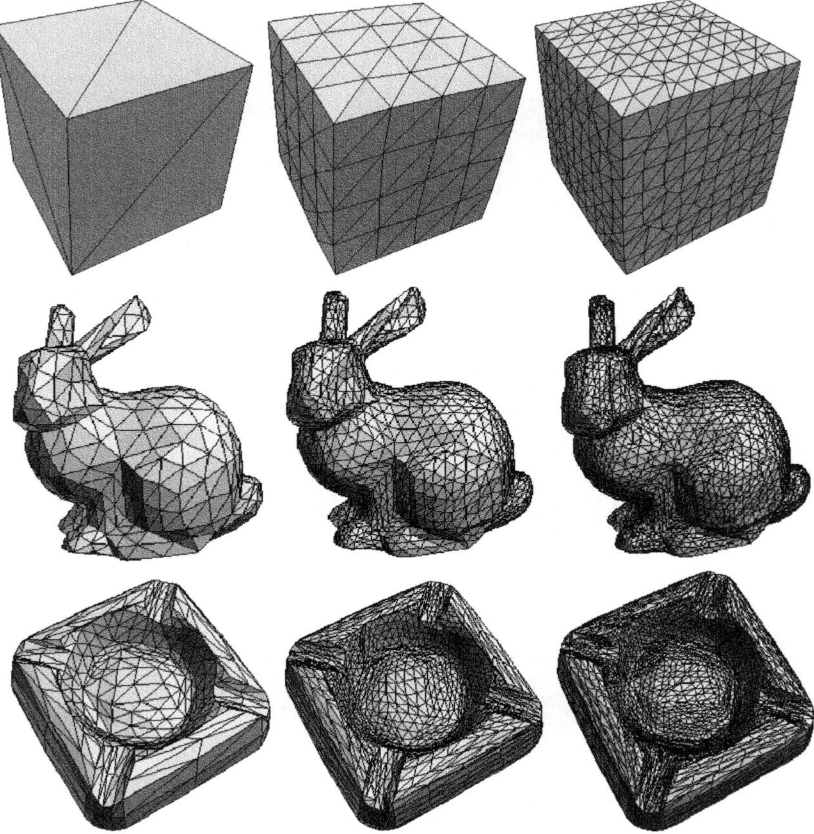

Fig. 10. Multiresolution models created by our algorithm with subdivision connectivity

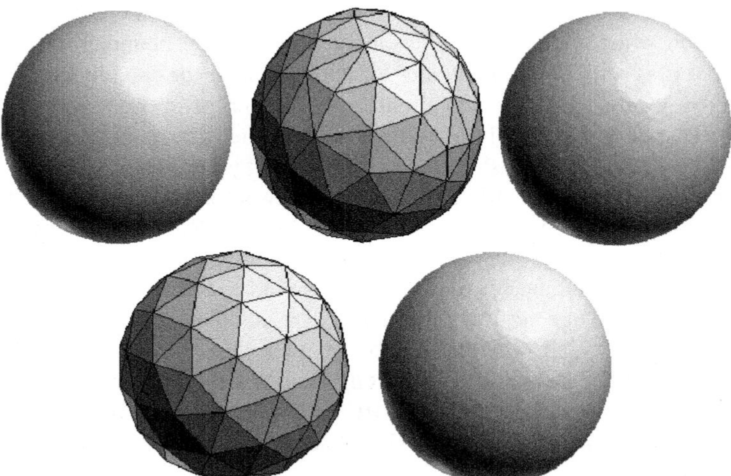

Fig. 11. Sphere reconstruction comparison between using an automatically decimated base mesh and elaborately created base mesh

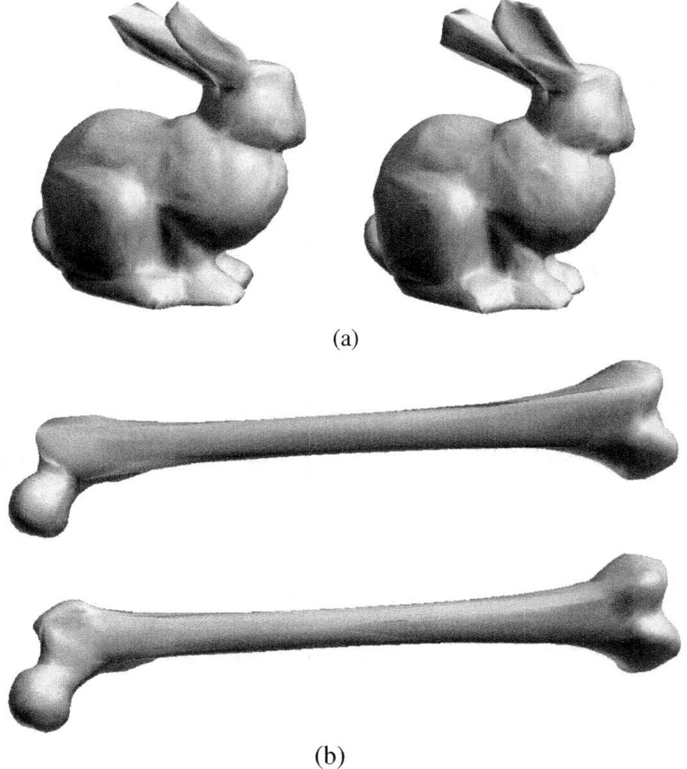

(a)

(b)

Fig. 12. Reconstructions of Stanford Bunny and femur bone models using different number of PDE patches: (a) Stanford Bunny model reconstructed from 180 and 250 PDE patches; (b) Femur bone model reconstructed from 60 and 132 PDE patches

Fig. 13. PDE shapes (highlighted) placed in a VRML environment

In Fig.12 the VRML shapes reconstructed from the original polygon meshes are displayed. The Stanford Bunny mesh has a size of 539 Kb while the derived coefficients file is only 86 Kb and 120 Kb for 180 and 250 PDE patches, respectively. The original femur bone mesh takes 1.26 Mb while the derived coefficients for 60 and 132 PDE patches take 28 Kb and 63 Kb, respectively. In Fig.13 these PDE shapes are displayed in a VRML environment.

In the 3D shared virtual world, efficient data exchange of 3D objects between servers and client computers plays a vital role. One of the advantages of using the patchwise PDE method is that it allows such data exchange via a few coefficents of the PDE spectral solution. As it was mentioned in Section 2, the efficiency and precision that the patchwise PDE method brings in part rely on proper selection of the number of PDE patches used to represent the original geometric shapes. We have developed an automatic deriving scheme of PDE boundary curves in this paper, based on the use of the developed automatic decimation algorithm. This decimation algorithm enables simplification of the large polygon meshes to a certain level, although this level, currently, has to be empirically governed by users through observations. For example, it is well known that human faces have variant geometries on the surface, different from person to person. Therefore, applying our decimation algorithm, which is geometry-based, to different faces with the same wireframe in an identical resolution may result in variant outputs after same iterations of operation due to the different facial geometries. To efficiently represent the face geometry, it is better to use less PDE patches produced with more iterations of decimation operation, but this may lead to an over-decimation. Fig. 14 shows a result of over-decimation with a 3D head model. One likely solution to resolve this problem is to introduce a generic facial mesh mask with a reasonable number of facets, which could be either automatically or manually superimposed onto the original face model. Once two meshes are aligned, the algorithms, except the decimation algorithm, described in Section 3 can be applied accordingly.

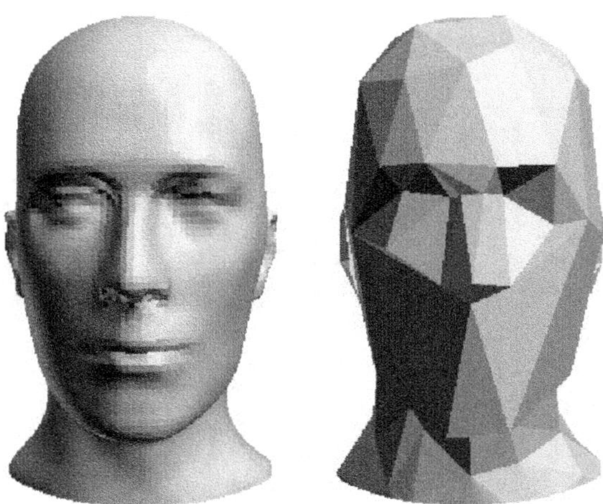

Fig. 13. Over-decimation of the head model and a simplified mask

5 Conclusion

We have proposed a fully automatic way of replacing large polygon meshes with much smaller PDE-based models approximating the original shapes. The precision of the approximation depends on the number of polygons in the original meshes as well as the number of PDE boundary conditions (PDE patches) which are reconstructed from the original meshes. Finally, PDE shapes are represented by small data files which are coefficients of the parametric PDE solution functions. These functions are then used for interactive visualization of the polygons approximating the original shapes. We have implemented the proposed algorithms within a framework of the FVRML/FX3D plug-in augmenting VRML and X3D with function defined shapes. Since the size of the PDE coefficients data is very small, it allows for an efficient web visualization and exchange of the PDE shapes in 3D Cyberworlds defined by VRML and X3D.

Acknowledgments

This project is supported by the Singapore National Research Foundation Interactive Digital Media R&D Program, under research Grant NRF2008IDM-IDM004-002 "Visual and Haptic Rendering in Co-Space", and by the UK Engineering and Physical Sciences Research Council Grant EP/G067732/1 "Function based Geometry Modeling within Visual Cyberworlds".

References

1. Sheng, Y., Sourin, A., Gonzalez Castro, G., Ugail, H.: A PDE Method for Patchwise Approximation of Large Polygon Meshes. The Visual Computer 26(6-8), 975–984 (2010)

2. Ugail, H., Sourin, A.: Partial Differential Equations for Function based Geometry Modelling within Visual Cyberworlds. In: 2008 Int Conf. on Cyberworlds, pp. 224–231. IEEE CS, Los Alamitos (2008)
3. Ugail, H., González Castro, G., Sourin, A., Sourina, O.: Towards a Definition of Virtual Objects with Partial Differential Equations. In: 2009 Int. Conf. on Cyberworlds, Bradford, September 7-11, pp. 138–145 (2009)
4. Bloor, M., Wilson, M.: Generating blend surface using partial differential equations. Computer Aided Design 21(3), 165–171 (1989)
5. Schroeder, W., Zarge, J., Lorensen, W.: Decimation of triangle meshes. Computer Graphics 26(2), 65–70 (1992)
6. Hoppe, H., DeRose, T., Duchamp, T., et al.: Mesh optimization. In: SIGGRAPH 1993, pp. 19–26 (1993)
7. Rossignac, J., Borrel, P.: Multi-resolution 3D approximations for rendering complex scenes. In: Falcidieno, B., Kunii, T. (eds.) Modeling in Computer Graphics: Methods Application, pp. 455–465 (1993)
8. Turk, G.: Re-tiling polygonal surfaces. In: SIGGRAPH 1992, pp. 55–64 (1992)
9. Lounsbery, M.: Multiresolution analysis for surfaces of arbitrary topological type [PhD dissertation], Department of Computer Science and Engineering, University of Washington (1994)
10. Garland, M., Heckbert, P.S.: Surface simplification using quadric error metrics. In: SIGGRAPH 1997, pp. 209–216 (1997)
11. Melia, S.: A simple, fast, effective polygon reduction algorithm. Game Developer 10, 44–49 (1998)
12. Levy, B.: Parameterization and deformation analysis on a manifold. Technical report, Alice (2007), `http://alice.loria.fr/publications`
13. Floater, M.: Mean value coordinates. Computer Aided Geometric Design 20(1), 19–27 (2003)
14. Tutte, W.: How to draw a graph. Proc of the London Mathematical Society, 743–768 (1963)
15. Chen, G.L., Pang, M.Y., Wang, J.D.: Calculating shortest path on edge-based data structure of graph. In: 2nd International Workshop on Digital Media and its Application in Museum and Heritage, pp. 416–421 (2007)
16. Chen, J., Han, Y.: Shortest paths on a polyhedron, Part I: Computing shortest paths. International Journal of Computer Geometry Application 6, 127–144 (1996)
17. Kaneva, B., O'Rourke, J.: An implementation of chen & han's shortest paths algorithm. In: 21st Canadian Conference on Computer Geometry, pp. 139–146 (2000)
18. Kapoor, S.: Efficient computation of geodesic shortest paths. In: 31st ACM Symposium on Theory of Computing, pp. 770–779 (1999)
19. Lanthier, M., Maheshwari, A., Sack, J.R.: Approximating weighted shortest paths on polyhedral surfaces. In: 13th Annual Symposium on Computational Geometry, pp. 274–283 (1997)
20. Martínez, D., Velho, L., Carvalho, P.: Geodesic Paths on Triangular Meshes. In: SIBGRAPI/SIACG, pp. 210–217 (2004)
21. Mitchell, J.: Geometric shortest paths and network optimization. In: Sack, J., Urrutia, J. (eds.) Handbook of Computational Geometry, Elsevier Science, pp. 633–702 (2000)
22. Surazhsky, V., Surazhsky, T., Kirsanov, D., et al.: Fast exact and approximate geodesics on meshes. ACM Transactions on Graphics 24(3), 553–560 (2005)
23. Sourin, A., Wei, L.: Visual Immersive Haptic Mathematics. Virtual Reality 13(4), 221–234 (2009)

On the Development of a Talking Head System Based on the Use of PDE-Based Parametic Surfaces

Michael Athanasopoulos, Hassan Ugail, and Gabriela González Castro

Centre for Visual Computing, University of Bradford,
Bradford BD7 1DP, United Kingdom
{mathana1,h.ugail,g.gonzalezcastro1}@bradford.ac.uk

Abstract. In this work we propose a talking head system based on animating facial expressions using a template face generated from a Partial Differential Equation (PDE). It uses a set of pre-configured curves (as boundary conditions for the chosen PDE) to calculate an internal template surface face. This surface is then used to associate various facial features with a given 3D face object. Motion retargeting is then used to transfer the deformations in these areas from the template to the target object. The procedure is continued until all the expressions in the database are calculated and transferred to the target 3D human face object. Additionally the system interacts with the user using an artificial intelligence (AI) chatterbot to generate response from a given text. Speech and facial animation are synchronized using the Microsoft Speech API, whereby the response from the AI bot is converted to speech.

Keywords: Facial animation, Speech animation, Motion re-targeting, PDE method, Parametric surface representation and Virtual interactive environments.

1 Introduction

A talking head system is an environment where a 3D human head is talking and interacting with a user. In such system, it is required to provide all the necessary tools for creating a human-computer interaction process. To this end, a text-to-speech (TTS) system is required to produce a sequence of phonemes from an input text. A phoneme is the basic unit of acoustic speech. A visual representation of the phoneme is called viseme. In speech, animation visemes are often used to represent the position of the lips, jaws and tongue in a given particular phoneme. Many phoneme sounds are visually ambiguous when pronounced. Therefore, one viseme can be used to represent several phonemes. This technique is very popular for generating realistic speech animation without having to manually set the key frame positions for a set of visemes [1] . The conventional lip synchronization technique initially consists of decomposing the speech into a set of phonemes. These phonemes will be visually represented in the system as a

M.L. Gavrilova et al. (Eds.): Trans. on Comput. Sci. XII, LNCS 6670, pp. 56–77, 2011.

set of visemes. A mapping between the phonemes in the speech signal and the visemes in the database is carried out to construct the appropriate lip shape.

In virtual worlds or environments such as cyberworlds, it is necessary to keep the size of data as small as possible for real time performance. Facial animation usually requires a large set of expressions to produce any given text. The facial data will need to be as small as possible to generate the speech animation without any lag or de-phase over the network. Additionally, the system will require a real time response to a user's input. Real-time dialog systems can include personality, emotions and interactive dialog in a human-computer or computer-computer situation. The process of human-computer interaction is facilitated with the use of chatterbots where actions such as personality, emotions and respond can be integrated within the talking head system [2]. Moreover, emotional tags embedded in the dialogue database can be used to generate facial expressions.

A new technique is introduced here that reuses facial animation to different 3D human face target models. The system uses a set of pre-configured viseme poses to generate a template human face model. Each new viseme pose will be then re-targetted to a given 3D target model for speech animation. The process can be repeated until all the required visemes are calculated. The 3D human head is synchronized with a TTS engine to generate voice, whereas a text-to-visemes function will return the current visemes of a given text in real time. Moreover, the system integrates an AI bot engine to calculate response from user input text. The engine used in this work is the Rebecca (Artificial Intelligence Markup Language) AIML library that implements the Alice bot language processing chatterbox. Chatterbots are computer systems that can produce a human-computer interaction in nature language. The user can enter a question or phrase and the AI bot will generate the appropriate answer to facilitate a real time conversation. Text response from the bot is then captured by the TTS engine and converted to a set of visemes to synchronize and animate the 3D human face. Animation is carried out by linearly interpolating any given set of visemes to generate the in-between transition of different visemes.

1.1 Related Work

A main problem that arises in facial animation consists of portraying the realism of the movement. The natural contact with facial expressions and the availability of better and more powerful hardware demand an ongoing improvement of the animation techniques for facial animation. Previous research in the area previously have introduced new techniques for achieving realistic motion. These include:

A Talking Head System for Korean Text animates the face of a speaking 3D avatar in such a way that it realistically pronounces the given Korean text [3]. The proposed system consists of SAPI compliant text-to-speech engine and MPEG-4 compliant face animation generator. The input to the engine is a Unicode text that is to be spoken with synchronized lip shape. The TTS engine generates a phoneme sequence with their duration and audio data. The TTS applies the co-articulation rules to the phoneme sequence and sends a mouth animation sequence to the face modeler.

Greta: A Simple Facial Animation Engine is a 3D facial animation engine compliant with MPEG-4 specifications [4]; the aim in this work was to simulate in a rapid and believable manner the dynamics aspect of the human face. Greta is a 3D proprietary facial model with the look of a young woman. The core of it is an MPEG-4 decoder and is compliant with the "Simple Facial Animation Object Profile" standard. The 3D model uses a pseudo-muscular approach to emulate the behaviour of face tissues and also includes particular features such as wrinkles and furrow to enhance its realism. Facial features such as wrinkles have been implemented using bump mapping which allows to create a high quality 3D facial model with a relative small polygonal complexity.
Real-time Lip Synchronization Based on Hidden Markov Model; A lip synchronization method that enables re-using of training videos when input voice is similar to training voice sequences [5]. The face sequences are clustered from video segments, then by making use of sub-sequence Hidden Markov Models, the system builds a correlation between speech signals and face shape sequences. This decreases the discontinuity between two consecutive output faces and obtains accurate and realistic synthesized animations. The system can synthesize faces from input audio in real-time without noticeable delay. Since acoustic feature data calculated from audio is directly used to drive the system without considering its phonemic representation, the method can adapt to any kind of voice, language or sound. Movement Realism in Computer Facial Animation is another work targeting realism in movement and behaviour of agents or avatars in virtual environments [6]. This system uses co-articulation rules for visual speech and facial tissue deformation producing expressions and wrinkles.

The aim of the work presented here consists of generating a template talking head system that will retarget realistic facial animation to a given human face model. It uses a pre-configured database of visemes curves that are used to compute a PDE solution for generating the template surfaces. Each generated viseme surface is then transferred to a different 3D human face object [7, 8]. Every template viseme surface is grouped into 4 facial areas; each of these areas is associated with the target model to transfer the facial features for the current viseme. This is a pre-processed procedure that takes place at the loading time of the application. The use of the PDE method for surface generation and representation provides several advantages for the manipulation of that surface. For instance, most of the information required to define a surface is contained at the boundary or the outline curves representing that object. The representation of the template human face expressions is controlled through the adjustments of parameters associated with these curves. Thus, the user can interactively transform the parameters and change the shape of the surface without having any knowledge of the mathematical theory behind the PDE method [9, 10].

2 Implementing a Talking Head System

In a Data-driven facial animation [11] system, the expressions are usually pre-configured and blended together in order to produce a sequence of letters; whereas in a 3D human head, the animation is synchronized with a text-to-speech engine to generate speech according to a set of phonemes for a given word. The system is usually interactive and can incorporate changes of emotions dependant on the user's input. Thus, it is necessary to build a viseme-driven talking head system where a given human head mesh model with similar topology, can be animated without the need to generate all the necessary expressions for the operation. The overall process consists of generating a set of PDE-based expressions that are used internally as the template expressions for any given human head model. The PDE method has been used for representing the surface template expressions, thus utilizing the advantages of parametric surfaces. The PDE method provides us with tools to control the surface resolution or smoothness of the facial geometry by adjusting the u, v parameters. At the same time, the PDE method enables us to generate facial animation from a given complex face model by adjusting only a small set of boundary curves. This methodology enable us to produce a set of template mesh surfaces that are used to transfer a given motion sequence to a given human head mesh model. The following explains further details on the implementation of the talking head environment.

The PDE method produces a parametric surface, which in summary is the graphic representation of the solution to an elliptic PDE of the form,

$$\left(\frac{\partial^2}{\partial u^2} + a\frac{\partial^2}{\partial v^2}\right)^n \chi(u,v) = 0\,, \tag{1}$$

where a represents an intrinsic parameter and n determines the order of the PDE. This work focuses on the use of the biharmonic equation; that is the values of a and n had been set equal to 1.0 and 2 on each respective case. Moreover, this work restricts the use of the PDE method to periodic boundary conditions where by $0 \leq u \leq 1.0$ and $0 \leq v \leq 2\pi$. Thus an approximate analytic solution to Equation 1 is found and it is gven by,

$$\chi(u,v) = \mathbf{A_0}\,(u) + \sum_{n=1}^{n=N} \left[\mathbf{A_n}\,(u)\; cos\,(nv) + \mathbf{B_n}\,(u)\; sin\,(nv)\right] + \mathbf{R_n}\,(u,v)\ , \tag{2}$$

where $\mathbf{A_0}(u)$ is a cubic polynomial, and, $\mathbf{A_0}(u)$ and $\mathbf{A_0}(u)$ are functions resulting form linear combination of exponentials functions and powers u. The term $\mathbf{R_n}\,(u,v)$ is known as the remainder term and it is responsible for exactly satisfying the original boundary conditions imposed to Equation 1. For further details on the mathematical foundations of the the PDE method, the reader is referred to [10] where additional details are given.

2.1 Generating Expressions

Viseme-driven speech animation approaches [11] often require manual design of key mouth poses in order to generate realistic speech animations. The first step for building the facial animation system consists of generating the facial expression data [12]. The template face, Figure 1.a, is represented as a set of

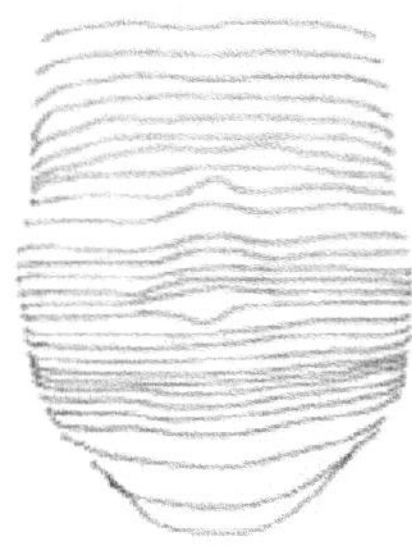
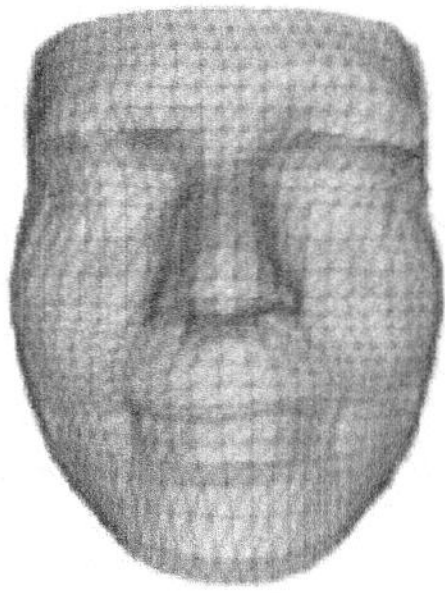

Fig. 1. The neutral expression template curve set (left). The resulting surface (right).

28 boundary curves acquired from a laser-scanned 3D face model in its neutral configuration.

The curve-set covers the entire face area, describing the most basic facial features. The template face surface is then reconstructed using a combination of nine different fourth orders PDEs that guarantee surface continuity. Note that here the u, v resolution of the PDE surface is set to 35x35. As mentioned before, a set of pre-configured expressions is required for animating speech cite13. A database of 22 visemes is used to identify the corresponding viseme of a given word or phrase within the TTS engine as shown in Figure 3. The talking head database consists of 15 viseme expressions. Some visemes can be repeated to save loading time e.g. viseme AW and OY can be represented using the same viseme.

Each viseme is represented as a curve-set derived from a face mesh model as seen in Figure 1.b. The viseme curve-set poses have been obtained from the work presented in [14]. The authors have developed a PDE-based facial deformation technique that uses the boundary curves of a PDE surface to deform a face model using the MPEG-4 compliant facial feature points. The idea behind this approach is that it utilizes the boundary curves for complex facial deformations rather than using conventional control points and surface interpolation techniques. A group of boundary points anatomically related to facial features, such as right corner of left eyebrow, left corner of inner lip contour, are selected as feature points according to the MPEG-4 definition. Animation of the face model is achieved by adjusting the position of the boundary curves before each calculation of the PDE surface. The deformations have been computed using a series of weighted sinusoids to parameterize the FAP-driven facial animation with the feature points. Additional boundary points have also been selected and used as weights in areas around the feature points in order to guarantee surface smoothness. The mouth and eyes area is achieved using sinusoidal animation of the corresponding feature points. For stretching the lip corners, linear interpolation is applied to the boundary curve points representing the lips.

The resulted curve-sets are stored in an internal viseme database and they are used for computing a PDE surface that passes though the control points of each

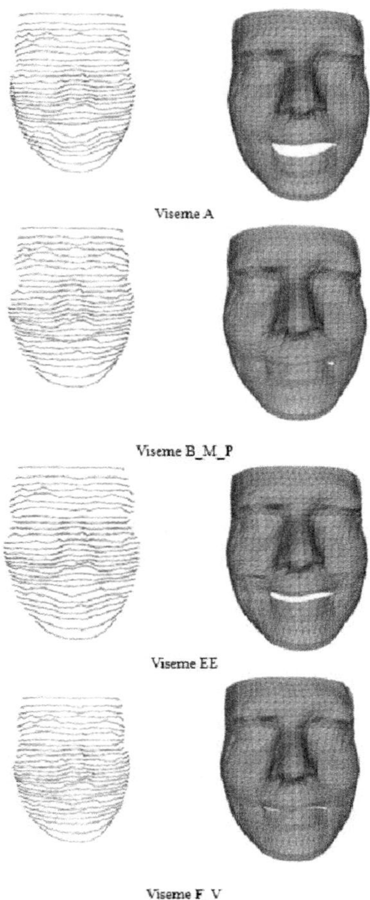

Fig. 2. Curved-based visemes (left). Template surfaces for each corresponding curve set (right).

curve to generate a surface representation. The process can be seen in Figure 3. Each curve-set viseme is used as the boundary conditions required for calculating the PDE method. The new surface is stored and used later for re-targeting the facial deformations of the corresponding viseme to a different face mesh model. The pre-configured viseme curve-sets contain all the necessary information to generate the PDE-based speech expressions for the animation system. Storing only a set of curves rather than a mesh object for each required viseme pose gives us the advantage in keeping the storage requirements to a minimum. This technology can be exploited to reduce network transmission bit-rates, by sending only animation parameters rather than the video sequence. Figure 2 contains various template viseme poses for calculating the required template expressions. The curve sets seen in Figure 2 (right), are used to generate the human face mesh for the given viseme and apply motion retargeting to transfer the deformations

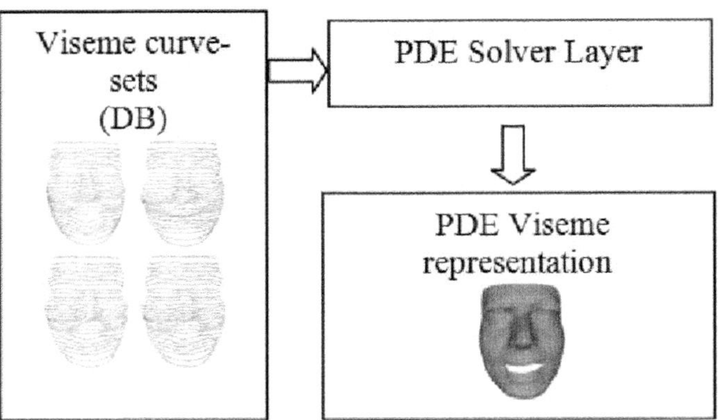

Fig. 3. PDE-based viseme generation process

to a different target human face model. Visemes A, B/M/P, EE and F/V used for the speech animation are shown in both configurations, curve and mesh representation.

3 Text-to-Speech Engine Integration

Various important developments in speech synthesis and natural language processing techniques resulted in the concept of text-to-speech synthesis. A text-to-speech synthesis can be defined as automatic production of speech, through a grapheme-to-phoneme transcription of the sentences to utter [15]. Figure 4 shows the general functional diagram of a synthesizer. It consists of a Natural Language Processing module (NLP), capable of producing a phonetic conversion of the text read together with the desired voice tone and rhythm (also called prosody) and a Digital Signal Processing module (DSP), which transforms the symbolic information it receives into speech. Phonetic and prosody information together can generate the symbolic linguistic representation that is output to the front-end application. The back-end application is usually provided from the API, converts the symbolic linguistic representation into sound usually referred as the synthesizer. Further details of the architecture of a TTS system is outside the scope of this paper and the interested reader can find additional information in [15].

The text-to-speech system converts normal language text into speech; it recognizes the input text and using a synthesized voice, chosen from several pre-generated voices, speaks the written text. The TTS engine used for this work is the Microsoft TTS which is fully programmable from the Speech API 5.0. The SAPI acts as a software layer that allows speech-enabled applications to

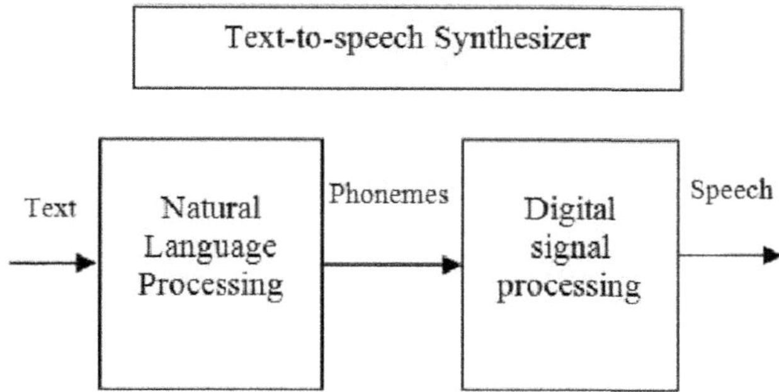

Fig. 4. General function diagram of the TTS system used in this work

communicate with both speech recognition and TTS engines. Additionally, the SAPI is responsible for a number of functions in a speech system, such as:

- Controlling audio input, whether from a microphone, files, or custom audio source.
- Converting audio data.
- Storing audio and serializing results for later analysis.
- Using SAPI grammar interfaces and loading dictation.
- Performing speech recognition.

The speech API provides a high-level interface between an application and the speech engine. All the low-level details needed to control the real-time operations of various speech engines are implemented in a collection of libraries and classes. With the help of the API, the TTS operation, which is required to convert a given text into a set of viseme, can be performed real time. The text-to-viseme process is used to identify which letter from the database is required to animate a given text and synchronize it with the voice. Using events, the application synchronizes the output speech with real time actions such as the phonemes or visemes generated from the input text.

In this case, the generated viseme is used to query from the database the current viseme index that is required for the speech animation. Table 1 contains the first 10 indexes used to identify which visemes are needed to simulate the speech animation. The TTS engine contains 22 visemes in total used for reproducing the sounds from any given input text. Note that many phoneme sounds are visually ambiguous when pronounced. Therefore, one viseme can be used to represent several phonemes. For example letters F and V can be reproduced by using the same viseme. Consonant letters such as B, M and P can represented as well with the same viseme. To that extent, the animation system can link various similar visemes together to minimize loading and processing time.

A total of 15 visemes are used in the talking head database to simulate speech. The remaining 7 visemes from the TTS engine have been associated with other similar visemes. For example, SP_VISEME_1, 2 and 3 from Table 1 can be represented as SP_VISEME_1.

Table 1. List containing the the first 10 viseme indexes association

SP_VISEME_1	ae, ax, a
SP_VISEME_2	aa
SP_VISEME_3	ao
SP_VISEME_4	ey, eh, uh
SP_VISEME_5	er
SP_VISEME_6	y
SP_VISEME_7	w, uw
SP_VISEME_8	ow
SP_VISEME_9	aw
SP_VISEME_10	oy

The integration of the TTS system in the talking head environment was developed using the .NET framework. Using the Component Object Model (COM) ISpVoice interface applications can control the TTS functionality. Initialization of the engine is achieved simply by creating a spVoice object, whereas for text to speech synthesis a single call to ISpVoice::Speak is required. Additional functionality for manipulation of voice and synthesis is provided; various function calls can control the speaking rate, the output speech volume and the current speaking voice. The speak method can be operated synchronously or asynchronously. An important feature for synchronizing the speech output with the rendering API as well as adjusting speech properties in real time. The SAPI communicates with the applications by sending events using standard windows callback mechanisms. Applications can then sync to real-time actions such as word boundaries, phoneme or viseme boundaries. Events are used for synchronizing the output speech. Each audio stream generated from the SAPI engine contains an event id where the application can identify position and status of the stream. A viseme event id is associated with the list in Table 1 to find the current viseme. This information will help calculate the interpolation time between previous and current viseme. Additionally, the TTS engine will playback the audio stream from the input text.

Communication between the application and the SAPI is processed as a two step operation. The application first receives a general window message from the SAPI. This message is similar to other window messages used by the operating system, such as mouse, keyboard, and window events. The second step in the communication process is determining which action occurred. The application needs to determine the exact action that is taking place. A list of all the SAPI defined action can be found using the **SPEVENTENUM** list. Using SPEVENT and GetEvents method the SAPI can identify specific information about the current event. Actions such as **START_INPUT_STREAM**, **END_INPUT_STREAM** and **VISEME** are used to determine the current activity that is taking place. Once the current event is determined, the application needs to take the necessary action. When a **START_INPUT_STREAM** event is capture, the input stream from a Speak call has begun synthesizing to the output, whereas an **END_INPUT_STREAM** is identified as the end of the

Speak event. In the case of the VISEME event id, the SAPI returns the relevant viseme with some additional information for the current Speech operation.

4 A.L.I.C.E's Brain

Once the speech engine is configured to capture and process real time events, the following step consists integrating an Artificial Intelligence Markup Language or AIML chatterbot to generate response from a given text. A.L.I.C.E. (Artificial Linguistic Internet Computer Entity), also referred to as Alicebot, or Alice, is a language processing chatterbox based on an experiment specified by Alan M. Turing in 1950. A chatterbox is a program that engages in a conversation with a user by applying some heuristical pattern matching rules to the human's input [11, 16], A.L.I.C.E. software utilizes AIML, an XML based language used for creating chat robots.

It contains a class of data objects defined in the XML specification [17] called AIML objects and describes the behaviour of computer programs that process them. Various Alicebot clones have been created based upon the original implementation of the program and its AIML knowledge base [18]. Some of these AIML interpretersare:

- RebeccaAIML(C++, Java, .NET/C#, Python)
- Program D(Java)
- Program R(Ruby)
- Program O(PHP)

This work uses the Rebecca AIML library for generating real-time responses. The system establishes a local connection with the Rebecca AIML bot to process and generate response to a given input text. The response is then handled as an input text for the SAPI text-to-speech engine.

Persona-AIML [19] presents the Persona-AIML architecture for the creation of chatterbots in AIML (Artificial Intelligence Markup Language) with personality. Computational models of personality are in general adapted from some psychology model or theory. The Personality Component defines the beliefs, the personality elements, and the rules that determine chatterbots behaviour.

The work presented in [20] demonstrates an emotional MPEG-4 compliant talking head system based on AIML. It uses Alicebot to generate response and emotion from given input text. Emotions are embedded in the AIML database as a set of predefined emotion tags. These emotional tags are passed to the personality model to simulate believable behaviours. The personality model, depending upon the current mood and the input emotional tags, updates the mood. Depending upon the output of the personality model, mood processing is done to determine the next emotional state. This processing determines the probabilities of the possible emotional states. Additionally, the system generates lip sync from the visemes generated from the Text-To-Speech engine.

TQ-Bot in [21], presents an Intelligent Tutoring System using AIML whose aim is in providing personalized instruction to students. The authors have developed an open e-Learning platform for helping the students during their learning process and for supporting the activities of the teacher. The bot is able to analyze the requests made by the learners in written natural language and to provide adequate and domain specific answers orienting the student to the right course contents. Additionally, TQ-Bot is able to track and supervise the student progress by means of personalized questionnaires. The brain of A.L.I.C.E. consists of 41,000 elements called categories [2]. Each category contains a question and answer, called the pattern" and template". The patterns are stored in a tree structure managed by an object called the Graphmaster, which implements a matching algorithm. Graphmaster matching is a special case of backtracking, depth-first search. In most cases matching is handled by a linear traversal of the graph from the root to a terminal node.

The AIML architecture allows the use of different models of personality in the construction of chatterbots. It implements various tags to introduce randomness in answers, and to keep track of small dialogue history. Although it does not use any syntactic or semantic language analysis techniques to generate the response, the content embedded in AIML is enough to engage the user in believable conversation to a certain degree. A.L.I.C.E contains a learning mode, called supervised learning since a botmaster is required to create and manage the content. The botmaster will monitor the bot conversations and can create new AIML content to make the bot responses more believable, accurate or human like.

Moreover, every AIML object has both a logical and a physical structure. The physical structure consists of units called topics and categories, while the logical structure is composed of elements and character references.

4.1 AIML Elements

The basic unit of knowledge in AIML is called a category [2]. The term category was borrowed from pattern recognition theory. Each category consists of an input question, an output answer, and an optional context. The question is called pattern, the response is called template and lastly, the optional context is divided into two types called that and topic. The AIML pattern tag consists only of words, spaces, and the wildcard symbols. Words must contain only letters and numbers separated by a single space whereas the wildcard characters can function like words. A pattern elementmust appear in each category and it must always be the first child element of that category element. Apatterndoes not have any attributes.

Additionally, AIML supports interaction with other languages and systems. For example the <system> tag can be used to execute any program or command accessible from the operating system and insert the result in the reply. Alternatively, the <javascript> tag can be used to allow scripting inside the templates. The <template> tag it is the most basic AIML tag and is always paired with a <pattern> tag. It always appears within<category>elements and it does not contain any attributes. The <template> must follow the <that>element or follow the <pattern>element. The majority of AIML content is within thetemplate. Thetemplatemay contain zero or more AIML template elements mixed with character data.

Figure 5.b shows an AIML example code, the pattern defines the input and the template tag defines the bots response to that input. The syntax of an AIML category is:

```
<aiml>
   <category>
      <pattern> InputText </
pattern>
   <template>Response </
template>
  </category>
</aiml>
```

(a)

```
<aiml>
  <category>
    <pattern> Hello </pattern>
      <template> Hi!   How are
you? </template>
   </category>
</aiml>
```

(b)

Fig. 5. General syntax of an AiML category tag (a). Simple example of input and output response (b).

The above AIML code matches the client text input, in this case the word "Hello" and sends back to the client the response "Hi! How are you?".

The optional context in the category tag consists of two elements, called <that> and <topic>. The <that> element appears inside the category, and its pattern must match the bots last response. The <that> element is a special type of pattern element used for context matching. Itis optional in a category, but if it exists it must occur no more than once, and must follow the<pattern>and <template> element. Remembering the last response is important for creating a more believable conversation. In the example below, Figure 6.a, the <category> element is activated when the client says yes. The bot must find out what was the question to that answer. If the bot asked, "Do you like AIML?", the category matches the <that> element and it continues the conversation using the <template> response element.

The <topic> is an optional element that might appear outside a <category> element, and it is used to group together categories. The <topic> element allows the bot to store duplicate patterns in different topics, this way the bot can generate different responses to the same input patterns depending on the topic. A<topic> element has a requirednameattribute that must contain a simple pattern expression. A <topic>element may contain one or morecategoryelements. The botmaster uses the <set_topic> tags to set the topic of current <category> element. Once the topic is set, for any new query from the client the bot will start looking for a response in the categories that match the current <topic> tag. If there is not a category defined in the current topic, then any categories that are not defined in topic tags are searched.

```
<category>
  <pattern>YES</pattern>
   <that>DO YOU LIKE
AIML?</that>
  <template>WHAT IS YOUR
FAVORITE INTERPRETER?
</template>
</category>
```

(a)

```
<topic name="AIML">
 <category>
    <pattern> * </pattern>
   <template> MY FAVORITE
INTERPRETER IS
RebeccaAIML
   </template>
</category>
```

(b)

Fig. 6. <That> element example (a). <Topic> element example (b).

Figure 6.(b) presents an example of using the <that> element. In this case, if the client says something that the bot does not have a specific response for, it could still respond within the current topic. For example, the response for any undefined pattern element under the AIML topic will be "My favourite interpreter is RebeccaAIML". AIML consists of various elements offering leaning capabilities and intelligent response to achieve realistic human computer interaction. Moreover, the <random> element tag can generate different responses to the same input text. Each possible template response for the current pattern element needs to be separated with the <li> tag element. AIML is extensible; the botmaster can include an infinite number of new tags for application specific properties. Predicate tags can be used according to a client based "set" and "get" method to generate an endless variety of responses. Recursive categories can be used to map one input to other inputs, either for language simplification or to identify similar patterns. The AIML implementation for recursion is the tag <srai>. Figure 7.(a) presents a basic recursion example, if the user says "Hi", "Hello", "Hi there", etc., the response template will be the same as for the "Hi" pattern element.

Recursions are useful for a variety of tasks, some of these include:

- Symbolic Reduction: Reduction and simplification of complex input patterns.
- Divide and Conquer: Split an input into two or more parts, and combine the responses to each.
- Synonyms: Generate different ways of saying the same thing to the same reply.
- Spelling or grammar corrections.
- Detecting keywords anywhere in the input.
- Conditionals: Branching can be implemented with the <srai> operator.

A common application for recursive categories is simplification and reduction of complex input patterns. A combination of the <star> tag and <srai> recursive

```
<category>
    <pattern>HI THERE!
</pattern>
   <template>
   <srai>HI</srai>
</template>
</category>
```

(a)

```
<category>
<pattern>WHAT IS  *  </
pattern>
  <template>
        <srai>WHAT IS<star/>
</srai>
  </template>
</category>
```

(b)

Fig. 7. Basic recursion example (a). Simplification of input pattern using recursion and the <star> element (b).

calls can be used to produce endless combinations. For instance, Figure 7.(b) shows an example of recursion using the <star> tag. The bot will match a pattern starting with "What is" and use the "*" value to define a recursion call to find the best match for the transformed input.

4.2 Emotions

One of the many advantages of using AIML is that it is fully customizable. The bot master can include new AIML content with custom handling of input text. As discussed in the previous paragraph, implementing custom AIML elements can add intelligence to the bot and make the human-computer interaction more believable, accurate and human like. Additionally, an infinite number of new tags can be added to extend the functionality of the bot. The selection of current mood in the talking head system is an example of AIML customization. The emotions have been embedded in the AIML files by introducing a new tag called <emotion> and a "name" property.

The property name will hold the mood name of the current AIML pattern element. This information has been added manually in a Custom AIML file to support a range of different input pattern elements. Figure 8 shows an example of emotion handling in AIML. When the input text matches the pattern element, the template element contains additional information. In this example, the response is produced by a <random> tag that contains two different responses and the emotion that is set for current template is happy. However, the new <emotion> tag requires additional processing since the AIMLBot interpreter does not recognize custom tags. A simple XML node processing function searches the <category> node element for each new response to check if it contains a valid emotion element. If an emotion element is found, it is parsed to the Blend Shape layer to produce the new expression. Alternatively, if the current template does not contain any emotion element, then the current mood is set to neutral.

```
<pattern>I LIKE YOU</pattern>
  <template>
   <emotion name="happy" />
    <random>
    <li> I Like you too.</li>
    <li> You are very kind. Thank you!.</li>
   </random>
 </template>
```

Fig. 8. An AIML example containing emotion data

5 Blend Shape Visemes

Once the curve-sets have used to computed the PDE method, the generated PDE surfaces for each viseme in the database are stored and represented in the system as a group of polygon meshes. This is a pre-processed procedure that takes place at the loading time of the application and it is repeated for all the required visemes in the database. The next step consists of generating animation for any two given facial expressions. The talking head system must be able to blend any required viseme poses in real time. For this task shape interpolation between two key-frames is used, where two or more facial expressions are captured and in between frames are computed by linear interpolation.

Linear interpolation is used to approximate a value using two known visemes; these indexes are computed from the text-to-speech engine and assigned a value according to the viseme mapping in the database. The interpolation is calculated for all the vertices in the coordinates of each viseme. Additionally, the normals of the new surface are calculated using the same technique.

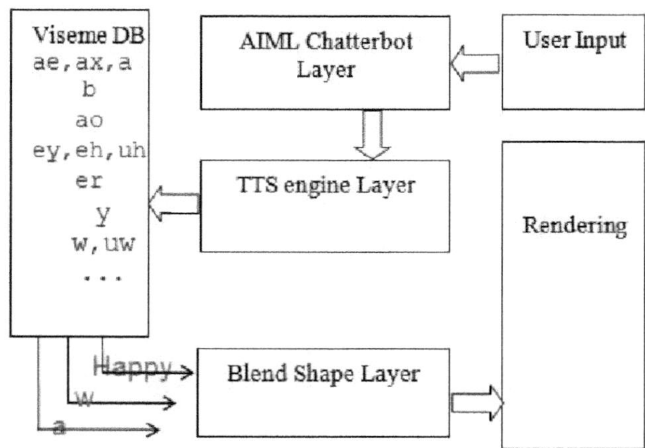

Fig. 9. The speech animation process layers

The talking head system presented in this work integrates emotions that are produced from the AIML chatterbox engine according to the current user input. This requires the shape interpolation of an additional expression of the overall speech animation. In this case, the additional expression will be included in the interpolation function with an additional variable to control the amount or intensity of the blend shape. The intensity variable is passed to the interpolation in real time during the calculation to produce a realistic conversation during the speech animation.

The process of speech animation is explained in the diagram shown in Figure 9, which shows the communication between the various layers of the speech animation. Animating two viseme poses is dependant to the user input. Input text is send to the chatterbot process to generate a response; response from the AIML engine is then processed by the TTS engine to produce the appropriate speech and viseme information used for the animation. The corresponding visemes will be parsed to the Blend Shape process to produce the interpolated shape at any given time. Additionally, a third expression is parsed to the shape interpolator indicating the mood change. This information is obtained from the AIML engine in real time.

6 Motion Retargetting

Finally, once the system is initialized with the PDE-based viseme expressions and synchronized with the AIML engine, it is required to retarget [22] the deformations to a different target face mesh model. The process of transferring the animation from a set of PDE-based template representations to a given face mesh model is carried out as follows:

- Alignment between the neutral viseme surface and the target mesh model in the same initial position. Key features of the model have to be positioned so that they nearly overlap. This will ensure a correct correspondence between the two surfaces.
- Mapping correspondence between models. This process consists of associating each point of the mesh model with the nearest point of the template surface in 'their initial configurations. This way, each point in the mesh model is represented on the template surface and it is assumed that this point remain as the closest one to the same particular point in every viseme surface in the database.
- Animation of the mesh model is carried out by finding the difference between the corresponding viseme pose and the original neutral surface for each point. This difference is then added to each point of the original mesh model according to the mapping correspondence previously found.

The above procedure is repeated for every expression in the viseme database. There are cases where a better mapping correspondence is required, e.g. mouth regions between the two objects, to achieve a more detailed representation. To solve the problem, the template face and the target object are split into face

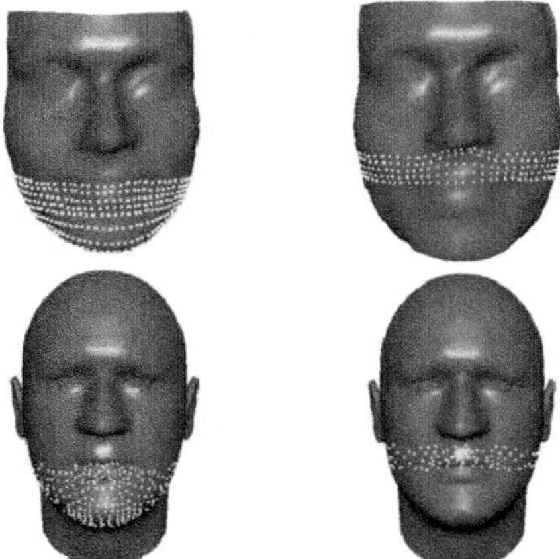

Fig. 10. Region maps used for motion re-target. PDE-based template (top). Mesh model target (bottom).

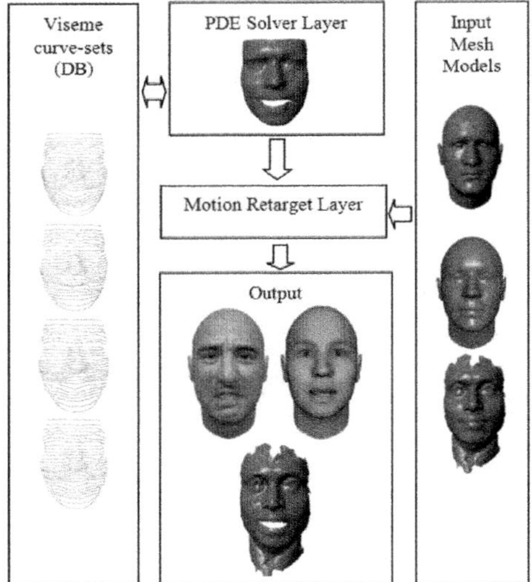

Fig. 11. Motion re-target process for each viseme in the database

region maps, such as bottom and top mouth, nose and eyes. Each map contains the index of each vertex that is included in that region. The collection of these points is handled using the Autodesk Maya environment, where the selection and output of the correspondent vertices is performed using MEL scripts. The correct selection of these points is very important since the motion transfer is based on the vertex indexes each map contains.

Figure 10, contains the top and bottom mouth region maps for the template face and the target human face object. Motion retarget is applied separately to each region map until all the required facial areas are processed. Once the motion transfer is complete, the resulting object can be associated with the expression it represents in the viseme database. Note that the quality of the mapping correspondence between the template viseme and the output face model depends also on the resolution of the grid used to compute template surface representation. Thus, using a grid with a similar number of points to the number of vertices in the original mesh will produce a better mapping correspondence. As seen in the diagram in Figure 11, the motion re-targeting process can be visualized as an additional layer in the talking head system that communicates between the PDE surface layer and the Input Mesh model.

7 Examples

Figure 12, contains a sequence of viseme expressions that are transferred from the template mesh Figure 12 (a, b, c) to the different target face mesh model Figure 12 (d, e, f). The motion re-targeting technique employed in this work requires a mapping correspondence between the two objects, such as each point of the target mesh model is associated with the nearest point of the template surface. This way each point of the target mesh will be represented on the template model. Finally, motion re-targeting is carried out by adding the difference between each point in the source model and the corresponding point in the target model.

Next, sequence of expressions in Figure 13, shows several examples of motion retargeting on expressions that are used within the talking head system to simulate change of mood during the speech. These expressions are included in the blend shape process to adjust the current viseme according to the current mood expression. The current Mood selection can be controlled by certain input

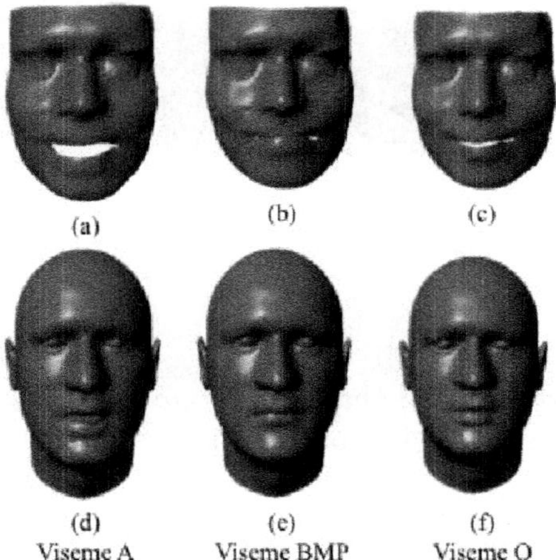

Fig. 12. Viseme templates. PDE-based visemes templates (a, b, c). Visemes retargetted to human head mesh model (d, e, f).

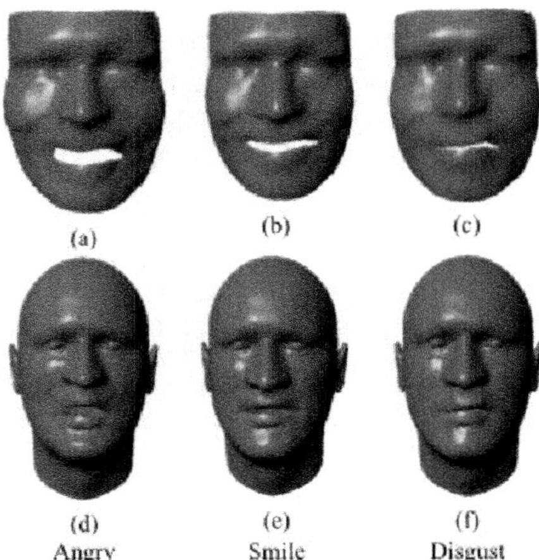

Fig. 13. Emotion templates. PDE-based emotion templates (a, b, c). Emotions re-targetted to human head mesh model (d, e, f).

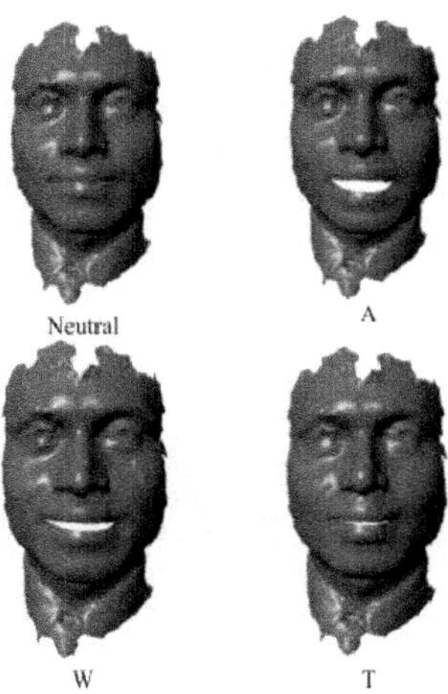

Fig. 14. Motion re-target to a 3D scan mesh model

text, duration or from the AIML engine [19]. AIML elements can encapsulate the response and certain mood that is generated from each parsed input text. This way, the response from the bot can also contain the appropriate mood expression.

Another example of motion re-targetting is shown in Figure 14; the target 3D human face has been replaced with a model acquired from a 3D scanner. The initial alignment of the two models plays a very important role in the correct motion transferring between the facial region maps. There are cases where facial regions of the target mesh need to be removed from the motion retarget process.

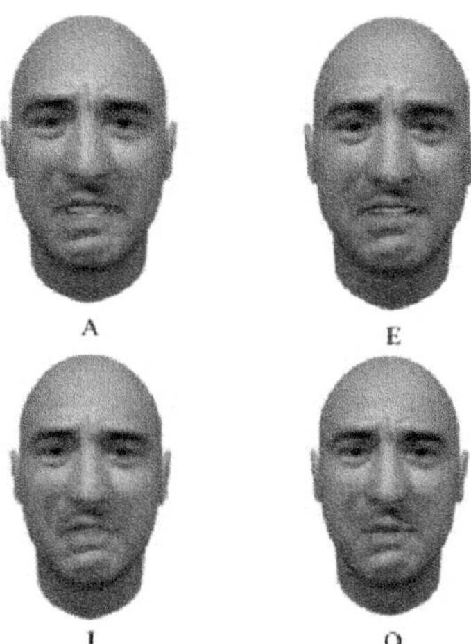

Fig. 15. Final rendering of the talking head system 3D face. The face model contains mouth, teeth and tongue animated according to each viseme.

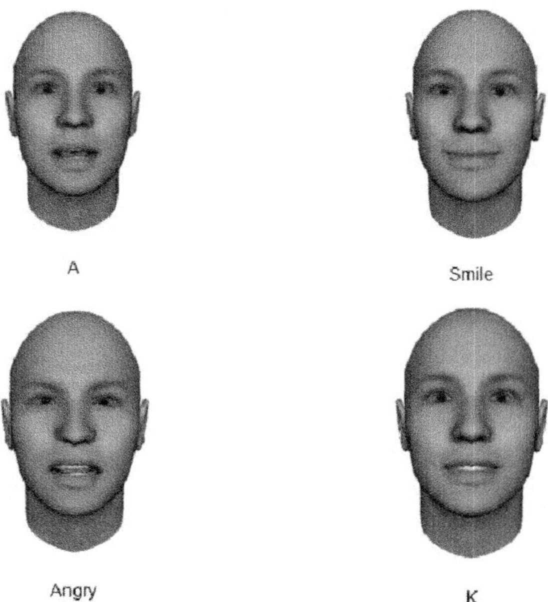

Fig. 16. Final rendering of the different face model retargeted to the talking head system. The face model contains mouth, teeth and tongue animated according to each viseme.

The generic face template used in this work contains only 4 basic regions maps. For more realistic facial deformations the number of face area maps can be increased according to the facial features of the target objects.

The final textured version of the animated target object is shown in Figures 15 and 16. The animated face contains teeth, tongue and mouth to increase the realism of the speech animation. The position of the teeth and tongue is interpolated alongside with the viseme blend shape process.

8 Conclusions

This work presents a technique for animating facial expressions within a talking head system. It uses a set of pre-configured curves representing various viseme poses to calculate a template PDE-based surface. The resulting surface is used to associate various facial feature areas with a different target face mesh model. Motion re-target is then applied to transfer the deformations in these areas from the template to the target model. This technique offers animation re-usage, since all the necessary viseme poses for the animation are pre-calculated and re-used for a different target mesh model. Additionally, it minimizes the storage requirements by storing only a small set of curves for each expression. The system interacts with the user using an AIML chatterbot to generate response from input text.

The user can enter a question or phrase and the AI bot will generate the appropriate answer to facilitate a real time conversation. The response is then captured and converted to speech from the text-to-speech engine; each word is split into a sequence of viseme poses that are used to synchronize the facial animation. The animation is carried out by linearly interpolating a given set of visemes to generate the in between transition of different visemes.

An improvement to be included in this technique is the development of a more generic template model representation that could enable the animation to a larger variety of character models. Future work could also be undertaken in automating the facial map extraction process. This process is required to associate various facial areas between the two objects for a seamless motion re-target.

References

1. Lee, R., Terzopoulos, D., Waters, K.: Realistic Modeling for Facial Animation. In: Proceedings of the 22nd Annual Conference on Computer Graphics and Interactive Technique, pp. 55–62 (1995)
2. Wallace, S.R.: The Anatomy of A.L.I.C.E.
3. Kim, S.W., et al.: A Talking Head System for Korean Text. World Academy of Science. Engineering and Technology 50 (2005)
4. Pasquariello, R., Pelachaud, C.: Greta: A Simple Facial Animation Engine. In: Proceedings of the 6th Online World Conference on Soft Computing in Industrial Applications (2001)
5. Huangm, Y., et al.: Real-time Lip Synchronization Based on Hidden Markov Models. In: The 5th Asian Conference on Computer Vision, Melbourne, Australia (2002)
6. Maddock, S., Edge, J., Sanchez, M.I.: Movement Realism in Computer Facial Animation. In: Workshop on Human-Animated Characters Interaction, vol. 4 (2005)

7. Fedorov, A., et al.: Talking Head: Synthetic Video Facial Animation in MPEG-4. In: International Conference Graphicon Moscow, Russia (2003)
8. Balcõ, K.: Xface: MPEG4 Based Open Source Toolkit for 3D Facial. In: Proceedings of the Working Conference on Advanced Visual Interfaces, pp. 399–402 (2004)
9. González Castro, G., et al.: A Survey of Partial Differential Equations in Geometric Design. The Visual Computer 24(3), 213–225 (2008)
10. Ugail, H., Bloor, M.I.G., Wilson, M.J.: Manipulation of PDE surfaces using an interactively defined parameterization. Computers and Graphics 23(4), 525–534 (1999)
11. Deng, Z., Noh, J.: Computer Facial Animation: A Survey in Data- driven 3D facial animation. Springer, Heidelberg (2007)
12. Marschner, S.R., Guenter, B., Raghupathy, R.: Modeling and Rendering for Realistic Facial Animation. In: Proceedings of the Eurographics Workshop on Rendering Techniques, pp. 231–242 (2000)
13. Haber, J., et al.: Face to Face: From Real Humans to Realistic Facial Animation. In: Proceedings Israel-Korea Binational Conference on Geometrical Modeling and Computer Graphics, pp. 73–82 (2001)
14. Sheng, Y. et al.: PDE-Based Facial Animation: Making the Complex Simple. In: Proceedings of the 4th International Symposium on Advances in Visual Computing, pp. 723–732 (2008)
15. Dutoit, T.: High-quality text-to-speech synthesis. Springer, Heidelberg (2001)
16. Galvão, A.M., Barros, F.A., Neves, A.M.M., Ramalho, G.L.: Adding personality to chatterbots using the persona-AIML architecture. In: Lemaître, C., Reyes, C.A., González, J.A. (eds.) IBERAMIA 2004. LNCS (LNAI), vol. 3315, pp. 963–973. Springer, Heidelberg (2004)
17. Wallace, R.: Artificial Intelligence Markup Language (AIML) v1.0.1, A.L.I.C.E. AI Foundation Working Draft (2001)
18. Mana, M., Pianesi, F.: HMM-based Synthesis of Emotional Facial Expressions during Speech in Synthetic Talking Heads. In: Proceedings of the 8th International Conference on Multimodal Interface, pp. 380–387 (2006)
19. Galvo, A.M., et al.: Persona-AIML: An Architecture for Developing Chatterbots with Personality. In: Third International Joint Conference on Autonomous Agents and Multiagent Systems, New York, USA, vol. 3 (2004)
20. Giacomo, T.D., Garchery, S., Thalmann, N.M.: Expressive Visual Speech Generation in Data-driven 3D facial animation. Springer, Heidelberg (2007)
21. Fonte, A.M.: TQ-Bot: An AIML-based Tutor and Evaluator Bot Fernando. Journal of Universal Computer Science 15(7)
22. Pighin, F., et al.: Synthesizing Realistic Facial Expressions from Photographs. In: Proceedings of the 25th Annual Conference on Computer Graphics and Interactive Techniques, pp. 75–84 (1998)

Real-Time Spatial and Depth Upsampling for Range Data

Xueqin Xiang, Guangxia Li, Jing Tong, Mingmin Zhang, and Zhigeng Pan

State key Lab of Computer Aided Design and Computer Graphics,
Zhejiang University, Hangzhou, 310058, Zhejiang Province, China
{xiangxueqin,lgx,tongjing,zmm,zgpan}@cad.zju.edu.cn

Abstract. Current active 3D range sensors, such as time-of-flight cameras, enable acquiring of range maps at video frame rate. Unfortunately, the resolution of the range maps is quite limited and the captured data are typically contaminated by noise. We therefore present a simple pipeline to enhance the quality as well as improve the spatial and depth resolution of range data in real time. To improve the spatial resolution of range data, we first upsample the depth information with the data from high resolution video camera. And then, a new strategy is utilized to increase the sub-pixel accuracy. We show that these techniques can greatly improve the reconstruction quality, boost the resolution of the range data to that of video sensor while achieving high computational efficiency for a real-time application.

Keywords: Time-of-Flight Camera, Super Resolution, Fast Multi-Lateral Filter, Sub-Pixel Estimation.

1 Introduction

In recent years, a variety of range measuring devices have been developed for 3D data acquisition.For example, by using extremely faster shutter (on the order of nanosecond),time-of-flight(TOF) sensors [1] measure time delay between transmission of a light pulse and detection of the reflected signals on an entire frame at once which are best suited for dynamic scene. This process is largely independent of the scene texture and full frame real-time depth estimates are possible.On the other hand, the main contender to TOF sensor- stereo vision [3] - is rather limited: it is known to be quite fragile in practice (e.g. due to lack of texture).

Unfortunately, being a relatively young technology, TOF sensors have not enjoyed the same advances, with respect to image resolution, quality and photo speed, that have been made in traditional 2D intensity imaging sensors. As a result, in current generation, these sensors provide range data of comparably low image resolution (e.g. only up to 176×144 for MESA SwissRanger^{TM} SR4000 [2]) that are heavily contaminated with noise in the distance measurement.

To overcome this issue, this paper proposes a simple framework to substantially enhance the spatial and depth resolution of range data, e.g., those from the

M.L. Gavrilova et al. (Eds.): Trans. on Comput. Sci. XII, LNCS 6670, pp. 78–97, 2011.

Mesa imaging sensor. To achieve this goal, firstly, we propose a new fast multi-lateral filter, termed as FMLF, to adaptively upsample the low resolution range data in real time by taking advantage of the significant information provided by registered high resolution video camera. To enhance the depth resolution and reduce the discontinuities caused by quantization in the depth map initiation process, a sub-pixel estimation algorithm then is formulated as a Markov Random Field (MRF) and treated it as a Maximum A Posteriori (MAP) problem which can be solved via the gradient descent method.

The main contribution of our method is to present a simple pipeline to enhance the spatial and depth resolution of range data while obtaining real time performance. We also extend our method in a new realm: combined with the low resolution intensity image generated by TOF sensor itself, the quality of range data can be greatly improved.

The rest of the paper is organized as follows. Section 2 introduces the previous works. Section 3 describes the multi-sensor setup of our system. The complete description of the proposed fast and simple super resolution technique is presented in Section 4. The extension is given in Section 5.The experimental results are given in Section 6. Finally the conclusions are outlined in Section 7.

2 Previous Work

There are many approaches that exploit additional information to improve the resolution of range data combining TOF sensor with one or two high resolution video cameras. The main assumption is that depth discontinuities are often related to color changes in the corresponding color image.

Prior researchers often use a probabilistic approach: In [5], MRF is first designed based on the low resolution depth maps and the high resolution camera images and solved via conjugate gradient. Unfortunately, this method gives promising spatial resolution enhancement only up to 10×. Yang et al. [6] then present a method modals a cost volume of depth probability and iteratively applies bilateral filter [7] to refine the cost volume, providing a spatial resolution enhancement of 100× (10× width and 10× height). However, they do not use a joint bilateral filter [8] to link the two images and even with GPU (Graphics Processing Unit) [9] optimization, their effective runtime would be very large due to the number of cost slices and the iterative scheme.

Another recent method [10] utilizes exclusively depth maps, without color image aid: a sequence of low resolution depth maps of same scene is aligned and then merged together to obtain a single depth map with improved resolution. But this method is restricted to static scenes' acquisition. Lindner et al. [23] apply noise and edge aware upsampling for range data. However, using a pure upsampling method, they do not to recover details which are beyond the depth sensor's resolution limit.

Key to our success is the use of multi-lateral filter, which is essential the extension of joint bilateral filter widely used in several state-of-the-art upsampling algorithms [11].Until recently, these edge preserving bilateral filters were too computationally intensive for real time applications. Several efficient methods [12] enable it to be computed at constant time or even video using GPU implementation [13]. Yang et al. [14] improve on this by not explicitly representing the entire space, but instead sweeping a plane through the intensity level, computing the output in intensity order. This low-memory, cache-friendly algorithm is the fastest known bilateral filter. Inspired by Yang's acceleration strategy, our multi-lateral filter is sliced into one bilateral filter and one joint bilateral filter that computed through discretization technology respectively. Therefore, the real-time performance can be eventually achieved via GPU implementation. What's more, compared with the work of Chan et al. [16], our multi-lateral filter method then allows for sub-pixel accuracy, in contrast with a potentially blocky range result.

TOF sensor also provides an intensity image that is perfectly registered with a depth map at each frame. Since a little interest has been put into this realm [15], we extend our algorithm for improving the quality of range data of TOF sensor by its own low resolution intensity image. Experimental results indicate that even with the help of low resolution intensity image, the quality of range data could be greatly improved by our algorithm.

3 Multi-sensor Setup

We combine a TOF sensor with a high resolution video camera (as shown in Fig.1). In our setup, it has a baseline about 100mm and two sensors are verged towards each other from the parallel setup.

The TOF sensor we have is a SwissRanger SR4000 [2], which can continuously emit a sin wave and detected its reflected signal to produce a depth map in 176 × 144 size. Its operational range is up to seven meters with the modulation frequency of 20MHZ. In addition, it will also produce an intensity image in the

Fig. 1. Multi-sensor setup

same resolution based on object reflectance. The video camera we have is a dragonfly2 video camera, providing color images with resolution up to 1024 × 768 pixels.

4 Algorithm

An overview of the framework of the approach is given out in Fig.2 and it has two main independent phases: First, up-sample the low resolution range image from TOF sensor to the same size as the high resolution camera image and fast multi-lateral filter (FMLF) is applied for the purpose of spatial super resolution and denoising afterwards. In contrast to Chan's method [16], our fast multi-lateral filter enables of *arbitrary spatial function and arbitrary range function.* To reduce the quantization effect of the depth map (i.e. for the enhancement of depth super resolution), then a sub-pixel refinement algorithm is proposed based on probabilistic model. We will explain the details below.

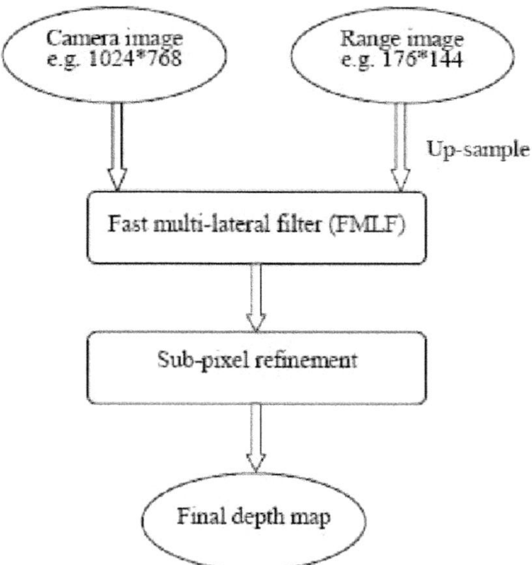

Fig. 2. Pipeline of our algorithm. The range image is up-sampled to the same size as the camera image, and two different images serves as the inputs of the fast multi-bilateral filter. The following is a sub-pixel refinement process.

4.1 FMLF for Depth Upsampling

To cope with the spatial super resolution requirement and meanwhile denosing for noisy real-time 3D sensors, like time-of-flight cameras, we propose a new fast multi-lateral filter for upsampling (FMLF). It is our goal to satisfy spatial super resolution and denosing requirement for real-time 3D sensors as fast as possible and to make our filter to be more flexible. Like [16], the FMLF filter takes the following form:

$$I_x^F = \frac{1}{K_p} \sum_{y \in N(x)} I(y) f_S(x,y)[(1-\phi(\sigma^2)) f_{\widetilde{R}}(D(x), D(y)) + \phi(\sigma^2) f_R(I(x), I(y))] \quad (1)$$

Where x is a pixel in low resolution range image and y is a pixel in the neighborhood of $N(x)$,$I(x)$ and $I(y)$ are the corresponding range values of pixel x and y, $D(x)$ and $D(y)$ denote the intensity values of pixel x and y in high resolution camera image respectively,f_S, $f_{\widetilde{R}}$ and f_R are all *arbitrary* functions, e.g. Gaussian function or Box function, $\phi(\sigma^2)$ represents a blend function, defining in the interval [0,1], σ^2 is the variance in pixel neighborhood N (e.g. 3×3 lattice) and K_p is a normalizing factor.

From the Equation (1), it is easy to conclude that a low weight ϕ makes our filter behave like a standard joint bilateral filter while a high weight ϕ gives higher influence to the latter range term $f_R(I(x), I(y))$ which makes the filter behave like an edge preserving bilateral filter that smoothes the 3D geometry independently of information from the high resolution camera image.

The main issue is to decide the value of weight ϕ since it controls the characteristic of our filter. We want the filter to switch to a bilateral filter in cases where the areas are actually smooth but heavily contaminated with random noise caused by range measure. Therefore, we intuitively define our blend function $\phi(\sigma^2)$ as follows:

$$\phi(\sigma^2) = \frac{\tau}{\sigma^2 + \tau} \quad (2)$$

Here, σ^2 is the variance in pixel neighborhood in N. We reason that if this variance is large, the local surface patch is most likely to be smooth and only noise-affected - thus former range term $f_{\widetilde{R}}(D(x), D(y))$ should be triggered. Once σ^2 being low, latter range term $f_R(I(x), I(y))$ will be triggered and our filter will act as a bilateral filter to ease the errors caused by range measurement. The unique parameter τ depends on the characteristic of the employed depth sensor and can be determined through experiments.

Please note that the computation of σ^2 on a low-pass filtered depth map is important, because it enables us to reliably disambiguate between isolated random noise peaks and actual depth edges. We have also found that it is better that the range term $f_{\widetilde{R}}$ takes the Gaussian form and cannot simply be set to a box filter in the range domain. With this design, we achieve much better preservation of depth discontinuities if ϕ lies in the transition zone. Finally, by choosing a small spatial support for our FMLF filter (3 × 3 lattice or 5 × 5 lattice), any form of texture copy around true depth edges can be reduced in practice while high-quality denoising is still feasible.

4.2 Acceleration Strategy

The complexity of Equation (1) makes direct compute could be time consuming and it is infeasible for real-time application. Several efficient numerical schemes

[14, 17], have been proposed to reduce the computational load of bilateral filter. Inspired by the fastest bilateral filter method [14] so far, our filter is sliced into one bilateral filter and one joint bilateral filter as follow:

$$I_x^F = \frac{1}{K_p} \sum_{y \in N(x)} f_S(x,y)(1-\phi(\sigma^2)) f_{\widetilde{R}}(D(x),D(y))I(y) + \frac{1}{K_p} \sum_{y \in N(x)} f_S(x,y)\phi(\sigma^2) f_R(I(x),I(y))I(y) \quad (3)$$

Here, the former is a joint bilateral filter while the latter is a bilateral filter. We then could take advantage of acceleration technology proposed by [14]: the range data of low resolution range image and the intensity data of high resolution camera image are discretized into a number of values, compute a linear filter for each such value respectively, the output of which is termed as PBFIC in [14] and get intermediate results by a linear interpolation between two closest PBFICs. The final result is obtained through adding operation between intermediate results. The details are given below:

In practice, we assume that the pixel intensity for an range image $I(x)$ is discrete with $I(x) \in \{0, \cdots, N-1\}$, where N is the total number of grayscale values. Letting $I(x) = k$, the latter term of Equation (3) $\frac{1}{K_p} \sum_{y \in N(x)} f_S(x,y) f_R(I(x), I(y))I(y)$ can be written as:

$$I^I(x) = \frac{\sum_{y \in N(x)} f_S(x,y) f_R(k, I(y))I(y)}{\sum_{y \in N(x)} f_S(x,y) f_R(k, I(y))} \quad (4)$$

For every pixel y and every intensity value $k \in \{0, \cdots, N-1\}$, we define:

$$W_k(y) = f_R(k, I(y)) \quad (5)$$

$$J_k(y) = W_k(y) * I(y) \quad (6)$$

Therefore, this bilateral filtering can then be decomposed into N sets of linear filter responses

$$J_k^I(x) = \frac{\sum_{y \in N(x)} f_S(x,y) J_k(y)}{\sum_{y \in N(x)} f_S(x,y) W_k(y)} \quad (7)$$

Thus, we have

$$I^I(x) = J_{I(x)}^I(x) \quad (8)$$

Where J_k^I is defined as Principle Bilateral Filtered Image Component (PBFIC) in [14]. In practice, only $N1$ out N PBFIC ($k \in \{0, \cdots, N1-1\}$) are used. Supposing x is $I(x) \in [L_k, L_{k+1}]$,therefore, the bilateral filtering value $I^I(x)$ can then be linearly interpolated [25] from $J_k^I(x)$ and $J_{k+1}^I(x)$ as following:

$$I^I(x) = (L_{k+1} - I(x))J_k^I(x) + (I(x) - L_k)J_{k+1}^I(x) \quad (9)$$

Note that, the range filter f_R is not constrained and any desired filter function can be chosen, but approximation can be poor if $N1$ is extremely small for some range filters, e.g., Box filter.

Similarly, the former term of Equation (3) $\frac{1}{K_p}\sum_{y\in N(x)} f_S(x,y)f_{\tilde{R}}(D(x),D(y))$ $I(y)$ can be reformulated as:

$$I^D(x) = \frac{\sum_{y\in N(x)} f_S(x,y)f_{\tilde{R}}(k,D(y))I(y)}{\sum_{y\in N(x)} f_S(x,y)f_{\tilde{R}}(k,D(y))} \tag{10}$$

Like Equation (9), it can be expressed as:

$$I^D(x) = (P_{k+1} - D(x))J_k^D(x) + (D(x) - P_k)J_{k+1}^D(x) \tag{11}$$

Where we assume that the intensity of pixel x in high resolution camera image is $D(x) \in [P_k, P_{k+1}]$, and $J_k^D(x)$ is defined according to:

$$J_k^D(x) = \frac{\sum_{y\in N(x)} f_S(x,y)Z_k(y)}{\sum_{y\in N(x)} f_S(x,y)U_k(y)} \tag{12}$$

Where $Z_k(y)$ and $U_k(y)$ are computed as:

$$U_k(y) = f_{\tilde{R}}(k, D(y)) \tag{13}$$

$$Z_k(y) = U_k(y) * I(y) \tag{14}$$

Finally, we get

$$I_x^F = (1 - \phi(\sigma^2))I^D(x) + \phi(\sigma^2)I^I(x) \tag{15}$$

These are the two main reasons why our approach outperforms the current state-of-the-art [12] for both accuracy and speed. The main storage required is six memory buffers with the same size as the input image for images. However, [12] requires a set of $N1$ image buffers to store the integral histogram during aggregation. Additionally, in our approach, image pixels are processed independently, allowing for parallel implementation.

Owing to the acceleration strategy discussed above, our GPU implementation of FMLF runs at about 35 frames per second using 8 PBFICs.

4.3 Sub-pixel Estimation

We obtain disparities of the range image on integer level after the process detailed in section above. However, unlike other methods, we also exploit the confidence of an established disparity value.

There has been a growing interest [6, 18] in obtaining accurate sub-pixel disparity since the parabola fitting approaches exhibit artifacts known as pixel-blocking [19]. With the help of Fourier analysis, Scharstein and Szeliski [20] have concluded that sinc interpolator is in theory the best interpolation to evaluate the disparity space image at fractional disparities.

Our approach performs a depth-edge-preserving smoothing on the disparity image, similar to [6] where bilateral filtering was used. Our sub-pixel estimation is similar to adaptive smoothing [26], however, unlike other methods we also further exploit the confidence of an established disparity value.

Our approach treats the sub-pixel estimation as energy minimization problem [27] with:

$$E_{tot} = \sum_{p \in V} E_p(d_p) + \sum_{(p,q) \in D} \alpha E_s(d_p, d_q) \tag{16}$$

Where data term E_p is the cost of assigning disparity d_p to pixel p, pairwise smoothness term E_s is the cost of assigning labels d_p and d_q to two neighboring pixels and α is a scale factor. The higher that α is chosen, the smoother is the resulting disparity map. One could also incorporate the image gradient or the gray value variance as a confidence measure to determine the value of α.

How can an appropriate data term be formulated? Let d_{int} be the integer disparity computed by our fast multi-lateral filter. The data cost of choosing d_p unequal to the former estimated d_{int} is formulated as following:

$$E_p(d_p) = (d_p - d_{int})^2 \tag{17}$$

From the Equation (17), we can conclude that the data cost tends to be small in low-textured regions, whereas it will be large in textured regions.

Let $\widetilde{d}$ be the average disparity within the considered patch D. The smoothness term E_s is defined according to:

$$E_s(d_p, d_q) = (d_p - \widetilde{d})^2 \tag{18}$$

Since our energy Equation (16) has a simple form, it is easy to compute the best solution for a certain point directly instead of to inference by belief propagation (BP) [4, 30] or graph cut (GC) [28, 31]. Partial derivation $\partial E_{tot} / \partial d_p = 0$ yields

$$d_p = \frac{d_{int} + \alpha(N-1)/N * \widetilde{d}}{1 + \alpha(N-1)/N} \approx \frac{d_{int} + \alpha * \widetilde{d}}{1 + \alpha} \tag{19}$$

Where N is the number of pixels within considered patch D. The higher that α is chosen, the smoother is the resulting disparity image.

In order to get close to the best solution of the above described problem, we need to iterate Equation (19) to propagate the update disparity values: d_{int} remains the origin value while d_p updated in every iteration.

This sub-pixel estimation favors solutions that are planar in 3D, i.e. fronto-parallel or slanted planes. This way, the algorithm is especially helpful for reconstructing flat object.

See Section 6.3 for results of our sub-pixel estimation.

5 Extended Range Data Super Resolution Based on a Single TOF Sensor

TOF camera robustly provides a range image of real world scenes at video frame rates that is perfectly registered with an intensity image. At this point, it looks like an ordinary color camera plus additional range information. We extend our range data super resolution *only* with a single TOF sensor, based on the

(a) Low-res depth map (b) Low-res intensity image

(c) Our refined result (d) Raw depth map (zoomed)

(e) Refined result (zoomed)

Fig. 3. From a low resolution depth map (a) and a low resolution grayscale intensity map (b) we create a depth map at a higher level quality. The significantly higher quality of our refined result (e) as opposed to the raw depth (d) is obvious.

insight that range measurement may be improved according to the low resolution grayscale intensity image of TOF sensor itself.

Unlike [10], our method relies on one frame and it does not require the setup or calibration process as literature [21] did previously. Therefore, it is available for real-time application, especially within dynamic environment.

Assume the $\widetilde{I}$ denote the low resolution grayscale intensity image obtained from TOF sensor. The fast multi-lateral filter we used is changed into following:

$$I_x^F = \frac{1}{K_p} \sum_{y \in N(x)} I(y) f_S(x,y)[(1-\phi(\sigma^2)) f_{\widetilde{R}}(\widetilde{I}(x), \widetilde{I}(y)) \\ + \phi(\sigma^2) f_R(I(x), I(y))] \quad (20)$$

This is almost identical to Equation (1) with the exceptions that the high resolution camera image is substituted with the low resolution grayscale intensity image.

Equation (20) can be also sliced into one bilateral filter and one joint bilateral filter as follows:

$$I_x^F = \frac{1}{K_p} \sum_{y \in N(x)} f_S(x,y)(1-\phi(\sigma^2)) f_{\widetilde{R}}(\widetilde{I}(x), \widetilde{I}(y)) I(y) \\ + \frac{1}{K_p} \sum_{y \in N(x)} f_S(x,y)\phi(\sigma^2) f_R(I(x), I(y)) I(y) \quad (21)$$

Therefore, the proposed acceleration strategy is utilized for speed up. The sub-pixel refinement strategy detailed in section 4.3 is also utilized to reduce quantization effect.

At the low integration times required for scene capture at 25 fps, the depth data provided by the SR4000 are severely noise-affected. As shown in Fig.3, our method successfully improves the quality of the low resolution depth maps and the resolution of the depth data can be effectively raised to the level of the video camera. True geometry detailed in data, such as discontinuities, are preserved and enhanced, the random noise level is greatly reduced.

Note that the generic artifacts that arise from the sensitivity of TOF sensor to object reflectance [21] are also prevented. By exploiting the GPU as a fast stream processor, real time performance is feasible. In a word, our design successfully handles the data produced by state-of-the-art time-of-flight sensors which exhibit significantly higher random noise levels than most active scanning devices.

6 Experimental Results

To demonstrate the effectiveness of our approach we applied our technique to various scenes including our own recorded sequences as well as scenes from the Middlebury stereo benchmark datasets [29]. In the following, we discuss implementation details in Section 6.1. Then, we analyze two main aspects of our approach more closely. We first demonstrate the visual superiority of our spatial super resolution results to results obtained with previous upsampling methods, Section 6.2. Thereafter, our depth super resolution results are described in Section 6.3. Finally, we discuss the advantageous run time properties of our algorithm, and discuss practical performance gains in comparison to optimization based upsampling methods, Section 6.4. We end the section by noting some overall limitations from our results and by discussing some possible future directions of investigation for our work, Section 6.5.

6.1 Implementation

Our experimental system consists of a Mesa Swissranger^{TM} SR4000 time-of-flight camera and a Point Gray^{TM} dragonfly2 video camera. The two cameras are placed side-by-side (as closely as possible) and are frame-synchronized. The Swissranger can produce range images with size up to 176 × 144 pixels and the dragonfly2 can provide color images with resolution up to 1024 × 768 pixels. To align the range and video images, we resort to the gray-scale intensity images that the Swissranger sensor provides in addition to range images. For the purpose of image registration, the approach proposed in [22] is applied. A better setup would be to use an beam-splitter to align the optical axes of both sensors to guarantee image alignment.

The approach is implemented on a state-of-the-art graphics card. Since our approach operations on individual pixels can be carried out independently, we can capitalize on the massive stream processing power of modern GPUs which by now feature up to 256 individual stream processors. In particular, we employ Nvidia's CUDA programming framework [9] to port our algorithm onto graphics hardware. Overall, we thereby achieve real-time up-sampling and denoising performance.

The simplicity of our method lies in that, Compared with previous work [16, 21], only two main parameters are involved in it, they are τ and α. τ is the constant used in Equation (2) which in essential denotes the expected variance due to noise, it is set to 50 experimentally. α is the magnification ratio of smooth term in Equation (16) and is set to 2 in this paper.

6.2 Spatial Super Resolution

Let the f_S in Equation (1) be Box function and $f_{\widetilde{R}}$ and f_R in Equation (1) be all Gaussian functions, we evaluate our algorithms on a real scene where a checkerboard and a cup are involved as shown in Fig.4. It is clear that our method successfully upsamples the low resolution depth maps to high resolution and with respect to the raw 3D data, the visual appearance of depth detail of the checkerboard is improved, especially on textured regions and around boundaries.

Our approach is also superior to the Chan's method proposed by [16]. Please pay attention to the details, indicated by the red boxes and arrows, for further comparisons. Furthermore, evaluated on Nvidia Geforce 9800 GT platform, the GPU implementation of our approach averages 31ms which is faster than that of the Chan's method (37 ms).

A visual comparison of the depth maps of the Middlebury datasets are provided in Fig.5.The original depth map is down-sample by 8 (2^3) from the ground truth.Currently, f_S is chosen to be Box functions, $f_{\widetilde{R}}$ and f_R in Equation (1) are all chosen to be Gaussian functions. The MRF approach in [5] also improves the stereo quality, but the improvement is relatively small compared to our approach. Cleary, the results using our approach have more clean edges than the input depth maps and the result using MRF approach [5]. According

(a) Low-res depth map

(b) High-res camera image

(c) Raw 3D data

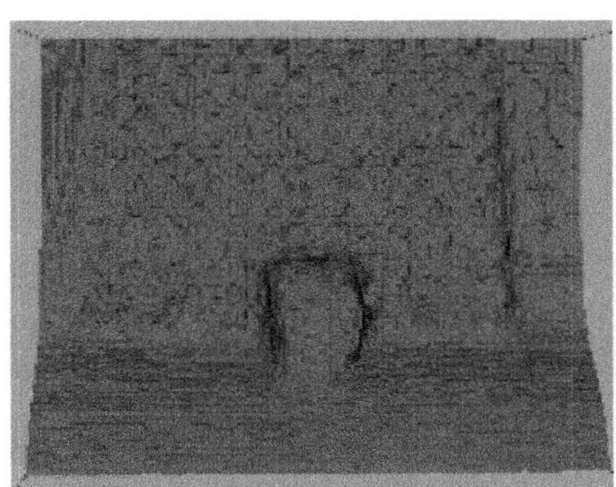

(d) Refined 3D data using Chan's approach

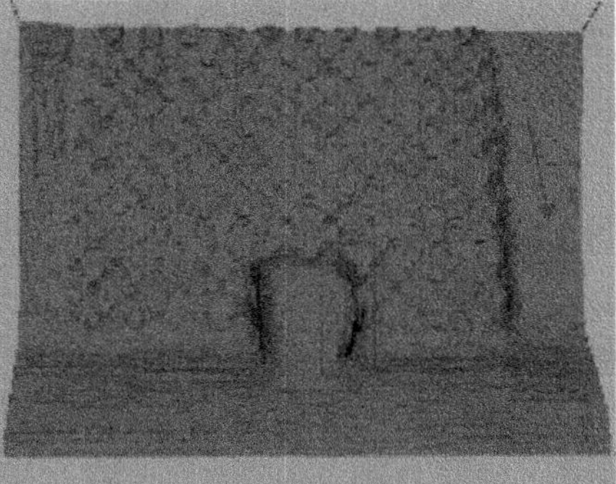

(e) Refined 3D data using our approach

Fig. 4. By using of the high-res camera image (b), our technique upsamples a low-res depth map (a) reconstructed as 3D geometry(c) to a high-res depth map which can be reconstructed as 3D geometry (e) with the comparison of Chan's results (d)

to Fig.5, it is faithfully acknowledged that our results are inferior to the results using Yang's method [6]. However, our approach is designed based on fast and simple pipeline whereas the Yang's method relies on iterations which make it impossible for real-time application.

By visual comparison, our approach outperforms the MRF approach as the resolution of the range sensor keeps on decreasing. In Fig.6, we show that even

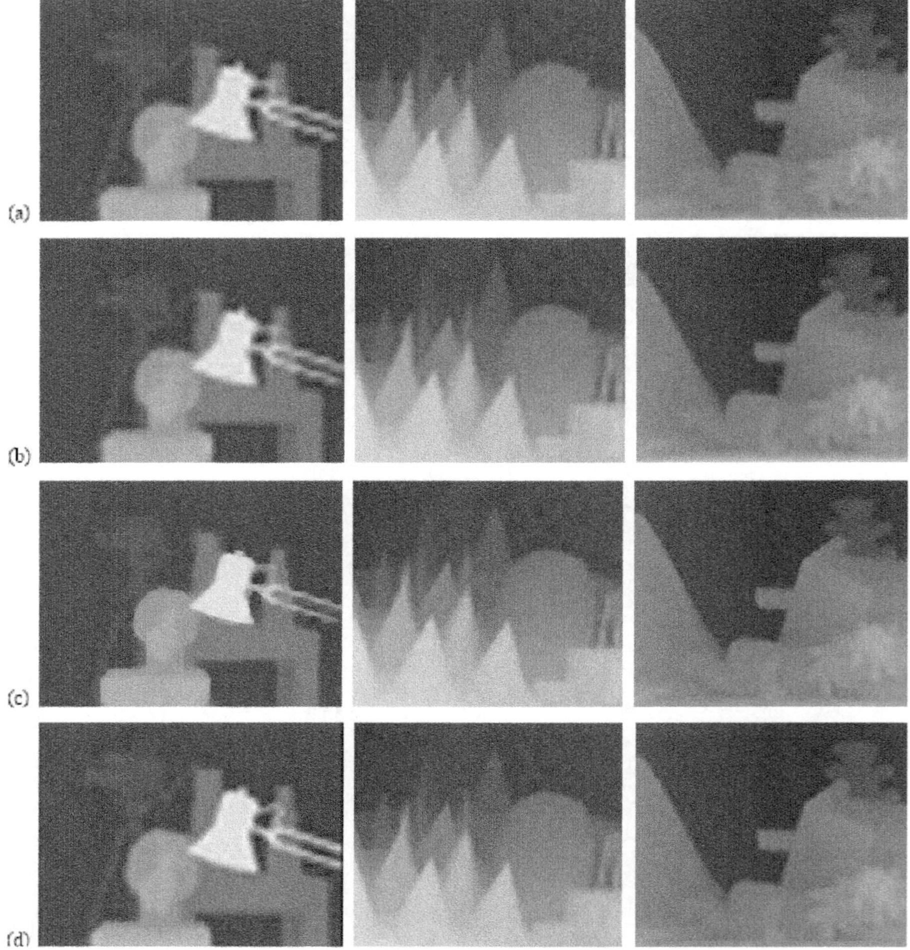

Fig. 5. Super resolution result on Middlebury datasets. (a) Before refinement. (b) Using Diebel's approach [5].(c) Using Yang's approach [6]. (d) Using our approach.

with tiny sensors (down-sample by 16, 2^4), we can still produce decent high-resolution range images.

Fig.7 show the performance of our algorithm on Middlebury datasets when f_S is chosen to be Box function or Gaussian function. Obviously, the two curves (corresponding to f_S is set to be Box function or Gaussian function) are almost coincidence in this experiment. A reasonable explanation may be that test images in Middlebury datasets are well taken under an ideal environment. However, practical experiments have proven that Gaussian function is more robust, especially for noisy cases.

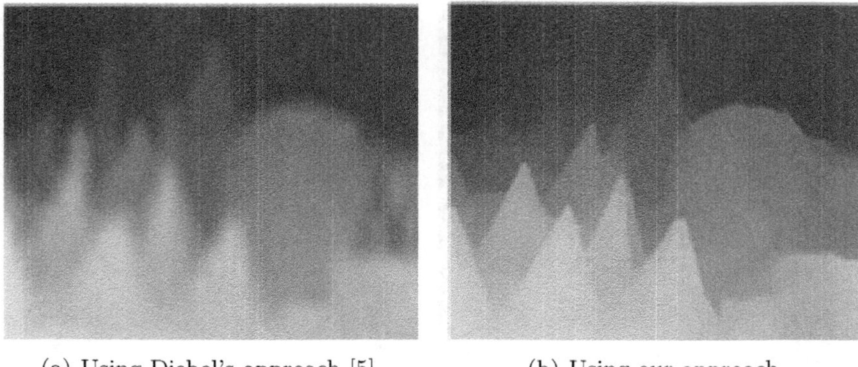

(a) Using Diebel's approach [5] (b) Using our approach

Fig. 6. Super resolution result on Cones dataset (down-sample by 16)

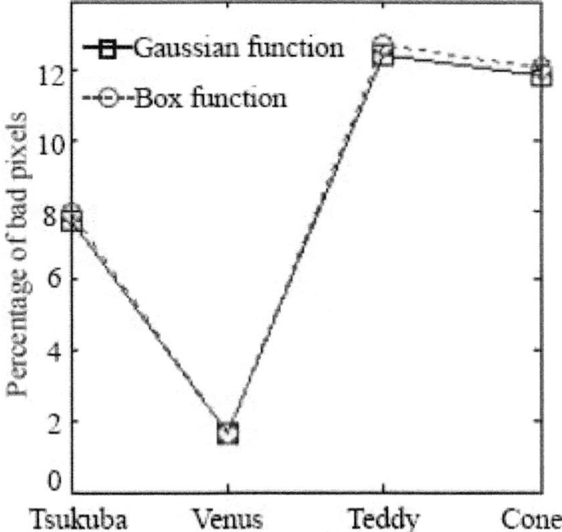

Fig. 7. The performance of our algorithm on Middlebury datasets with regard to f_R being Box or Gaussian function (with error threshold 1)

6.3 Depth Super Resolution

Besides the enhancement of the spatial resolution of range images, our approach also provides sub-pixel estimation for the further enhancement of the depth resolution of range images. A set of synthesized views are shown in Fig.8, providing a visual comparison of the algorithms with and without sub-pixel refinement. The enhancement of the depth resolution is clear: As shown in Fig.8(a), Fig.8(c) and Fig.8(e), the results are quantized into discrete number of planes. After sub-pixel estimation, the quantization effect is removed, as it is shown in Fig.8(b), Fig.8(d) and Fig.8(f).

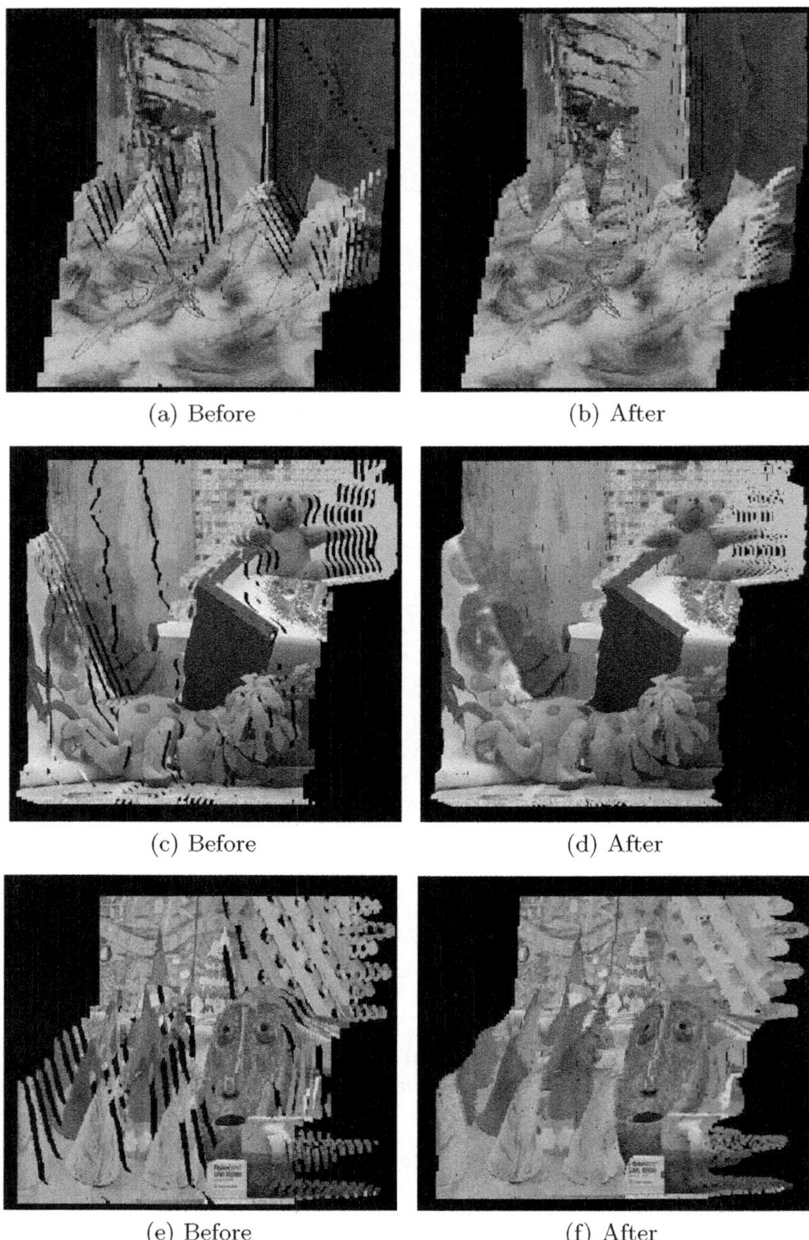

(a) Before (b) After

(c) Before (d) After

(e) Before (f) After

Fig. 8. Synthesized views produced by our approach before or after sub-pixel estimation

Table 1 evaluates the performance of our approach with and without sub-pixel estimation on Middlebury datasets. The original depth map is down-sample by 8 (2^3) from the ground truth. By comparing bad pixel percentages with

Table 1. Comparison of the results on Middlebury datasets with or without sub-pixel refinement (with error threshold 1, down-sample by 8)

	Tsukuba	*Venus*	*Teddy*	*Cone*
Without sub-pixel refinement	8.23%	1.73%	13.5%	12.9%
With sub-pixel refinement	7.71%	1.62%	12.4%	11.9%

Table 2. Comparison of the results on Middlebury datasets with or without sub-pixel refinement (with error threshold 1, down-sample by 2)

	Tsukuba	*Venus*	*Teddy*	*Cone*
Without sub-pixel refinement	2.45%	0.52%	2.66%	3.25%
With sub-pixel refinement	2.12%	0.43%	2.54%	2.98%

Table 3. Comparison of the results on Middlebury datasets with or without sub-pixel refinement (with error threshold 1, down-sample by 4)

	Tsukuba	*Venus*	*Teddy*	*Cone*
Without sub-pixel refinement	4.93%	1.02%	7.64%	7.42%
With sub-pixel refinement	4.06%	0.58%	6.90%	6.32%

and without sub-pixel estimation, we can conclude that sub-pixel refinement improves the performance of our approach for all data sets.

For further comparison, Table 2 and Table 3 list the performance of our approach with and without sub-pixel estimation on other two scales, i.e., down-sample by 2 (2^1) or 4 (2^2). Clearly, sub-pixel refinement improves the performance of our approach throughout all scales.

6.4 Runtime Analysis

In [5], Diebel et al. use an iterative solver to find the MAP upsampled depth values based on a MRF. Chan et al. [16] report their runtime analysis on Diebel's method: They used an implementation of the error metric and gradient computation on a cpu solved with an iterative L-BFGS-B solver [32] to derive depth results.

In [16], Chan et al. have shown that, to find an optimal set of MRF parameters, it consistently took over 150 iterations to converge to a solution when ran upon upsampling several scenes from the Middlebury dataset. When performing the error metric and gradient computation on the GPU along with appropriate loading of data to and from the GPU, it took on average 755 ms to compute 150 iterations. In the event of a full GPU solution, the iterative runtime of the gradient computation is 279 ms. This value is also absent of the time required for a GPU based solver to find a new gradient direction. In contrast, the GPU implementation of our approach averages 38 ms, which includes transfering data to and from the GPU.

Our approach is implemented with CUDA technique on a Geforce 9800 GT graphics card (512 MB video memory) GPU, together with a 2.7GHz CPU with dual core. In our experiments, the spatial super resolution phase contributes to major time spent in our algorithm while the run-time of depth super resolution is negligible since the Equation (19) is well suited for parallel execution. Generally speaking, processing an entire video camera image with large size (1024 $\times$ 768 pixels), our approach takes around 40 ms that makes it applausible for real-time applications. The runtime of some experiments is listed in Table4.

Table 4. Runtime on some experiments

	Figure3	*Figure4*	*Figure6*
Runtime	19ms	42ms	30ms

Finally, we would like to note that our approach performs much more efficiently than the multi-plane bilateral filtering and upsampling method of Yang et al. [6]. Although their method reportedly produces higher quality results, it will requires many iterations of a bilateral filter at each time step. Therefore, it is infeasible for real-time applications.

6.5 Discussions and Future Work

Although our approach improves depth details well, it does poorly in some cases. Fig.9 depicts such cases, e.g., the transparent and textureless glass and the high specular head statue. This is because the complimentary nature of the TOF sensor and color camera is invalid. Our formulation (color camera with the TOF sensor) cannot deal with this problem.

However, 3D shape of specular and transparent objects could be recovered by three viewpoints if incoming light undergoes two reflections or refractions [24]. Although we have not implemented this method, we can image that shape by light path may provide the depth for transparency and textureless objects which is currently not addressed in this paper.

In previous sections we have demonstrated that our approach allows us to produce high-resolution noise-reduced depth data in real-time even with a highly noisy TOF camera. The real-time requirement, however, makes some necessary for which we would like to contrive improved alternatives in the future. First, our FMLF switches between its two operating modes using a fundamentally heuristic model that requires manual parameter setting. In future, we plan to investigate how to learn the correct blending function from real data. Although our approach is defined in a local 3D domain which enables further improvement of our spatial super resolution results, also, we plan to research if it is feasible to improve accuracy of depth super resolution results.

(a) a transparent and textureless glass

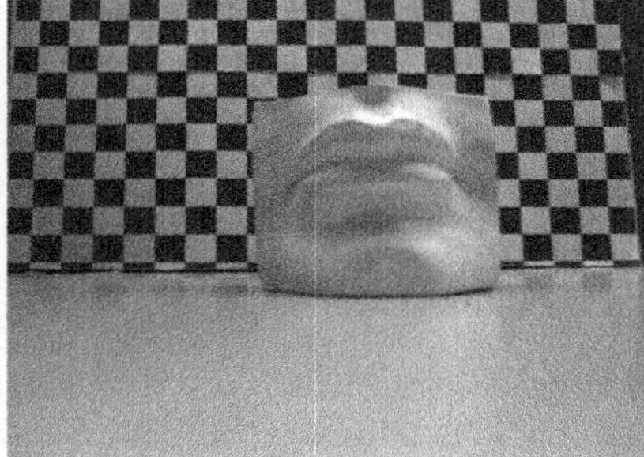

(b) a high specular head statue

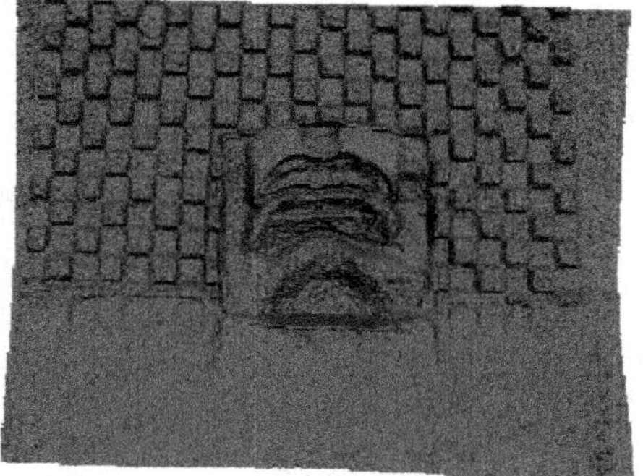

(c) Raw 3D data from the high specular head statue

Fig. 9. Some problematic cases for our approach

Furthermore, some stepping artifacts in the results are due to 8-PBFICs quantization which we will resolve in future. Finally, we would like to note that our current warping-based alignment, completed before starting experiments, may lead to non-exact depth and video registration in some areas, which may explain remaining blur in our results around some actually sharp depth edges. A feasible hardware solution would have the video and depth sensors record through the same optics which would greatly facilitate alignment.

7 Conclusions

In this paper, we present a fast and simple framework that enable us substantially enhance the spatial and depth resolution of range data in real-time while preserving features, reducing random noise and eliminating artifacts like texture copying phenomenon. We have shown that the results of our approach exceed the reconstruction quality obtainable with related methods from the previous literature. Adapting the fastest acceleration strategy ever known and using the parallel processing power of a modern graphics processor, the construction of dynamic scene with a high resolution is feasible. In addition, the super resolution method is extended to one single TOF sensor case. Look into future, there are still rooms for improvement. For instance, some constraints and priors (e.g. gradient profile prior) are hoped to be incorporated into our algorithm for further improvement.We also want to investigate how to tackle some difficult cases, such as specular and transparent objects.

Acknowledgements. This work was supported by the National Natural Science Foundation of China (Grant No. 60970076) and the National High Technology Research and Development Program of China (Grant No. 2009AA062704).

References

[1] Ogger, T., Griesbach, K., et al.: 3D-Imaging in real-time with miniaturized optical range camera. In: Proc. OPTO, pp. 89–94 (2004)
[2] SwissRanger^{TM} SR-4000, MESA Imaging inc., `http://www.mesa-imaging.ch`
[3] Yang, Q., Wang, L., Yang, R., Stewnius, H., Nistr, D.: Stereo matching with color-weighted correlation, hierarchical belief propagation, and Occlusion Handling. IEEE Trans. PAMI 31(3), 492–504 (2009)
[4] Liang, C.-K., Cheng, C.-C., Lai, Y.-C., Chen, L.-G., Chen, H.: Hardware efficient belief propagation. In: Proc. CVPR, pp. 80–87 (2009)
[5] Diebel, J., Thrun, S.: An application of markov random fields to range sensing. In: Proc. NIPS (2005)
[6] Yang, Q., Yang, R., Davis, J.: Spatial-depth super resolution for range images. In: Proc. CVPR, pp. 1–8 (2007)
[7] Tomasi, C., Manduchi, R.: Bilateral ltering for gray and color images. In: Proc. ICCV, pp. 839–846 (1998)
[8] Kopf, J., Cohen, M., Lischinski, D., Uyttendaele, M.: Joint bilateral upsampling. ACM Transactions on Graphics (TOG) 26(3), 96(1–5) (2007)
[9] Nvidia Corporation. CUDA: compute unified device architecture programming guide. Technical report (2008)
[10] Schuon, S., Theobalt, C., Davis, J.: Thrun. S.: High-quality scanning using time-of-flight depth superresolution. In: Proc. CVPRW 2008, pp. 1–7 (2008)
[11] Riemens, A.K., Gangwal, O.P., Barenbrug, B., Berretty, R.-P.M.: Multistep joint bilateral depth upsampling. In: SPIE 7257: Proc. VCIP (2009)
[12] Porikli, F.: Constant time O(1) bilateral ltering. In: Proc. CVPR, pp. 1–8 (2008)
[13] Chen, J., Paris, S., Durand, F.: Real-time edge-aware image processing with the bilateral grid. ACM Transactions on Graphics (TOG) 26(3), 103:1–10 (2007)
[14] Yang, Q., Tan, H.-H., Ahuja, N.: Real-time O(1) bilateral ltering. In: Proc. CVPR, pp. 557–564 (2009)
[15] Bhme, M., Haker, M., Martinetz, T., Barth, E.: Shading constraint improves accuracy of time-of-flight measurements. In: Proc. CVPR 2008 (2008)
[16] Chan, D., Buisman, H., Theobalt, C., Thrun, S.: A noise-aware lter for real-time depth upsampling. In: M2SFA 208: Workshop on Multi-camera and Multi-modal Sensor Fusion Algorithms and Applications (2008)
[17] Weiss, B.: Fast median and bilateral ltering. In: Siggraph., vol. 25, pp. 519–526 (2006)

[18] Gehrig, S.K., Franke, U.: Improving stereo sub-pixel accuracy for long range stereo. In: Proc. ICCV, pp. 1–7 (2007)
[19] Shimizu, M., Okutomi, M.: Precise sub-pixel estimation on area-based matching. In: Proc. ICCV, pp. 90–97 (2001)
[20] Szeliski, R., Scharstein, D.: Sampling the disparity space image. IEEE Trans. PAMI 26(3), 419–425 (2004)
[21] Zhu, J., Wang, L., Yang, R., Davis, J.: Fusion of time-of-flight depth and stereo for high accuracy depth maps. In: Proc. CVPR (2008)
[22] Guizar-Sicairos, M., Thurman, S.T., Fienup, J.R.: Efficient subpixel image registration algorithms. Opt. Lett. 33, 156–158 (2008)
[23] Lindner, M., Lambers, M., Kolb, A.: Data fusion and edge-enhanced distance fefinement for 2D RGB and 3D range images. IJISTA, Special Issue on Dynamic 3D Imaging 5(1), 344–354 (2008)
[24] Kutulakos, K.N., Steger, E.: A theory of refractive and specular 3D shape by light-path triangulation. In: Proc. ICCV, pp. 1448–1455 (2005)
[25] Durand, F., Dorsey, J.: Fast bilateral filtering for the display of high-dynamic-range images. In: Siggraph., vol. 21, pp. 1–7 (2002)
[26] Barash, D., Camiciu, D.: A common framework for nonlinear diffusion, adaptive smoothing, bilateral filtering and mean shift. Image and Vision Computing 22(1), 73–81 (2004)
[27] Alvarez, L., Deriche, R., Sanchez, J., Weickert, J.: Dense disparity map estimation respecting image discontinuities: A pde and scale-space based approach. Technical report. Research Report 3874, INRIA Sophia Antipolis, France (January 2000), 2,3
[28] Freeman, W.T., Pasztor, E.C., Carmichael, O.T.: Learning low-level vision. IJCV 40(1), 25–47 (2000)
[29] Middlebury datasets, http://vision.middlebury.edu/stereo
[30] Tseng, Y.C., Chang, N., Chang, T.S.: Low memory cost block-based belief propagation for stereo correspondence. In: Proc. ICME, pp. 1415–1418 (2007)
[31] Tappen, M.F., Freeman, W.T.: Comparison of graph cuts with belief propagation for stereo, using identical MRF parameters. In: Proc. ICCV, pp. 900–906 (2003)
[32] Byrd, R., Lu, P., Nocedal, J., Zhu. C.: A limited memory algorithm for bound constrained optimization. SIAM J. Sci. Comp. 16(5), 1190–1208 (1995)

Six Degree-of-Freedom Haptic Rendering for Biomolecular Docking

Xiyuan Hou and Olga Sourina

Nanyang Technological University,
50 Nanyang Avenue, Singapore
{houx0003,eosourina}@ntu.edu.sg

Abstract. Haptic device enable the user to manipulate the molecules and feel interactions during the docking process in virtual environment on the computer. Implementation of force-torque feedback allows the user to have more realistic experience during force simulation and find the optimum docking positions faster. In this paper, we propose a haptic rendering algorithm for biomolecular docking with force-torque feedback. It enables the user to experience six degree-of-freedom (DOF) haptic manipulation in molecular docking process. The linear smoothing method was proposed to improve stability of the haptic rendering during molecular docking. Collaborative docking with two devices was implemented.

Keywords: haptic rendering, biomolecular docking, torque feedback, stable algorithm, collaboration.

1 Introduction

Cyberworlds can integrate visual, audio and haptic tools for both research and e-learning applications. Biomolecular docking is a new research area which includes development of software system with both visual and haptic interfaces for rational drug design. In previous papers [33,32], we proposed a visual haptic-based biomolecular docking system for helix-helix docking research and implemented the application of this system in e-learning. In this paper, we propose an improved haptic rendering algorithm for biomolecular docking with torque force. The user can experience 6-DOF haptic force-torque feedback during the process of molecular docking. Moreover, we developed one application for collaborative molecular docking system which can be used for different haptic devices.

The molecular docking process of drug design can be simulated in a three-dimensional space where a ligand can be docked on to a receptor. By using computer-aided design system, the manipulation of molecules can be realized with real-time interactive visualization in virtual environment. Especially, it has been proved to be helpful for users to understand the interactions between molecules in e-learning applications. As described in [6], Cooper et al. developed a multiplayer online molecular docking game Foldit for non-scientists to engage them in protein prediction problem solving.

M.L. Gavrilova et al. (Eds.): Trans. on Comput. Sci. XII, LNCS 6670, pp. 98–117, 2011.

Beside the basic visual technology for molecules, the haptic interface appears to be another effective tool to improve the immersion and interaction during the molecular docking process in virtual environment. Haptic technology provides interactivity between real and virtual environment through force and torque feedbacks transmitted by the haptic devices. This makes it possible to manipulate molecules and transform biomolecular interactions into sensory experiences during a virtual experiment. Therefore, haptic-based visual biomolecular docking allows developing more interactive systems that could be used in rational drug design and molecular medicine.

Biomolecular docking is an assembling process for molecular structures to predict the preferred complimentary molecular shapes that can bind molecules to form a stable complex. Since there is an exponential increase in conformations as the number of atoms increases, the simulation of docking task with an automatic conformation search algorithm could be difficult. By using visual haptic-based molecular docking system, the user can manually explore the conformational molecular space to find an optimal conformation within the minimum time.

Some previous studies have proved that the force display can provide a better understanding of the molecular docking process compared with traditional visual display methods [23]. As the haptic device provides realistic force feedbacks to users, in recent years, more and more researchers tend to explore and analyze the molecular docking process with haptic interfaces [35,15,11]. In earlier work [25], Ouh-young et al. proposed a real-time system for interactive molecular docking that allows the user to manipulate the position of ligand and feel the interactive forces between molecules. A force smoothing method was presented in [20] where forces from the Lennard-Jones (LJ) force field were calculated, and instability was eliminated when two atoms are in contact. Besides traditional 3-DOF haptic force feedback, the torque force also plays an important role in molecular docking process. Persson and Cooper [26] developed a force-torque haptic molecular interaction system. They also conducted an evaluation to test the importance of force feedback in learning and understanding the interactions between molecules. Their results proved that both force and torque forces play important roles in helping students to understand the concepts of molecular interaction. In [19], a 5-DOF haptic device and computational engine were developed for computer-aided molecular docking (CAMD) which provided both force and torque feedback.

This paper is organized as follows. Section 2 introduces the research background in biomolecular docking. Section 3 describes basic concepts of our haptic rendering algorithm and the method that we used to improve the stability of haptic force and torque display. Simulation results and analysis are given in section 4. Section 5 introduces one collaborative molecular docking application developed with our 6-DOF haptic rendering algorithm. In section 6, conclusion and future work are discussed.

2 Related Work

Modern molecular visualization systems such as RasMol [30], PyMol [10], JMol [2], MDVQS [31], etc allow visualize and analyze complex molecular structures.

On the other hand, haptic-based technology allows molecular docking with force feedback in such way that the user could "feel" force field of bimolecular interactions. There are haptic-based systems that also enable users to feel an electrostatic force of the explored molecule.

Lai-Yuen and Lee [18] developed a computer-aided design system with the lab-built 5-DOF haptic device for molecular docking and nanoscale assembly. Nagata and Mizushima [24] developed a prototype for protein-ligand docking simulation with the total potential energy (Van der Waals potential energy, electrostatic potential energy and hydrogen bond potential energy) calculation between atoms of ligand and protein. In [29], a grid map was used to generate the electrostatic field data around the molecular structure. The haptic forces at any position were calculated using tri-linear interpolation of the potential energy. Stocks and Hayward developed a haptic system HaptiMol ISAS [34]. It allows the user to interact with the biomolecular solvent accessible surface through the haptic device. A navigation cube is used to visualize the explored surface region, and the cube can be automatically scaled to fit the workspace of haptic device. The cube approach allows choosing a limited interaction area of very large molecules. In another approach proposed by Subasi and Basdogan [36], the user can insert a rigid ligand molecule into the cavities of protein to search for binding cavity. Similarly to the cube approach, an Active Haptic Workspace (AHW) was implemented for the efficient haptic-based exploration of large protein-protein docking in high resolution. In the system, the user could feel a tunneling effect when the ligand molecule is pulled towards the binding cavity. In [14], an Interactive Global Docking (IGD) approach was presented. An immersive environment for docking interface with both visual and haptic rendering was implemented. The commercial Falcon haptic device from Novint Technologies was used to provide 3-DOF force feedback. In [35], an Interactive Molecular Dynamics (IMD) system was implemented. A real-time force feedback based virtual reality system Steered Molecular Dynamics (SMD) [27] was developed for dynamic simulation of bimolecular interaction.

For the 6-DOF haptic-based molecular docking, Daunay and Micaelli [8] developed a haptic-based molecular docking system which enables the feeling of both force and torque. The system provides haptic feedback for a flexible ligand-protein docking. However, the system has limitations on the size of protein molecules. Project CoRSAIRe [12], is an example of a multisensory virtual reality system with 6-DOF haptic device provided by the Virtuous haptic interface of Haption Company. It was designed for a study of protein-protein docking. To create an immersive virtual environment, the visual, audio and haptic feedbacks were combined to enhance the process of exploration.

Haptic-enable web-based molecular docking is a new research direction in development of molecular docking simulation. In [9], Davies implemented a prototype of Molecular Visualiser (MV) system with Web3D standards adding an extension to support haptic interaction. MV provides the following features: visualizations of molecular systems, visualization of potential energy surfaces, and implementation of wavepacket dynamics. The system can run in a web browser using VRML, or be delivered to a virtual environment in which haptic properties are assigned based on the molecular dynamics of the system. Applications of MV for both research and teaching were also discussed. The authors mainly focused on the visualization of the molecular models and did not study molecular docking problem. A new multiplayer online game Foldit has been developed by Cooper et al. [6]. It is a novel approach for solving the protein prediction problems. The Fodit has proved that the accurate protein structure model can be produced through the game play. As predicted by the Fodit system, human search process has advantages in complexity, variation and creativity. Liu and Sourin [21] proposed a functional approach for modeling geometry of objects. The function-based objects were added into VRML. The user is able to define function of any shape with implicit functions. Later, Wei et al. [37,38] extended the system proposed in [21] by incorporating haptic based features to new FVRML nodes. A new density node was proposed for haptic implementation. It allows exploring an electrostatic force of molecule with a probe in VRML.

In the previous works [33,32], we described our work on haptic-based visual molecular docking targeting on the research of helix-helix docking. In [32], we proposed an e-learning scenario with our system. In paper [16], we described the system interface and further development of a prototype of biomolecular docking system Haptic-based Molecular Docking (HMolDock). In this paper, we describe our 6-DOF haptic rendering which allows the user having more real experience in virtual molecular docking.

3 Basic Concepts and Algorithm Description

We developed the prototype of biomolecular docking system HMolDock using the haptic device PHANTOM 1.5/6DOF. A Protein Data Bank [3] format file of molecular structure is used as an input of the molecular model. Atoms coordinates are got from the input PDB files. The radius, position, and corresponding color are also determined based on the atom type and its belonging residue which are extracted from PDB data. An interaction force including torque force between ligand and receptor is calculated and displayed in real time, so that the user could feel an attractive/repulsive force and rotation torque through the 6-DOF haptic device simultaneously.

3.1 Lennard-Jones Potential

The algorithm is based on the Lennard-Jones Potential which is usually used to describe an interaction between a pair of atoms or molecules.

The common Lennard-Jones Potential can be expressed as follows:

$$V_{(r)} = 4\varepsilon[(\frac{\sigma}{r})^{12} - (\frac{\sigma}{r})^{6}] \tag{1}$$

where r is the distance between the atom pair. ε is the depth of potential well, σ represents the specific distance where the inter-particle potential is zero. The values of these two parameters will be different for different interacting particles.

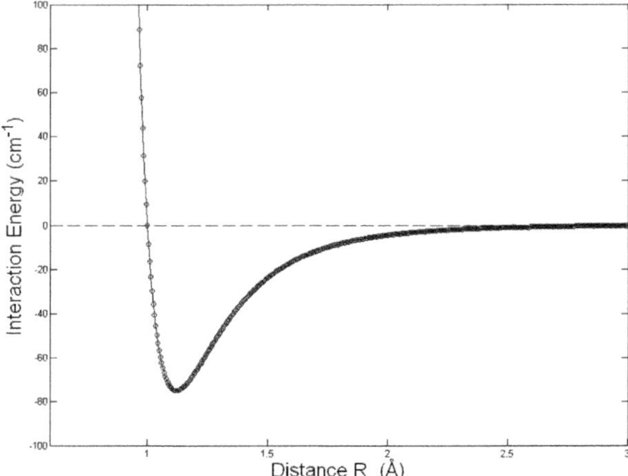

Fig. 1. The simulation of Lennard-Jones potential

The Lennard-Jones potential is an efficient mathematical model used to calculate the interaction force according to the distance between two atoms or molecules. As shown in Fig. 1, the potential is expressed as mildly attractive when one molecule is approaching to another from a far distance. However, when two molecules are close enough, the potential will be strongly repulsive.

The aim of bimolecular docking is to achieve an optimized conformation and orientation between two molecules such that the potential overall energy is minimized. Through the calculation of Lennard-Jones potential and real-time force torque feedback of 6-DOF haptic device, the user can find the optimal docking positions faster and more efficiently. It can totally overcome disadvantages of automatic simulation methods which take much longer time to calculate the best binding position for exploring a large energy landscape.

3.2 Force Calculation

In calculation of interaction forces between ligand and receptor, we use the method of Lennard-Jones potential which has been considered as the most important factor in tansmemebrane a-helix interaction [4].

The interaction forces between molecules are well approximated by LJ potential which changes according to the distance between molecules.

Therefore, a potential energy can be used to represent the interactions between two large molecules. Assume that there are M atoms in receptor and N atoms in ligand, the LJ potential between two molecules is calculated as follows:

$$V = \sum_{i=1}^{M} \sum_{j=1}^{N} 4\varepsilon_{ij} [(\frac{\sigma_{ij}}{r_{ij}})^{12} - (\frac{\sigma_{ij}}{r_{ij}})^{6}] \quad (2)$$

where ε_{ij} and σ_{ij} are LJ parameters for atom i in receptor and atom j in ligand, and r_{ij} is the distance between the atom pair.

During the interactive force calculation between ligand and receptor, first, we calculate the sum force that one atom of ligand is suffered from all atoms of receptor. As shown in Fig. 2, the interactive force that the yellow atom suffered comes from the force composition of all possible atom pairs between yellow atom and green atoms. In practice, the ligand consists of more than one atom, thus composition of forces of all ligand atoms is the final interaction force of the ligand. Also this force composition will be directly exerted to the hand of the user.

From the LJ potential equation, the force calculation function can be derived as follows:

$$F = \nabla V(r) = \frac{d(V(r)\tilde{r})}{dr} \tag{3}$$

$$F = \sum_{i=1}^{M}\sum_{j=1}^{N} 24\varepsilon_{ij}[2(\frac{\sigma_{ij}^{12}}{r_{ij}^{13}}) - (\frac{\sigma_{ij}^{6}}{r_{ij}^{7}})]\tilde{r} \tag{4}$$

where $\tilde{r}$ is the distance unit vector.

The algorithm described in Algorithm 1 is sensitive to the distance r_{ij}. Depending on the distance between two molecules, the interaction force can be classified into two opposite force sections: short-range repulsion force and long-range attraction force.

In (4), the term $(\frac{1}{r_{ij}})^{13}$ describes the short-range repulsive force. When the distance between a pair of atoms or molecules is very close, the dominator distance is relatively small. Therefore, the magnitude of repulsion force is strongly large.

The term $(\frac{1}{r_{ij}})^{7}$ describes the long-range attractive force. As the separation between a pair of atoms or molecules increases, the 7th parameter plays the dominate role. Correspondingly, the interaction between two particles will change

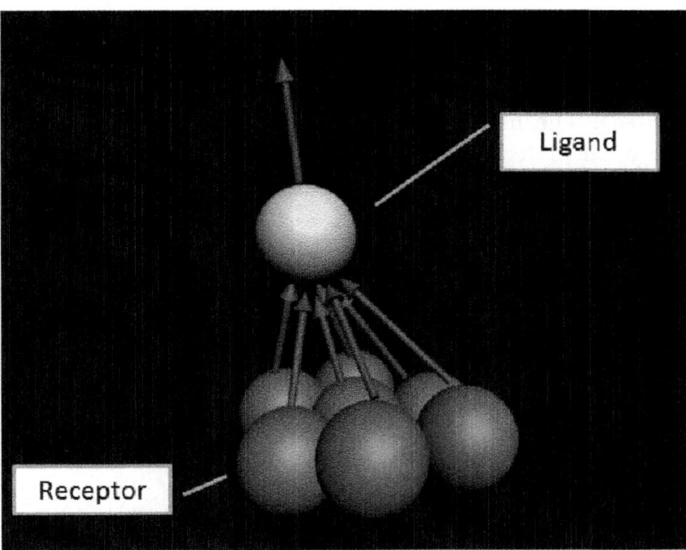

Fig. 2. The interaction forces between one atom and other atoms

Algorithm 1. Force Calculation

```
Input: Postition of ligand P_l, Position of receptor P_r
Output: Interactive forece F_device
{The ligand is grasped by the haptic device}
if grasp object is true then
   {Initialize variables}
   F ← 0 {Interactive force}
   D_ij ← 0 {Distance between one atom pair}
   {Retrieve all atoms of ligand}
   for all i such that 1 ≤ i ≤ M do
      {Retrieve all atoms of receptor}
      for all j such that 1 ≤ j ≤ N do
         S ← σ_ij
         E ← ε_ij
         D_ij ← P_rj − P_li
         r ← |D_ij| {Get the magnitude of distance vector}
         r_dir ←normalize(D_ij) {Get the direction of distance vector}
         F ← f(r) {Calculate the interactive force between this atom pair}
         F_device ← Scale(F) {scale the force magnitude to fit the haptic device}
      end for
   end for
end if
return F_device
```

form intensively repulsive to mildly attractive. Finally, the interaction force tends to the zero as the distance tends to infinity. In practice, these two opposite force factors play the dominant role in force calculation.

There are many different force field models that can be used to simulate proteins and other molecules. The model used and described in here is OPLS-aa [17,7,28] which is parameterized for small organic molecules in protein simulation.

For the homo-atomic pairs, there are published LJ parameters available (i.e. [1] for OPLS-aa). The interaction of hetero-atomic pairs, the effective values of ε and σ are calculated from those for the homo-atomic pairs. This way of calculation is called mixing rule. OPLS-aa uses the same non-bonded functional forms as AMBER [39], and the Lennard-Jones terms between unlike atoms are computed using the mixing rule [22] as follows:

$$\sigma_{ij} = \sqrt{\sigma_{ii}\sigma_{jj}} \tag{5}$$

$$\varepsilon_{ij} = \sqrt{\varepsilon_{ii}\varepsilon_{jj}} \tag{6}$$

3.3 Torque Calculation

Most of the haptic rendering algorithms for molecular docking are based on 3-DOF haptic devices which can only simulate force effects by moving along three coordinates. However, with 6-DOF haptic devices, the torque feedback of ligand can be calculated to produce the torque effects which could add rotation force around three axes in virtual environment.

For some previous torque calculations, the torque is only calculated around the position of haptic device's attachment point which is set to be the closest atom to the ligand's center of mass [26]. In our algorithm, as shown in Algorithm 2, we propose a more flexible way: the torque force can be calculated at any surface position of the ligand. As long as a haptic interface point (HIP) is attached to the ligand and the docking process started, the position of HIP is stored and updated to calculate the torque feedback. In this way, the user has more freedom in manipulation of the ligand. Especially in e-leaning, the users could have a better understanding of interactions between molecules with changing position of HIP.

The calculation of the torque force requires atom forces derived from (4) and the position of the haptic interface point x_{HIP} on the ligand as well. Like the computation of force, the torque feedback is the sum of all torque effects from atom pairs between ligand and receptor. The haptic torque T is calculated as follows:

$$T = \sum_{j=1}^{N} [(x_j - x_{HIP}) \times \sum_{i=1}^{M} F_{ij}] \tag{7}$$

where x_j is the position of ligand atom j, and x_{HIP} is the position of contact point between haptic interface point and ligand. For the specific ligand atom j, $\sum_{i=1}^{M} F_{ij}$ represents the interaction force between one ligand atom and all atoms of the receptor. The torque is calculated by the cross product between a force vector and a displacement vector (vector from the point that torque is measured to the point where Van der Waals force is applied).

Algorithm 2. Torque Calculation

```
Input: Postition of ligand P_l, Position of receptor P_r, Grasp position P_HIP
Output: Torque T_device
if grasp object is true then
  {Initialize variables}
  T ← 0 {Torque vector}
  T_d ← 0 {Distance vector from ligand center to grasp position}
  {Retrieve all atoms of ligand}
  for all i such that 1 ≤ i ≤ M, i ∈ ligand do
    {Retrieve all atoms of receptor}
    for all j such that 1 ≤ j ≤ N, j ∈ receptor do
      F_ij ← f(r) {Calculate the force on this ligand atom}
      T_d ← P_i − P_HIP {Calculate direction vector for torque}
      T_ij ← T_d × F_ij {Calculate the torque using cross product}
    end for
    T ← Σ_{i=1}^{M} F_i {Accumulate the torque value on one ligand atom}
  end for
  T_device ← Scale(F) {Scale the torque magnitude to fit the haptic device}
end if
return T_device
```

3.4 Stability of Haptic Rendering

The force and torque are calculated from the sum of interaction forces of all possible atom pairs between two molecules. It is well known that this force is extremely sensitive to the distance. There would be a sudden change on the force direction and magnitude when the Van der Waals force changes from attraction to repulsion. This sharp change of the force vector can cause sudden jump effect in successive force and torque frames of haptic devices.

To achieve the smooth Van der Waals force, we use a linear smoothing method similar with [13] to avoid this kicking phenomenon which is caused by the sudden change of force display during molecular docking process. In Algorithm 3, we set a maximum step size F_{step} for the successive force magnitudes. To achieve a smooth haptic update, the setting of step size depends on the type of haptic device and computer configuration. The force calculated in a previous frame is F_{t-1} and the current force is set to be F_t. The kicking problem is often happened when $F_t > F_{t-1}$. Due to the high update rate of haptic pipeline, a simple algorithm is used for the force display. When $F_t - F_{t-1} > 2F_{step}$, the current force is set to be $F_{t-1} + F_{step}$. When $2F_{step} > F_t - F_{t-1} > F_{step}$, the current force is set to be $F_{t-1} + \frac{F_{step}}{2}$. When $F_{step} > F_t > F_{t-1}$, the force does not change. In addition, this algorithm is also used for the rendering of torque update. Although there are some other methods that could be used to improve stability of the haptic rendering, this method is proved to be computationally efficient in practice.

The purpose of bimolecular docking is to find the position with the minimal potential energy. However, in practice, this position is hard to be captured manually since there is a great change in the force magnitude from attraction force to repulsion force. As a result, it is difficult for the user to stay with a haptic device in this minimal energy position.

To solve the problem of potential unstable factor and to produce an accurate manipulation, we set a zero area when the energy of molecular docking force is very small, as it is shown in the red area of Fig. 3. When the force changes from

Algorithm 3. Stabilization Method

Input: Previous force values F_{t-1}, F_t
Output: Optimized force F_o
$F_{step} \leftarrow stepsize$ {Initialize force step size}
$F_{dif} \leftarrow |F_t - F_{t-1}|$ {Calculate the force step between previous two frames}
if $F_{dif} > 2F_{step}$ **then**
  {Compare the force difference with defined force step size}
  $F_o \leftarrow F_{t-1} + F_{step}$
else if $F_{dif} > F_{step}$ **and** $F_{dif} < 2F_{step}$ **then**
  $F_o \leftarrow F_{t-1} + \frac{F_{step}}{2}$
else
  $F_o \leftarrow F_t$
end if
return F_o

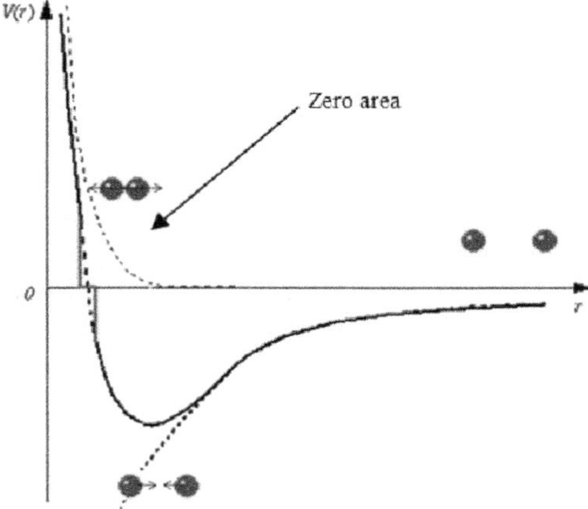

Fig. 3. The zero area (in red) is set when the potential energy approximates to zero

attraction to repulsion, the user can maintain position in this area. Therefore, the user can have a stable control of ligand and maintain it in the position of the minimal energy area. Otherwise, there would be an obvious vibration around this area.

4 System Implementation and Performance

In this section, we describe the HMolDock system and the results of molecular docking interaction. To display the force and torque feedback, the PHANToM Premium 1.5/6DOF designed by SensAble Technologies is used in our system.

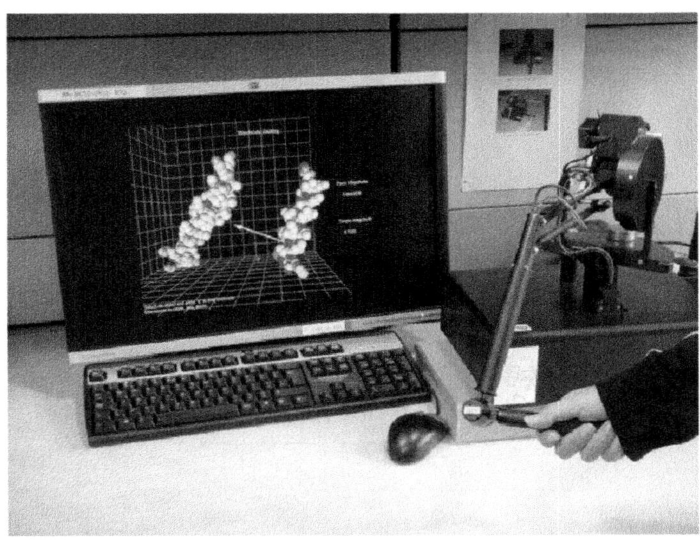

Fig. 4. HMolDock system with PHANToM Premium 1.5/6DOF haptic device

4.1 Implementation

The proposed haptic force-torque rendering algorithm has been implemented in our prototype HMolDock system. The system is developed and tested on a dual 1.86GHz CPU workstation using OpenHaptics Toolkit, OpenGL library and Visual C++ programming language. Fig. 4 shows the set-up of the HMolDock system with 6-DOF force-torque feedback haptic device in the laboratory.

Although there are different file formats for molecular structures, we use a Protein Data Bank (PDB) format for the input. Two molecules could be visualized on the screen as shown in Fig. 4. The user could assign a haptic interface point to one of the molecules and move this molecule towards/around of another one to feel both force and torque feedback around three axes.

The haptic rendering pipeline mainly consists of molecule transformation, force and torque calculation, and stable force algorithm implementation. An

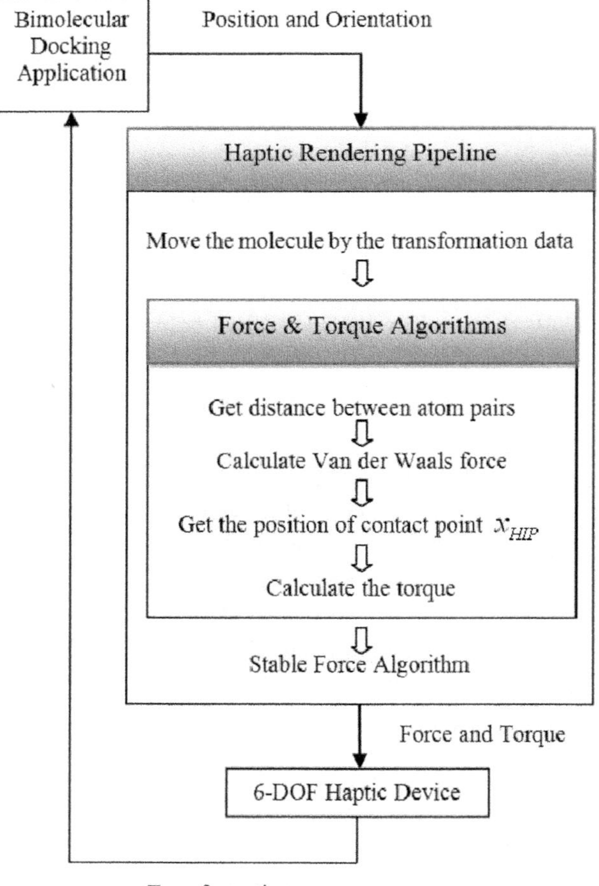

Fig. 5. The structure of the haptic rendering pipeline with force-torque feedback

overall structure of the haptic rendering pipeline for bimolecular docking process is shown in Fig. 5.

At first, when the user controls the haptic device to manipulate the ligand around the receptor, the system transfers the molecule according to the position and orientation data from the device. During the manipulation process, the force algorithm calculates an interaction force depending on the distances between all atom pairs of two molecules. Meanwhile, the algorithm calculates a vector between every atom of the ligand and the contact point x_{HIP}. This vector will be used for the torque computation. Finally, both force and torque values are optimized by the stable algorithm to display continuous and smooth force-torque feedback to the user.

4.2 System Performance

Fig. 6 shows a bimolecular docking process between one αIIb helix (with 154 atoms) and one designed antibody-like complementary peptide anti-αIIb(with 266 atoms). The coordinate grid is used to help the user to improve the accuracy of manipulation in 3D environment. After molecules are loaded into the system, the user can assign a haptic interface point to probe and grab the ligand which can be moved towards/around the receptor.

The position of contact point between haptic probe and ligand is used to calculate the torque force caused by all atom pairs between ligand and receptor. The resulting attraction/repulsion forces and rotation torques forces can be displayed through the 6-DOF haptic device. Therefore, the molecule can be selected by the haptic mouse and moved around to let the user feel the force-torque feedback in our bimolecular system. Values of force and torque magnitudes are shown on the right side of the screen to help the user to find the optimum docking position. In addition, the force direction and magnitude are visualized as a yellow vector, and the cyan vector indicates the change of torque vector. Here, the direction and length of the arrow indicates the attraction/repulsion force and its magnitude. Fig. 6a shows an attraction force between two separated molecules. As the distance becomes smaller, the repulsion force will appear and increase rapidly. Fig. 6b shows the force and torque directions and magnitudes when two molecules contact with each other. Both directions of force and toque change to the opposite, and the increased magnitudes can be read from the value shown on the right side of the screen.

With the intuitive vector representation and force-torque feedback, the user can experience more realistic force feeling during the docking process and analyze the optimal positions of minimal potential energy.

4.3 Force and Torque Results

Fig. 7 represents both force and torque response in the molecular docking process. During this process, the haptic device manipulates the ligand approach to the receptor from a far apart distance to a contact status. We divide this molecular docking process into the following three time intervals:

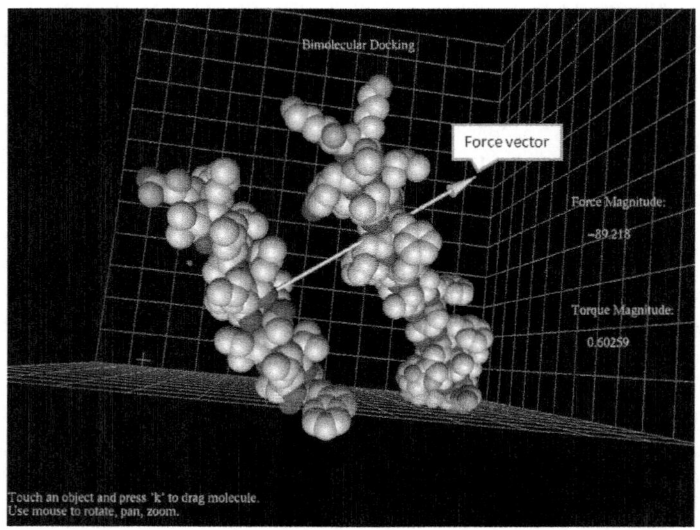

(a) Separate

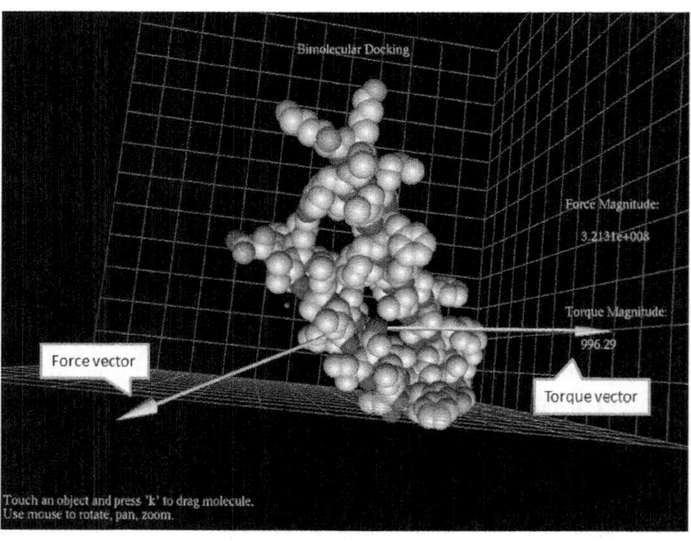

(b) Contact

Fig. 6. An interaction between two α-helices: the yellow and cyan arrows represent the force and torque vectors. In (a), two molecules are separated in a distance, and the force vector indicates the attractive force. In (b), two molecules contact with each other, and the force vector changed to the inverse direction as a repulsive force.

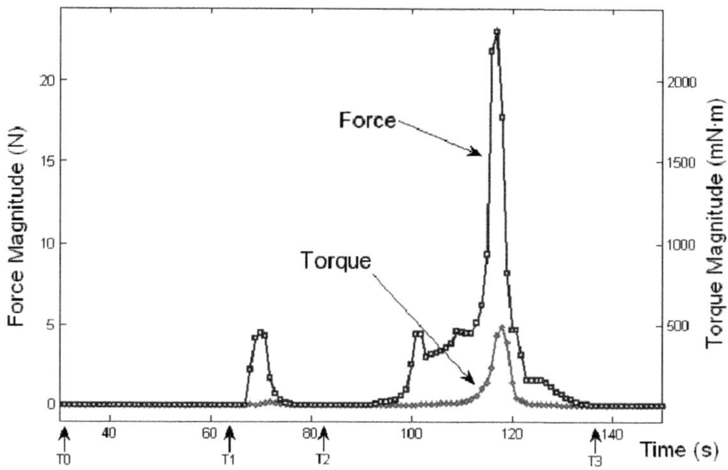

Fig. 7. Force and torque magnitudes in a molecular docking process

1. $T_0 < T < T_1$ (Separate): two molecules are far apart, the force and torque magnitudes are very weak.
2. $T_1 < T < T_2$ (Approach): two molecules are separated with a limited distance.
3. $T_2 < T < T_3$ (Contact): the ligand contacts with receptor, the repulsion force increases greatly.

As shown in Fig. 7, both Van der Waals force magnitude and torque magnitude change according to the distance between two molecules with the same trend. In addition, the force magnitude is more sensitive than the torque magnitude in the process of molecular docking. Through our biomolecular docking system HMolDock, the force and torque change from attraction to repulsion can be directly experienced by the user while performing drug design or molecular docking simulation. Therefore, the optimum docking position can be found by the intuitive haptic feeling instead of complex computation of docking algorithms.

4.4 Analysis of Stabilization Methods

To solve the instability and vibration problems appeared during the molecular docking process, we applied a linear smoothing method to improve the stability of docking manipulation. Since these changes of force and torque feedbacks have similar trends in docking process, and the force is much more sensitive to the distance between molecules, we use the force feedback to analyse how stabilization methods could improve stability of molecular docking manipulation.

Fig. 8a shows the simulation results of the complete molecular docking process. The stabilization method were used in the docking process shown in Fig. 8a. Correspondingly, the docking process without stabilization methods is shown in 8b. The same ligand and receptor were used in these two molecular docking

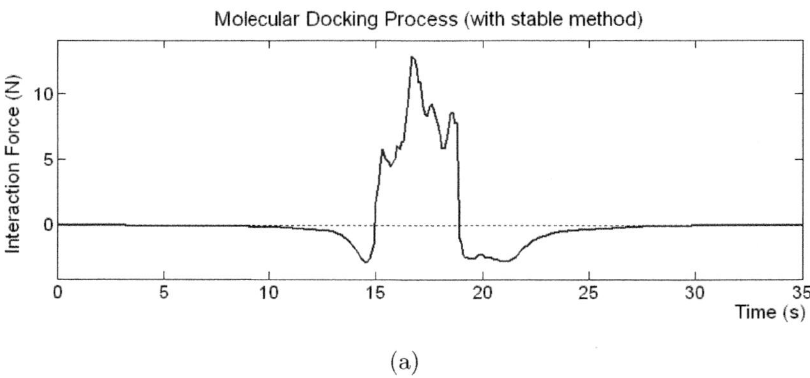

(a)

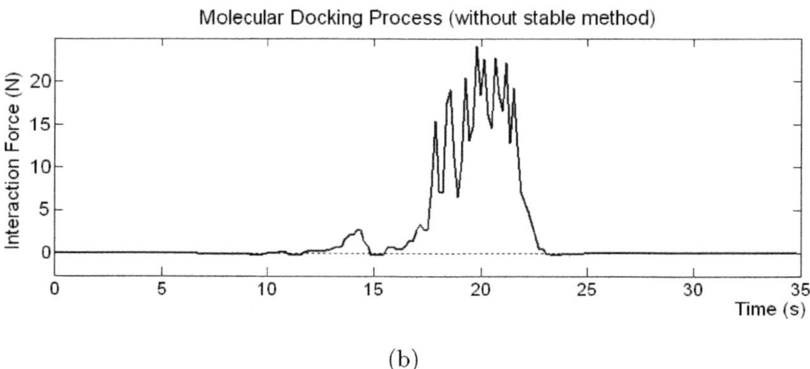

(b)

Fig. 8. The simulation results of a molecular docking process using same ligand and receptor. (a) The optimized interaction force feedback during a molecular docking manipulation with stabilization method. (b) The interaction force feedback without stabilization method during a docking process. All the experiments were done in the same manipulation and docking positions within 35 seconds.

experiments. The docking manipulations of the experiments were also completed with the same docking path and docking position within the certain time range.

When the ligand approaches to the receptor from a far distance, the interaction should be mildly attractive force. In the stabilization method, this force is represented by the smoothly degressive curve from the 10th to 15th second in Fig. 8a. In contrast, the approaching process without stabilization method contains tiny vibrations during the same process. Because of device vibration and hand movement, the original weak attractive force could increase to become repulsion force in far distance in this process at the 14th second, shown in Fig. 8b.

As the distance between molecules gets smaller, the interaction force would have a sudden change from attractive to repulsion force. With stabilization method, although this change is fast after the attraction reaches its maximum value, the process of force change is relatively smooth at the 15th second in Fig. 8a. However, in the simulation without stabilization method, the maximum attraction cannot be felt because of the continuous vibrations in the previous frames. Moreover, these vibrations will not disappear during the process. Therefore, for the user with a haptic device, a sudden kick phenomenon could happen at the 17th second in Fig. 8b.

When the distance between visual representations of two molecules was small enough and molecules contact with each other, the interaction force will be strongly increased. Although there were some tiny vibrations appeared in the simulation with stabilization method, the user can feel a continuous increase of interaction force and a point of maximum repulsion force. In contrast, the simulation without stabilization method has strong vibrations around the 20th second with great vibration amplitude, shown in Fig. 8b. In practice, such vibration could be so strong that the user felt hard to control the haptic device during manipulation. As a result, the maximum repulsion force cannot be clearly felt by the user.

Stabilization method has obvious advantages in improving the speed of docking and accuracy of manipulation.

5 Collaborative Molecular Docking

In addition to implementing both force and torque feedback in the molecular docking system, we developed a collaborative virtual environment for multi-user molecular docking manipulation with two haptic devices. In this collaborative molecular docking system, we used the CHAI 3D library for computer haptics and real-time simulation. The CHAI 3D library supports most of commercially available desktop haptic devices [5].

Haptic rendering is performed on a quad processor to update position and orientation information, and to calculate force and torque results. Haptic feedback is provided by two different kinds of haptic devices. As shown in Fig. 9, the left 3-DOF device is a Novint Falcon haptic interface made by Novint Technologies, Inc. And a PHANToM Premium 1.5/ 6-DOF haptic device made by SensAble Technologies, Inc is on the right side of screen. In this virtual environment, the receptor is controlled by the Phantom haptic device and the ligand is controlled by the Falcon haptic device. During the molecular docking process, both users can feel the interaction force between ligand and receptor.

It proved that the multi-user games can be used to produce accurate protein structure model [6]. Compared with previous molecular docking methods, multi-user human search allows users come up with creative solutions to docking of complex molecular structure. In this collaborative system, we combine the virtual environment and various haptic devices to improve the problem-solving ability for molecular structure exploration.

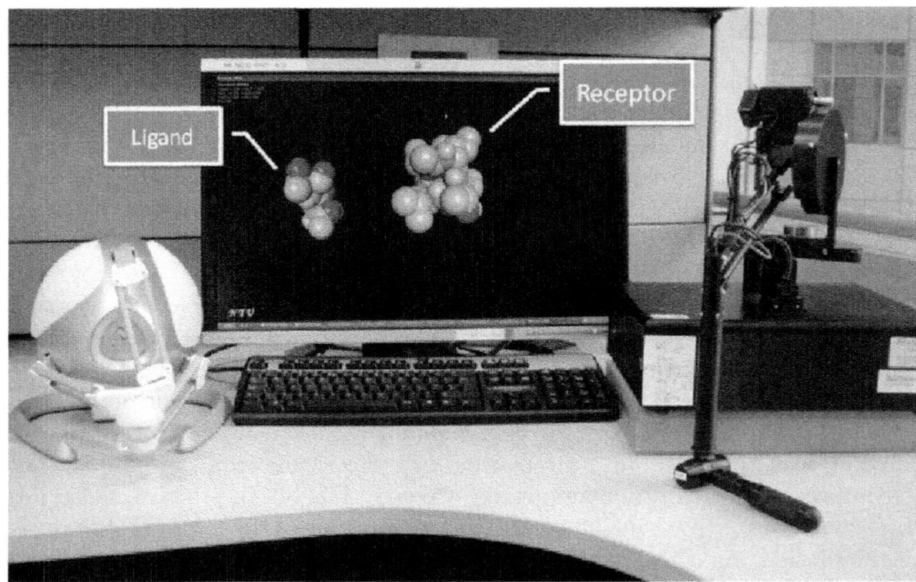

Fig. 9. Collaborative molecular docking with two different haptic devices

Since haptic devices from different companies were used in our system, the stability problem became more complex than the system with a single device. In order to maintain stability for different devices, our program could recognize the type of device automatically and set the corresponding parameters to optimize force results.

6 Conclusion and Future Work

In this paper, first, we introduced research background of molecular docking and reviewed visual and haptic based molecular docking systems. Most of the systems use devices with 3-DOF that lack of providing the user "real" feeling of molecular interaction with force-torque feedback during the molecular docking. In this paper, we proposed and described the 6-DOF haptic rendering algorithm with torque force for molecular docking system. To provide an additional flexibility of the torque display in the molecular docking process, the user can change the attachment position where to apply device to move a ligand. Also we introduced stabilization method to provide smooth force and toque feedback. Finally, we developed a multi-user molecular docking application for general haptic devices.

Currently, the system is implemented as a standalone application. There are many directions for future research. We are planning to use this collaborative biomolecular docking system in e-learning and implement the haptic rendering for more complex molecular models.

Acknowledgment. This project is supported by IDM Grant NRF2008IDM-IDM004-002 "Visual and Haptic Rendering in Co-Space" of National Research Fund of Singapore and by MOE NTU grant RG10/06 "Visual and Force Feedback Simulation in Nanoengineering and Application to Docking of Transmembrane - Helices".

References

1. Opls-aa force field parameter (2007), http://egad.berkeley.edu/EGAD_manual/EGAD/examples/energy_function/ligands/oplsaa.txt
2. Jmol: an open-source java viewer for chemical structures in 3d (2010), http://jmol.sourceforge.net
3. Pdb - protein data bank (2010), http://www.rcsb.org/pdb
4. Bowie, J.: Helix packing angle preferences. Nature Structural Biology 4(11), 915–917 (1997)
5. Conti, F., Barbagli, F., Morris, D., Sewell, C.: Chai 3d: An open-source library for the rapid development of haptic scenes. In: IEEE World Haptics (2005)
6. Cooper, S., Khatib, F., Treuille, A., Barbero, J., Lee, J., Beenen, M., Leaver-Fay, A., Baker, D., Popovic, Z., Players, F.: Predicting protein structures with a multiplayer online game. Nature 466(7307), 756–760 (2010)
7. Damm, W., Frontera, A., Tirado-Rives, J., Jorgensen, W.: Opls all-atom force field for carbohydrates. Journal of Computational Chemistry 18(16), 1955–1970 (1997)
8. Daunay, B., Micaelli, A., Regnier, S.: Energy-field reconstruction for haptic-based molecular docking using energy minimization processes. In: IEEE International Conference on Intelligent Robots and Systems, pp. 2704–2709 (2007)
9. Davies, R., John, N., MacDonald, J., Hughes, K.: Visualization of molecular quantum dynamics - a molecular visualization tool with integrated web3d and haptics. In: Web3D Symposium Proceedings, pp. 143–150 (2005)
10. DeLano, W.L.: The pymol molecular graphics system (2010), http://www.pymol.org
11. Férey, N., Bouyer, G., Martin, C., Bourdot, P., Nelson, J., Burkhardt, J.M.: User needs analysis to design a 3d multimodal protein-docking interface. In: 3DUI - IEEE Symposium on 3D User Interfaces 2008, pp. 125–132 (2008)
12. Férey, N., Nelson, J., Martin, C., Picinali, L., Bouyer, G., Tek, A., Bourdot, P., Burkhardt, J., Katz, B., Ammi, M., Etchebest, C., Autin, L.: Multisensory vr interaction for protein-docking in the corsaire project. Virtual Reality 13(4), 273–293 (2009)
13. Gregory, A., Mascarenhas, A., Ehmann, S., Lin, M., Manocha, D.: Six degree-of-freedom haptic display of polygonal models. In: Proceedings of the IEEE Visualization Conference, pp. 139–145+549 (2000)
14. Heyd, J., Birmanns, S.: Global interactive docking and hessian filtering for multi-resolution fitting of biomolecular assemblies. Microscopy and Microanalysis 14(suppl. 2), 130–131 (2008)
15. Heyd, J., Birmanns, S.: Immersive structural biology: a new approach to hybrid modeling of macromolecular assemblies. Virtual Reality 13, 245–255 (2009)
16. Hou, X., Olga, S.: Haptic rendering algorithm for biomolecular docking with torque force. In: 2010 International Conference on Cyberworlds, CW 2010, pp. 25–31 (2010)

17. Jorgensen, W.L., Maxwell, D.S., Tirado-Rives, J.: Development and testing of the opls all-atom force field on conformational energetics and properties of organic liquids. Journal of the American Chemical Society 118(45), 11225–11236 (1996)
18. Lai-Yuen, S., Lee, Y.S.: Interactive computer-aided design for molecular docking and assembly. Computer-Aided Design and Applications 3(6), 701–709 (2006)
19. Lai-Yuen, S.K., Lee, Y.S.: Computer-aided molecular design (camd) with force-torque feedback. In: International Conference on Computer Aided Design and Computer Graphics, pp. 199–204 (2005)
20. Lee, Y.G., Lyons, K.: Smoothing haptic interaction using molecular force calculations. CAD Computer Aided Design 36(1), 75–90 (2004)
21. Liu, Q., Sourin, A.: Function-defined shape metamorphoses in visual cyberworlds. Visual Computer 22(12), 977–990 (2006)
22. Martin, M.: Comparison of the amber, charmm, compass, gromos, opls, trappe and uff force fields for prediction of vapor-liquid coexistence curves and liquid densities. Fluid Phase Equilibria 248(1), 50–55 (2006)
23. Ming, O.Y., Beard, D., Brooks, Jr., F.: Force display performs better than visual display in a simple 6-d docking task. In: IEEE International Conference on Robotics and Automation, 1989, vol. 3, pp. 1462–1466 (May 1989)
24. Nagata, H., Mizushima, H., Tanaka, H.: Concept and prototype of protein-ligand docking simulator with force feedback technology. Bioinformatics 18(1), 140–146 (2002)
25. Ouh-young, M., Pique, M., Hughes, J., Srinivasan, N., Brooks, Jr., F.P.: Using a manipulator for force display in molecular docking. In: 1988 IEEE International Conference on Robotics and Automation, vol. 3, pp. 1824–1829 (April 1988)
26. Persson, P., Cooper, M., Tibell, L., Ainsworth, S., Ynnerman, A., Jonsson, B.H.: Designing and evaluating a haptic system for biomolecular education. In: Proceedings - IEEE Virtual Reality, pp. 171–178 (2007)
27. Prins, J., Hermans, J., Mann, G., Nyland, L., Simons, M.: Virtual environment for steered molecular dynamics. Future Generation Computer Systems 15(4), 485–495 (1999)
28. Rizzo, R., Jorgensen, W.: Opls all-atom model for amines: Resolution of the amine hydration problem. Journal of the American Chemical Society 121(20), 4827–4836 (1999)
29. Sankaranarayanan, G., Weghorst, S., Sanner, M., Gillet, A., Olson, A.: Role of haptics in teaching structural molecular biology. In: 11th Symposium on Haptic Interfaces for Virtual Environment and Teleoperator Systems, HAPTICS 2003, pp. 363–366 (2003)
30. Sayle, R., Milner-White, E.: Rasmol: Biomolecular graphics for all. Trends in Biochemical Sciences 20(9), 374–376 (1995)
31. Sourina, O., Korolev, N.: Visual mining and spatio-temporal querying in molecular dynamics. Journal of Computational and Theoretical Nanoscience 2(4), 492–498 (2005)
32. Sourina, O., Torres, J., Wang, J.: Visual haptic-based biomolecular docking and its applications in E-learning. In: Pan, Z., Cheok, A.D., Müller, W., El Rhalibi, A. (eds.) Transactions on Edutainment II. LNCS, vol. 5660, pp. 105–118. Springer, Heidelberg (2009)
33. Sourina, O., Torres, J., Wang, J.: Visual haptic-based biomolecular docking. In: Proceedings of the 2008 International Conference on Cyberworlds, CW 2008, pp. 240–247 (2008)

34. Stocks, M., Hayward, S., Laycock, S.: Interacting with the biomolecular solvent accessible surface via a haptic feedback device. BMC Structural Biology 9 (2009)
35. Stone, J.E., Gullingsrud, J., Schulten, K.: A system for interactive molecular dynamics simulation, pp. 191–194 (2001)
36. Subasi, E., Basdogan, C.: A new haptic interaction and visualization approach for rigid molecular docking in virtual environments. Presence: Teleoperators and Virtual Environments 17(1), 73–90 (2008)
37. Wei, L., Sourin, A., Sourina, O.: Function-based haptic interaction in cyberworlds. In: Proceedings - 2007 International Conference on Cyberworlds, CW 2007, pp. 225–232 (2007)
38. Wei, L., Sourin, A., Sourina, O.: Function-based visualization and haptic rendering in shared virtual spaces. Visual Computer 24(10), 871–880 (2008)
39. Weiner, S.J., Kollman, P.A., Case, D.A., Singh, U.C., Ghio, C., Alagona, G., Profeta, S., Weiner, P.: A new force field for molecular mechanical simulation of nucleic acids and proteins. Journal of the American Chemical Society 106(3), 765–784 (1984)

Design of a Multiuser Virtual Trade Fair Using a Game Engine

I. Remolar, M. Chover, R. Quirós, J. Gumbau, P. Castelló, C. Rebollo, and F. Ramos

Institute of New Image Technologies,
Universitat Jaume I,
Castellón, Spain
{remolar,chover,quiros,jgumbau,castellp,rebollo,jromero}@uji.es

Abstract. The current world economic situation makes it necessary to develop new ways of establishing commercial relationships. One possible solution is to explore the advantages of virtual worlds, and for this reason online virtual trade fairs are becoming more popular in the business world. They enable companies to establish a trade relationship with their customers without the need to visit them in person. This is very attractive for exhibitors because it can save them money, which is a priority for many companies today. In this line, this article presents a multiuser virtual trade fair developed using 3D game engine technologys. Users represented by avatars can interact with each other while they are visiting the virtual fair, which has some interactive objects included in the stands to provide information about the exhibitors. This virtual world is accessible online, and visitors only require a plug-in on their computers to be able to enter the virtual world. The game technology makes it possible to obtain a high degree of realism: very real lighting, cast shadows, collision detection, etc. Moreover, the virtual world presented builds the 3D objects automatically. Participants in the trade fair can customize their virtual stand and the application will generate the code necessary for its inclusion in the rendered virtual world.

Keywords: virtual worlds, game engines, web application, business application, multiuser environment.

1 Introduction

Virtual business is currently an interesting way to increase sales and income. This is the main reason why virtual trade fairs are becoming so popular: *"make real money in a virtual world"* [1] [2]. In these digital events, customers and sellers can meet in a virtual scenario. The companies show their products and provide information about the services they offer [3] [4] [5]. Virtual trade fairs have actually many benefits more than regular trade shows [6]. They are more accessible than a physical exhibition, as they are available 24 hours a day. A huge benefit is the reduced cost of participating in a virtual trade fair as opposed

M.L. Gavrilova et al. (Eds.): Trans. on Comput. Sci. XII, LNCS 6670, pp. 118–139, 2011.

to a physical one. There are no travelling costs, nor any costs associated with employing staff for maintaining the trade show area.

The "Virtual Tradeshow Survey" [7] recently published, demonstrates how important is it for show organizers to have virtual events as part of their portfolio in order to truly engage an attendee in a year round environment. It also offers a comparison between physical events and virtual events from the collective standpoint of an event organizer, exhibitor and attendee and it explains how virtual events enhance other event attendance.

For all these reasons, the number of virtual trade fairs available on the web is increasing nowadays. Regarding the technology they use, most of them are not based on 3D models [8] [9]. Only very few virtual worlds are composed of 3D objects. These kinds of objects increase the realism considerably. The most popular technology used in these worlds is VRML or X3D [10]. Another widely used technology is the 3D engine of Adobe Shockwave. The virtual fair we present in this article has been implemented using the technology offered by game engines [11] [12] [13]. This type of software provides a suite of visual development tools that can produce very realistic scenes. Moreover, they allow us to design multiuser environments where some phisical effects are easily included. Moreover, game engines provide the necessary code to include in the virtual scenes players, that programmers can used to represent the users. Figure 1 shows a screenshot of the virtual fair. Up until now, game engines have been mainly used in the game industry. However, these tools make it easy to create virtual worlds, whose purpose is more serious than entertainment alone.

Our virtual fair has some characteristics that make it interesting to the business world. The most important one is the easy method implemented to manage this fair. Someone with no computer knowledge can build a scenario in it because the 3D objects appearing in the fair are automatically built [14] [15] . The participating companies can configure a virtual stand, adding pictures, contact information, and so on, and the application will generate the code necessary to create it, taking all this information into account. The main objective of the virtual fair is to provide publicity for the companies that have a stand within it. This has been taken into account and some selectable objects have been included in the stand of the trade fair. These objects can be configured by the companies to enable information about their business and the products they offer to be accessed by just clicking on these selectable objects.

The graphics hardware available in companies was a very significant obstacle. Not all businesses had the latest graphics cards installed in their computers. So this fact restricted the programming of our application. The latest technology in graphics programming could not be used because it would have prevented most companies from rendering the virtual fair. The virtual fair we are presenting can be displayed on the type of computers that most companies have.

In order to make a visit to the fair friendlier, users are represented by means of avatars, i.e., virtual characters. Users can customize these features: the gender, the clothing, etc. Moreover, various users can be connected at the same time and they can communicate with each other [16] [17]. All of them show their username

Fig. 1. The outdoor environment of the virtual fair

in order to be identified. Text-based messaging has been implemented in order to make this communication possible, as in [18].

The virtual fair is very easy to visit, because it is accessible using an Internet browser [19]. End users do not have to install specific software in order to connect to it. When they first visit the website, their browser will automatically download and install a plug-in, making the connection process transparent to the user.

In the next section related work is reviewed, then the system architecture of the virtual fair is explained in Section 3. Section 4 deals with the virtual trade fair management, and the automatic creation of the trade fair is detailed in Section 5. Sections 6 detail the web environment and the interaction in this virtual world. The client-server organization is explained in Section 7. Finally, some conclusions and future work on this application are presented in Section 7.

2 Related Work

The commercial potential of the new virtual worlds is impressive and makes them well worth a look. Currently, many new virtual trade fairs are appearing, even companys that offer to develop customized virtual trade fairs, such as BusinessGlobal [20]. Nowadays, most companies are looking for cost-cutting, so they are seeking new sales opportunities and new ways to show their products and services to possible customers using a global communication channel such as Internet. The main requirement of these virtual worlds is that they have to be remotely accessible by typical computer hardware configurations of companies not dedicated to computer graphics.

This requirement determines the technology used in building virtual worlds. Many of the virtual fairs that have appeared up until now are modeled in 2D technologies [8] [9]. However, this produces a lack of realism in the virtual world. Other projects model 3D objects [21] [20], but the best realism is obtained using a game engine [10]. For years now, the huge and growing computer games industry has helped drive the innovation of certain technologies, particularly graphics and user interfaces. Therefore the use of this technology to design virtual fairs makes these virtual markets more attractive to users.

Moreover, online gaming allows people to interact together in a computer game over a network. Huge virtual worlds are continually running on hundreds of servers for users to interact with each other and the virtual world they inhabit. This aids in building a virtual world where people can interact.

The most popular market in a 3D world is Second Life [1]. This is the most visited 3D virtual world inhabited by millions of residents from around the globe. As it is a community, albeit a virtual one, there are as many opportunities for innovation and profit in the Second Life world as there are in the real world. Thousands of residents are making part or all of their real life income from their Second Life businesses [2]. However, Secondlife is a very general and open virtual world. Real business are looking for some personalized virtual environment where the access is controlled by an administrator and has been designed according to its requirements, the same as a regular trade show. Summarizing, our objective with this virtual fair is to promote real business in a virtual world.

3 System Architecture

The main components have been organized into three different modules that have different tasks, as can be seen in Figure 2:

- *Virtual Fair Management*: this allows the fair administrator to manage the access of the companies to the virtual world. All the information about the exhibitors is stored in a database to allow the configuration of the virtual scenarios. In addition, this module enables the participant companies to configure their stands in the fair. They can easily register, choose the type of stand they prefer, add some multimedia content and choose the location and orientation of the stand in a pavilion.
- *Automatic Creation of the Fair*: this module automatically creates the pavilions from the information previously entered by the companies. Pavilions are

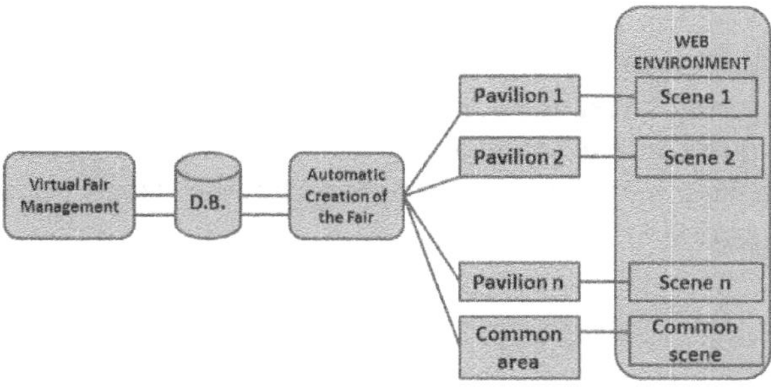

Fig. 2. The organization of the different modules that make up the developed application

the buildings where the virtual stands are located. They are configured in independent virtual worlds, which are usually called missions in game engine terminology.

- *Web environment*: this module provides Internet access. It enables the selection of avatars and manages the rendering of the exhibition and the clients through the Internet server. It is the module that makes it possible to access the virtual trade fair from a computer with an Internet connection.
- *Client-Server module*: this allows to implement a client-server model required by the game engine used in this application: Torque Game Engine (TGE) 1.5.2.

4 Virtual Fair Management

This part of the application makes it possible for users without specialized computer training to manage the 3D fair. Two kinds of users can access it: the fair administrator and the business administrator.

4.1 Fair Administrator User

As shown in Figure 3, after loggin in, the fair administrator can choose from a series of actions for efficient fair management. In brief, these actions include management of companies, assignment of areas in different pavilions for the companies and stand configurations. Figure 4 shows an example of the interface used for creating a company. It is easy to observe what type of fields are required for a company to be successfully registered.

Thus, the main role of the fair administrator is to enter general information about the participating businesses into the database and to provide them with a password so they can access the application. The other task of this user is to manage the location of the stands in the pavilions. A layout of available spaces in these pavilions has been designed as a square matrix. Each space has a virtual size of 16 m^2. This layout accommodates the three types of stands designed. The smallest one includes one square, i. e. 16 m^2, the medium one includes two squares, i.e. 32 m^2 and the largest one includes four squares, i.e. 64 m^2. The application shows the occupied stands colored red and the empty ones colored green. Figure 5 shows an example of available and occupied stands within a pavilion. It is important to note that the label "Access" shows where visitors will enter and leave the pavilion.

In order to position the stand for a business, the fair administrator simply selects as many empty squares as are needed for the size of the stand, and then checks its orientation.

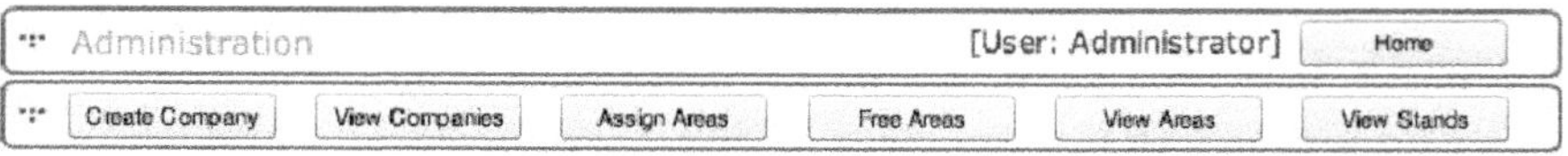

Fig. 3. Available options for fair administrators

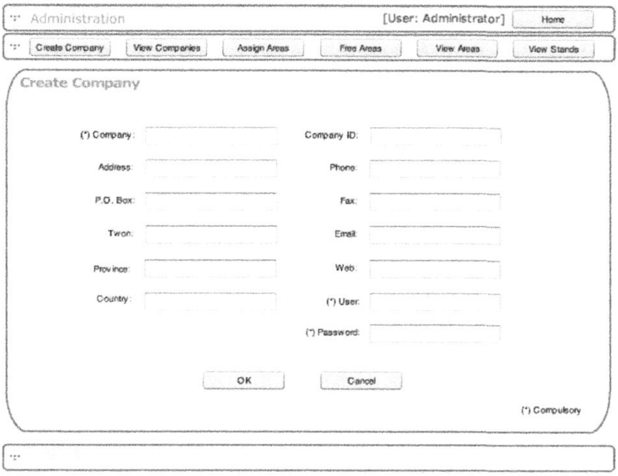

Fig. 4. Creating a company in the Virtual Fair Manager

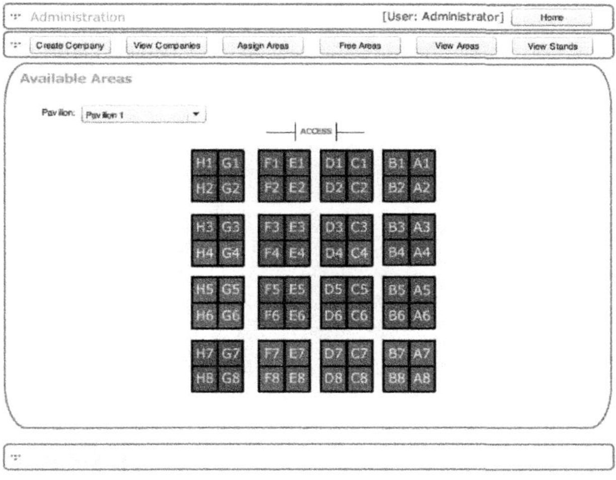

Fig. 5. Available and occupied areas within a pavilion

4.2 Business Administrator User

After the information about the participant is stored in the database, the fair administrator issues a username and password to the business administrator. The business administrator is responsible for configuring the final appearance of the stand and customizing the content displayed in the virtual environment. Once the size of the stand has been chosen, there are three different stand models available for selection. In this step, this user perform two important actions. In the first stage, companies select a color palette and a design from some predefined

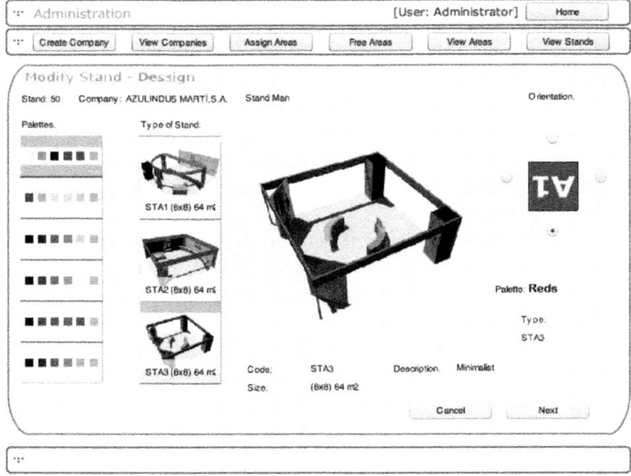

Fig. 6. User interface for basically configuring the stand of the company

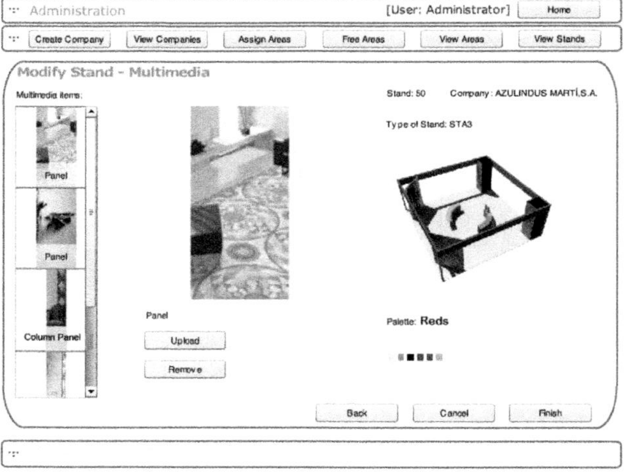

Fig. 7. User interfaz for updating some images or textures for the stand

models (Figure 6). Later, at the second stage, companies can freely customize their stands by combining different type of multimedia items (Figure 7).

Figure 8 shows an example configuration of each one of the different sizes, 64, 32 and 16 m^2 stands. The customization also enables different color combinations to be selected for the stand. Moreover, the business administrator can include some multimedia information for display in the fair. This is stored in the database and it may include pictures of products and/or services multimedia materials and Internet information, such as the website address of the business.

Fig. 8. Different virtual stand configurations, 64 m^2 on the left, 32 m^2 in the middle and 16 m^2 on the right

To implement this application, PHP and MySQL were employed. This framework has a series of advantages such as low cost, ease of use, HTML embedding, cross-platform compatibility, stability, speed, open source licensing, many extensions, fast feature development and strong user communities for supporting.

A MySQL database is used to store all the persistent information, PHP is used for efficient interaction between the database and client applications and a Flash movie embedded into a web page is used for administration. The combination of these technologies makes it possible to automatically manage the virtual world displayed in the game engine through a web environment. Both types of users are able to easily access the administration application, as this only requires a web browser with a Flash player.

5 Automatic Creation of the Trade Fair

The game engine employed in this project, Torque, has conditioned the organization of the information. Torque implements every game following a client-server model, even if there is a single player. This concept will be extended in Section 8. This client-server architecture conditions that the files of the project are organized in three main categories: files for the *server* module, files for the *client* module and the general *data* of the project. The first time a user connects to this virtual world, the folders *client* and *data* are downloaded to his/her computer. The final organization of the files involved in this virtual world is shown in Figure 9.

In the virtual trade fair, two main classes of 3D objects, following the requirements of Torque, have been designed. The first type is composed by the 3D objects where the player can move around, as buildings or outdoor environments. The other type is integrated by the rest of the 3D objects, objects that the designer can add to the virtual world in order to make it more attractive. In the first group, this virtual trade fair has two objects, the outside environment and the pavilion. They both have been modeled with a specific tool such as *Torque Constructor* [23]. These type of objects are called *interiors* in the Torque

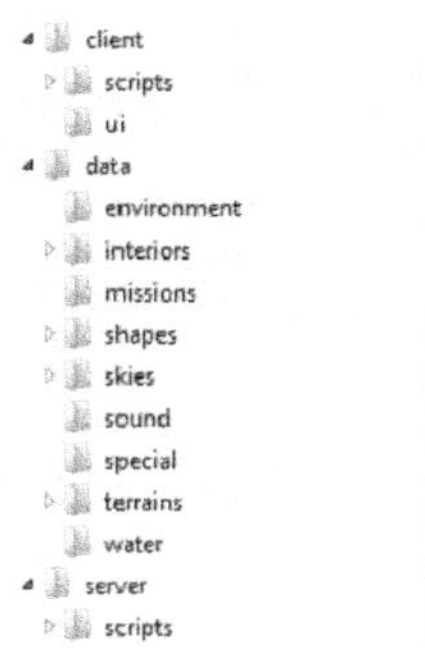

Fig. 9. Files organization in the virtual trade fair

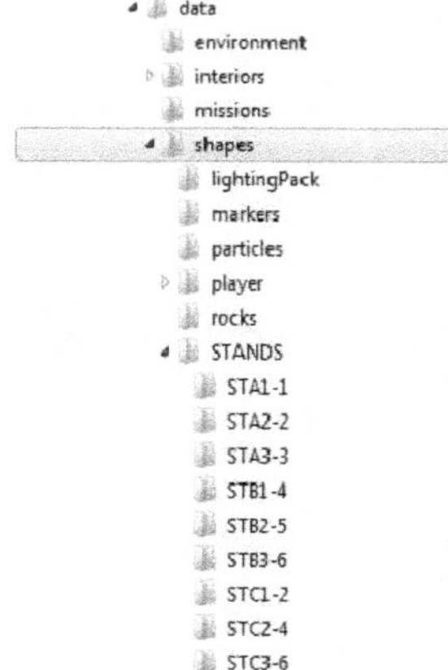

Fig. 10. Files organization in the virtual trade fair

terminology. The other kind of objects have been modeled with the 3D Studio Max 2009 [24]. They all are stored in the *shapes* folder.

As it is mentioned before, three different stands have been created for every size. All the geometry and the necessary files for their configuration are stored in the folder */shapes/stands*. Let A be the smallest stand, B the medium and one C the largest one. Three different versions of every size have been modeled. Every folder in */shapes/stands* stores all the data involved in the render of the type of stand specified by the letter and number that appear in the name of every folder. The organization is shown in Figure 10.

The scenarios are built from the information stored in the folder *missions*. Initially, one file is stored in this folder for every basic scenario. These files, also called missions in the game engine, have some typical components of an outdoor scene, such as the sky, the sun, the sea and the terrain, and the position where the player is going to appear in them (*SimGroup*). The basic mission of the outdoor environment has the six pavilions. Nevertheless, the initial mission of the pavilion is the same for all the pavilion scenarios, and later it is configured according to the one is going to be rendered. Briefly, this code is shown next:

```
 new InteriorInstance(Pavilion) {
   canSaveDynamicFields = "1";
   Enabled = "1";
   position = "0 0 0";
   rotation = "1 0 0 0";
   scale = "1 1 1";
   interiorFile = "/data/interiors/pavilion/PB0001.dif";
   showTerrainInside = "0";
   smoothLighting = "0";
};
new Sky(SkyObject) {
```

```
    ...
}
new Sun(SunObject) {
    ...
}
new TerrainBlock(Terrain) {
    ...
}
new SimGroup(PlayerDropPoints) {
    ...
}
new WaterBlock(WaterObject) {
    ...
}
```

The module in charge of the Automatic Creation of the virtual fair processes the database and automatically generates a mission for the scenario the avatar is going to appear. If he/she is going to be in a pavilion, the mission that generates this scenario is configured with all the stands and geometry that the fair and the businesses administrators have included in it. The generated code is added to the initial mission file in order to render the final scenario with all the changes that the users have included. This code establishes the type of stand, its position, its rotation and some details about the lighting of this stand. One example of code that renders this initial stand is briefly detailed next. The name assigned to it determines the pavilion and its position in it. In the following example, the stand is placed in the pavilion 1 (P1) and in the area labelled by F1.

```
new TSStatic(st-P1-F1) {
    canSaveDynamicFields = "1";
    Enabled = "1";
    position = "226.252 -0.353153 16.6096";
    rotation = "1 0 0 0";
    scale = "1 1 1";
    shapeName = "/data/shapes/STANDS/STA1-1/sta1-1.dts";
    receiveSunLight = "1";
    receiveLMLighting = "1";
    useCustomAmbientLighting = "0";
    customAmbientLighting = "0 0 0 1";
    usePolysoup = "1";
    allowPlayerStep = "1";
};
```

The next step is to customize the stand following the requirements established by the Business administrator. This is possible by adding some code to the current mission. One example of code is shown next: the stand named *st-P1-F1* will change the default texture *P01M02.png* by the established by the business user, *NewTex*.

```
nametoid("stA1").ResetDynamicSkin("st-P1-F1","P01M02.png","NewTex");
nametoid("stA1").UpdateDynamicSkins();
```

Every time a user connection is established, this module accesses the database to create the stands configured in a previous process and the code is generated to render the selected type of stand in the position and orientation configured for each business by the fair administrator. Moreover, multimedia objects in the stand are assigned a link to the business web page where the end user can obtain more information about their products and services.

Once this process ends, the necessary structure of files is created and it is possible to render the scene where the configured stand appears inside the pavilion. If something has been changed from the previous connection, the different textures and materials as configured by the business administrator are copied to the game engine file structure in the final user's computer, and assigned to the stand generated.

This part of the application has been implemented in PHP using the MySQL database.

6 Web Environment

All the modules described above are managed over the web, so the main function of the Web environment is to provide a website where the virtual world can be accessed by the virtual fair users. This module has been implemented using different web technologies such as HTML, JavaScript and PHP with a MySQL database and a web browser plug-in.

This module performs various tasks. One of them is to control the access of visitors to the 3D world. The module checks the data stored in the database in order to allow entrance to the virtual fair. All the users have to be registered in advance. If the user is a visitor, before connecting for the first time, s/he has to fill out a short questionnaire. Figure 11 depicts a screenshot of the user registration form. The application stores this information in a MySQL database and then allows the user to enter.

Each user in the virtual world is represented by an avatar. There are four different avatars available in the web environment: two male and two female, also shown in Figure 11. Moreover, users can choose the type and color of their avatar's clothing. This process allows users to complete their profile information, which can be changed later if necessary.

The final integration of the virtual world on the web is done through the client-server module. It is responsible for the communication between the client and the server of the game engine. As mentioned above, the game engine chosen was the Torque game engine, TGE, [22]. It is a modified version of a 3D computer game engine originally developed by Dynamix for the 2001 Tribes 2 FPS game. The Torque engine provides robust networking code, scripting, in-engine world editing and GUI creation. Moreover, the source code can be compiled on most popular platforms, including Windows, Macintosh, Linux, Wii, Xbox 360

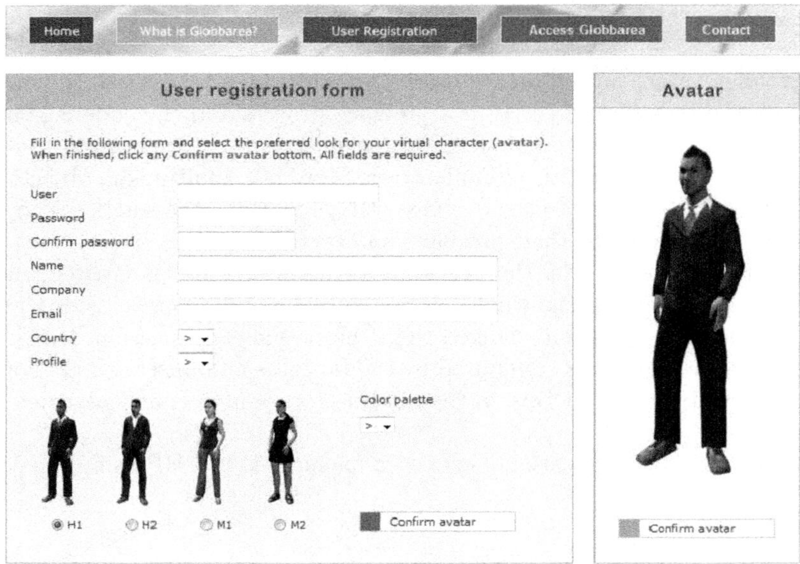

Fig. 11. A screenshot of the user registration form

and iPhone. This game engine also allows environment mapping, Gouraud shading, volumetric fog, and other effects such as decals that enable textures to be projected onto interiors in real time.

The version of the Torque game engine used in the implementation of the virtual fair is TGE 1.5.2. This version does not support web browser integration, so an external tool (plug-in) capable of integrating a game application into web browsers has been used. This tool is called *Igloader* [25]. *Igloader* is a cross-browser plug-in that allows virtually any Windows- based software to be delivered and operated within the browser window.

A light version of the Torque game engine (TGE 1.5.2) was used in order to lower the hardware requirements needed to run our virtual fair. We had to adapt the game engine code to make it compatible with the plug-in. The latest version of the game engine Torque (Torque 3D) can be integrated into some web browsers without any external plug-ins. However, this latest version is very complex, so the minimum hardware requirements are very high. This means that the graphics card installed on the user's computer system must support all recent programmable graphics processing units (GPUs) features such as shaders, which significantly limits the number of potential users in a web environment. This could be problematic for our application because even users with basic computer systems must be able to use the virtual trade fair over the Internet.

The *Igloader* plug-in can control the behavior of the application that runs in the browser by passing in parameters specified in an XML configuration file which can be created before the application starts running. In this way the web environment module is able to pass the type of avatar, its name, its position in the virtual world, etc. to the client game engine. The plug-in also has a powerful data streamer at its core that can be utilized to stream content in the background and patch current installations.

The web environment is hosted on a separate server to the one hosting the game engine. This prevents the server game engine from being affected by the load generated by the web environment. Each instance of the game engine can only run one mission, in our case pavilion or outdoor environment, thus several instances are running with a different TCP/UDP port and waiting for clients to connect to the virtual environment.

6.1 Interaction in the Virtual Fair

Different types of interactions have been implemented in the virtual fair. By default avatars can access the common area in the virtual fair. They are labeled with their username, as can be seen in Figure 12. The user can move around the virtual world with the avatar, using some on the keyboard to walk around and visit places of interest.

From the common area, the avatars can access the pavilions through the gateways (Figure 12). These gateways allow the user to leave the scenario s/he is visiting and to automatically appear in another one. The application quits the client's current mission of the client and downloads the new one. In the virtual fair, these gateways are usually placed at the entrance to the pavilions. Another way of getting to a stand quickly is through the trade fair directory.

Fig. 12. A screenshot of the application, showing where the gateways are located in the virtual world. In this screenshot, a user is approaching the entrance to a pavilion.

Fig. 13. A screenshot of the interior of a pavilion, showing text-based communication between users

The position of the stand and its direction are retrieved from the database, and then the avatar is placed near the selected stand and looking towards it. This information is passed from the web environment module to the client-server module of the 3D application, which is responsible for uploading the mission selected and placing the user in the correct position.

More interaction with the 3D world is possible through special objects located in the stands. These allow the user to view information about the company as well as its website address, e-mail, etc.

One of the main features of this 3D virtual business fair is the text-based communication it offers through chat. The chat system is part of the client-server architecture of the Torque game engine. It takes advantage of the multiuser connections to the virtual world and it allows the user to send and receive text messages. Figure 13 shows an example of chat. To send a message, the user only has to click in the dark area in the lower left corner of the window and type. The other users connected to the virtual fair will be able to read the message when the writer presses the enter key. Just above this area there is a display of the messages sent by users.

7 Client-Server Module

The 3D game engine used supports networked games over a LAN and the Internet with traditional client-server architecture. Server objects are "ghosted" on clients and updated periodically or upon events. This architecture is running regardless of whether the game is single-player or multi-player.

Torque implements the client-server model using a single executable. That is, whenever the engine is run, it contains both a server and a client. In the Torque implementation of a client-server architecture, the game world is controlled by the server. Game objects representing the game state are created and maintained on the server. For that reason, these objects are often called *server objects.*

During a game, individual clients are provided with copies of server objects. These copies are called *ghosts.* All game calculations are done using server objects. Thus, only server objects affect the game and its outcome. In order to keep clients up to date, the server will send information across the individual server-to-client connections, updating the ghost on each client.

Clients then render the game based on the state of their own ghosts. Figure 14 depicts the concept of server objects and copies of those objects being ghosted to a client. The server is aware of all server objects and all ghosts. Individual clients are only aware of their own ghosts. This provides a strong measure of security and prevents a number of game cheats involving clients having direct access to server objects and/or knowledge of other clients' objects.

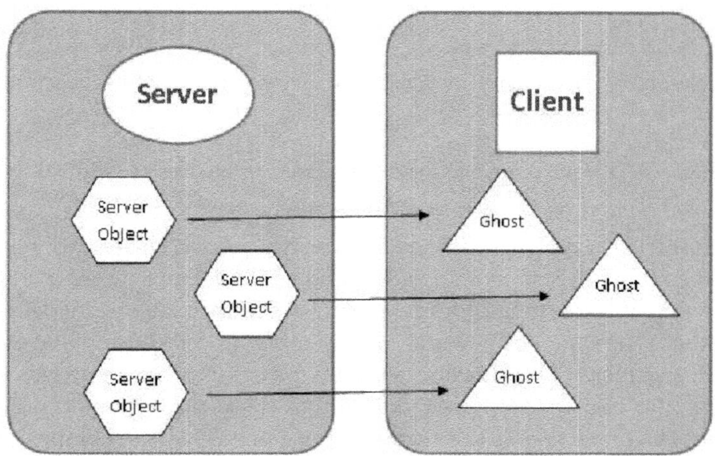

Fig. 14. Torque game engine server objects and ghosts

Torque game engine games operate in any of three modes: dedicated-client, dedicated-server, and client-server.

In dedicated-client mode, the engine is only executing client tasks. It parses user inputs, sends these inputs to the server, and does any rendering necessary to present the game. To participate in a game, the engine must connect to another copy of the engine running in either dedicated-server or client-server mode.

In dedicated-server mode, the engine is only executing server tasks. It receives user inputs (from clients), maintains the game state, and updates clients regarding that state. External copies of the engine running in dedicated-client mode may connect to this server to participate in a multiplayer game.

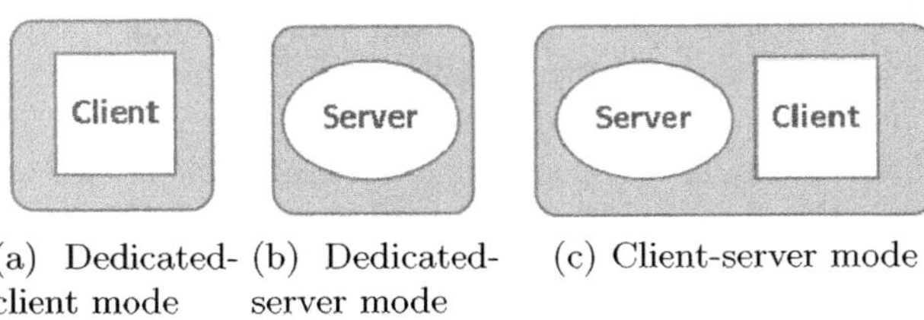

(a) Dedicated-client mode (b) Dedicated-server mode (c) Client-server mode

Fig. 15. Torque game engine modes

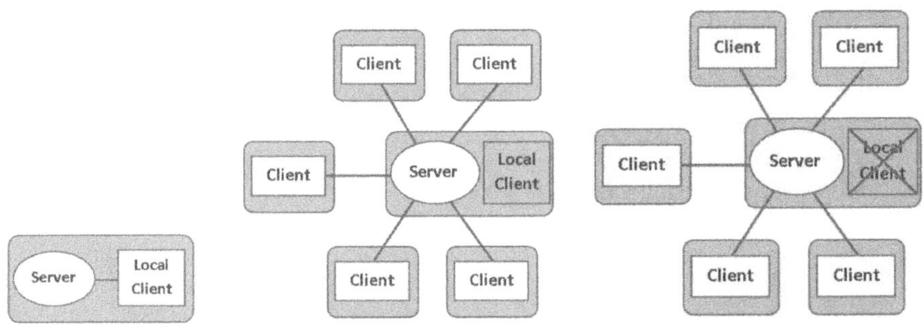

(a) Single-player connection (b) A multiplayer listen server connection (c) A multiplayer dedicated connection

Fig. 16. Torque game engine client-server interconnection diagrams

In client-server mode, the engine is executing both client and server tasks. An engine running in this mode can be used to implement either a single-player game or a multiplayer game. Figure 15 shows the three modes of Torque.

Given these three modes, a game can be interconnected using one of the three connection schemes. The connection scheme we select is based on the game type we wish to run. Figure 16 shows a diagram of these three game types.

The simplest game type is the single-player game. This is accomplished by running a single instance of the executable on one machine. In this case, the server and client connect via an internal (local) connection. When this connection is requested, the server becomes active. The second game type involves a single executable with an active client and an active server running on one machine as a listen server. One player uses the local client and a local connection. The remaining players use client-only executables, running on separate machines, and connect remotely to the listen server. This mode is appropriate for local-area network parties and other cases where a user wants to host a game while participating. The last game type involves a single executable running as a dedicated server. This means that only the server is active. Multiple client-only executables, running on separate machines, can then connect with this executable, again allowing for multiplayer games. Although this could be used for a local-area network party, it is more suited to a professional hosting setup, where your company hosts one or more sessions on a machine used only as a server. This last game

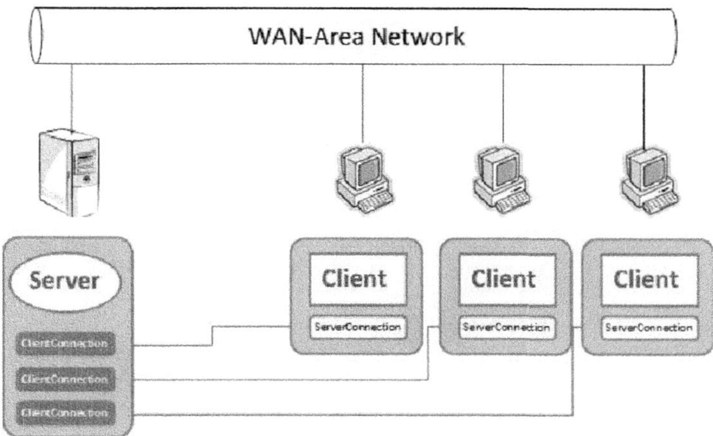

Fig. 17. Three-person wide-area network party

type was used in the implementation of the virtual fair. Figure 17 depicts an example of a three-person wide-area network party which uses six instances of the class *GameConnection*. This is, in fact, an example of the current settings for the virtual fair.

The Torque game engine is implemented in C++. It also allows the use of scripts written in its own scripting code similar to C++. The game engine also has its own integrated development environment (IDE) called *Torsion*. This facilitates the development of prototypes without the need for a constantly compilation. It also uses 3D graphics libraries, such as OpenGL and DirectX. The function of the server is to synchronize the position of clients and other animated objects and to detect possible collisions between them during the execution of a mission. In addition, it controls the messages sent among the clients and provides the infrastructure for the chat system. As mentioned above, currently the virtual fair consists of three scenarios: one for the common area, one for pavilion 1 and one for pavilion 2. Figure 18 shows an example of connection between multiple clients and servers.

A single machine can run multiple instances of the game engine server. This allows us to have multiple missions running on a single computer, each one assigned to a different TCP/UDP port. In a current virtual fair setup this means that we need three server game instances running. However, each server can be migrated to a separate machine depending on system load, thus making this architecture scalable.

Through the client-server module of the web environment, information is passed to the server game engine, including the username, avatar model, color combinations, position and rotation of the avatar in the world. An XML file is generated with this information, which is needed in order for the plug-in to run the client game engines, connecting them to the correct mission with the options selected by the user.

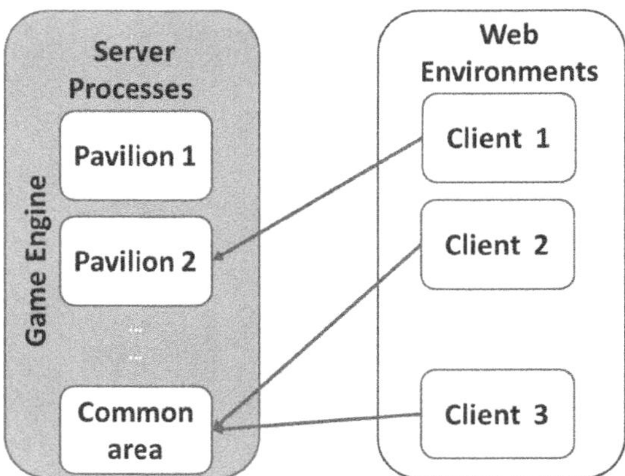

Fig. 18. The application organization of the client/server module

When a client connects to a game engine server for the first time, the web browser plug-in downloads a client module from the web server and installs it on the client machine. This client module contains the client version of the game engine and all the data needed to visualize the virtual world. This software is stored in multiple compressed and encrypted files. For subsequent connections, the files will not be downloaded again because they are already on the client machine. Even if an error occurs during the software download, the web browser plug-in is able to resume the download, reducing the data transferred between the client and the server.

8 Conclusions

In this paper, a multiuser 3D virtual world for business has been introduced (Figure 19). Firstly, we have taken advantage of the latest technology for the creation of 3D games. Secondly, the universality of the virtual fair was an important objective. To this end, we have used a client-server architecture on the Internet, a popular and widely used networked environment.

Our virtual fair offers an easy way for clients to connect. They just need a compatible Internet browser, which will connect to the virtual fail server using an plug-in. Once connected, the user can navigate though the world with the high realism offered by our solution. Moreover, we take into account client computer configurations and try to offer the maximum compatibility with standard graphics cards. Multiuser functionality is also present, offering companies the possibility of interacting or establishing communication with all the customers connected to the virtual fair. On the server side, fair content can be created with no specialized knowledge of graphics technology. That is, a virtual fair can

Fig. 19. This scene shows two users represented by avatars connected to the virtual world

be automatically created using a series of parameters such as the number of pavilions, stands and different style. In this way, the virtual fair administrator is able to easily design each stand and customize its content as desired. Finally, the virtual fair is generated and put online for the visitors.

Thus, our virtual fair is an interesting technology for companies and customers as it offers a highly realistic communication between them in a virtual world that is increasingly real. New functionalities may be incorporated into our virtual fair. Its architecture allows us to easily add new features by means of a plug-in system. In this sense, voice over IP support would be an important functionality to develop as it will enable us to offer better interaction between users. Thus, meeting or conference rooms, where conversations among the users could take place, would become essential in a virtual business world.

Moreover, support for video is another interesting area for exploration as the use of this technology to show new product features, company presentations, etc. would give to the customer a better view of the virtual fair content.

Acknowledgments

This work was supported by the Spanish Ministry of Science and Technology project TIN2010-21089-C03-03. And also by Bancaja, project P1 1B2007-56.

References

1. SecondLife, http://secondlife.com/whatis (retrieved January 12, 2010)
2. Guo, Y., Barnes, S.: Why people buy virtual items in virtual worlds with real money. SIGMIS Database 38(4), 69–76 (2007)
3. Berger, H., Dittenbach, M., Merkl, D., Bogdanovych, A., Simoff, S., Sierra, C.: Playing the e-business game in 3D virtual worlds. In: ACM, Proceedings of the 18th Australia Conference on Computer-Human Interaction, pp. 333–336. ACM, New York
4. Czerniawska, F., Potter, G.: Business in a Virtual World: Exploiting Information for Competitive Advantage. Palgrave Macmillan, Basingstoke (1998)
5. Greenfield, D.: Doing Business in the Virtual World. IT Infrastructure, http://www.eweek.com/c/a/IT-Infrastructure/Doing-Business-in-the-Virtual-World/ (retrieved January 14, 2010)
6. Virtual Trade Fair Shows, http://www.virtualtradefair.org (retrieved December 14, 2010)
7. TSSN, Onstream Media: 2010 Virtual Event Report: Insights & trends from Industry Insiders, http://www.marketplace365.com/marketing/virtualtradeshowsurvey/TSNN.pdf (retrieved December 20, 2010)
8. BusinessWomanFair, from http://www.businesswomanfair.com/information.htm (retrieved January 14, 2010)
9. VirtualRealEstate Fair, http://www.virtualrealestatefair.com/index_en.html (retrieved January 14, 2009)
10. Luebke, D.: The present & future of Web3D. In: Web3D 2008: Proceedings of the 13th International Symposium on 3D Web Technology, p. 6. ACM, Los Angeles (2008)
11. Eberly, D.H.: 3D game engine design: a practical approach to real-time computer graphics. Morgan Kaufmann Publishers Inc. address, San Francisco (2000)
12. Noh, S.S., Hong, S.D., Park, J.W.: Using a Game Engine Technique to Produce 3D Entertainment Contents. In: Pan, Z., Cheok, D.A.D., Haller, M., Lau, R., Saito, H., Liang, R. (eds.) ICAT 2006. LNCS, vol. 4282, pp. 246–251. Springer, Heidelberg (2006)
13. Trenholme, D., Smith, S.P.: Computer game engines for developing first-person virtual environments. Virtual Real 12(3), 181–187 (2008)
14. Cencioni, R., Bertolo, S.: From Intelligent Content to Actionable Knowledge: Research Directions and Opportunities under Framework Programme 7, http://www.3dexpositions.com/ (retrieved January 2009)
15. Groenewegen, S., Heinz, S., Fröhlich, B., Huckauf, A.: Virtual world interfaces for special needs education based on props on a board. Computers&Graphics 32(5), 589–596 (2008)
16. Maurina III, E.F.: Multiplayer Gaming and Engine Coding for the Torque Game Engine. A. K. Peters, Ltd. Press, Natick (2008)
17. Gehorsam, R.: The Coming Revolution in Massively Multiuser Persistent. Computer Journal 36(4), 93–95 (2003)

18. Belfore, L.A., Battula, S.: Virtual worlds: VRML clients linked through concurrent chat. In: Proceedings of the 34th Conference on Winter Simulation, pp. 518–524 (2002)
19. Holmberg, N., Wunsche, B., Tempero, E.: A framework for interactive web-based visualization. In: AUIC 2006: Proceedings of the 7th Australasian User Interface Conference, pp. 137–144. Australian Computer Society, Inc., Hobart (2006)
20. BusinessGlobal, http://www.businessglobal.com/ (retrieved December 14, 2010)
21. 3dexpositions, http://www.3dexpositions.com/ (retrieved January 14, 2010)
22. Torque 3D, http://www.torquepowered.com/ (retrieved January 12, 2010)
23. Torque Constructor, http://www.torquepowered.com/products/constructor (retrieved December 12, 2010)
24. 3D Studio Max, http://usa.autodesk.com (retrieved December 20, 2010)
25. Indiepath Ltd., http://www.indiepath.com/ (retrieved January, 12, 2010)

Applying Biometric Principles to Avatar Recognition

Marina L. Gavrilova[1] and Roman Yampolskiy[2]

[1] Dept. of Computer Science, University of Calgary, Canada
marina@cpsc.ucalgary.ca
[2] Dept. of Computer Engineering and Computer Science, University of Louisville, USA
roman.yampolskiy@louisville.edu

Abstract. Domestic and industrial robots, intelligent software agents, and virtual world avatars are quickly becoming a part of our society. Just like it is necessary to be able to accurately authenticate identity of human beings, it is becoming essential to be able to determine identities of the non-biological entities. This paper presents current state of the art in virtual reality security, focusing specifically on emerging methodologies for avatar authentication. It also makes a strong link between avatar recognition and current biometric research. Finally, future directions and potential applications for this high impact research field are discussed.

Keywords: biometric, avatar, recognition, robot, synthesis, artimetrics.

1 Introduction

Over the course of history, the greatest minds: scientists, philanthropists, educators, politicians, leaders, philosophers, were fascinated with the way human brain works. From Michelangelo to Lomonosov, from DaVinci to Einstein, there have been numerous attempts to uncover the mystery of human mind and to replicate its working first through simple mechanical devices an later, in the 20th century, through computing machines, software and robots.

In Alan Turings's 1950 work "Computing Machinery and Intelligence," Turing posed the question "can machines think?" In order to establish a credible criteria to answer this question, he proposed a test, now widely known as "The Turing Test" – to estimate a machine's ability to demonstrate intelligence. At the core of the test is conversation in a natural language between the human judge and the opponent, who can be either human or a machine. If the judge cannot reliably tell the machine from the human, the machine is said to have passed the test. In the light of recent developments, it can be viewed as the ultimate multimodal behavioural biometric, which can detect differences between a man and the machine. After the theoretical platform for an Automated Turing Test (ATT) was developed by Naor in 1996, the new generation of researchers continued to study the same concept of human / machine disambiguation. In addition to ATT, the new developed procedures were "reversed Turing test" (RTT); "human interactive proof" (HIP); "mandatory human participation" (MHP); and the "completely automated public Turing test to tell computers and humans apart" (CAPTCHA) [1].

M.L. Gavrilova et al. (Eds.): Trans. on Comput. Sci. XII, LNCS 6670, pp. 140–158, 2011.

Following Turing's work, another foundation of modern artificial intelligence was laid out by John von Neumann in the 1950's in his theory of automata and self-replicating machines. His theoretical concepts were based on those of Alan Turing. The main difference was that instead of being able to read and write data, a self-replicating system reads instructions and converts these into assembly commands that result in the assembly of replicas of the original machine. The vast majority of work in this area is in the form of non-physical self-replicating automata (e.g., computer viruses, the "game of life" computer program, etc.). The only physically based concepts that have been explored related to true self-replication pertain to self-assembling systems and robots.

Self-replication is an essential feature in the definition of living things. At the core of biological self-replication lies the fact that nucleic acids can produce copies of themselves when the required chemical building blocks and catalysts are present. This self-replication at the molecular level gives rise to reproduction in the natural world on length scales ranging the ten orders of magnitude. Self-replication in non-biological contexts has been investigated as well, but to a much lesser degree. These efforts have resulted in the field of "Artificial Life". This field is concerned with the sets of rules that, when in place, lead to patterns that self-replicate. The research has been fruitful in the past decade. Cornell University researchers have created a machine that can build copies of itself. Their robots are made up of a series of modular cubes -- called "molecubes" -- each containing identical machinery and the complete computer program for replication. The cubes have electromagnets on their faces that allow them to selectively attach to and detach from one another, and a complete robot consists of several cubes linked together.

However, the bigger question of authentication and labeling of such "self-replicating" robots and software (such as viruses) has rarely been posed, despite the growing concerns that uncontrollable development of self-replicating machines and machines with artificial intelligence can be somewhat harmful for the human society. And examples are plentiful. Domestic and industrial robots, intelligent software agents, virtual world avatars and other artificial entities are quickly becoming a part of our everyday life. Just like it is necessary to be able to accurately authenticate identity of human beings, it is becoming essential to be able to determine identity of the non-biological entities rapidly infiltrating all aspects of modern society. Military soldier-robots [27], robots museum guides [6], software office assistants [7], human-like biped robots [35], office robots [2], bots [44], robots with human-like faces [31], virtual world avatars [57] and thousands of other man-made entities all have something in common: a pressing need for a decentralized, affordable, automatic, fast, secure, reliable, and accurate means of identity authentication. To address these concerns, we proposed [62, 65, 64] the concept of *Artimetrics* – a field of study that will allow identifying, classifying and authenticating robots, software and virtual reality agents. In this paper, unless otherwise qualified, the word *robot (or agent)* refers to all of the above mentioned non-biological entities.

While the area of robot and agent authentication may seem a bit futuristic at first, careful analysis of recent news stories shows that the proposed research is years behind where it needs to be.

Fig. 1. Facial images of a humanoid robot-model, robot celebrity and a 3D-virtual avatar [41] [22] [43]

To give just some examples: Al-Qaeda terrorists have been reported recruiting and communicating in virtual communities such as Second Life [8]. Cybercrime, including identity theft, is rampant in virtual worlds populated by millions of avatars and operating multibillion dollar economies [40]. Security experts have testified to the US Senate that defenses are lacking when it comes to emerging threats to the nation's Cyberinfrastructure. International teams of hackers assisted by semiautomatic hacking software agents have perpetrated numerous attacks against the Pentagon and other government agencies' computers and networks [59].

A novel paradigm, unique to virtual communities, has appeared in recent years and was labeled "interreality". In the *Second Life* visitors are allowed to populate, build and exploit initially empty spaces. As a result "the new reality that is thus created is, remarkably enough not entirely 'virtual', but is becoming gradually more linked to our physical reality" [40]. Relationships between social, economical, and psychological status of game players and their respected avatars in the virtual environment are a subject of current research. Early results show that avatars for the most parts resemble their "owners" rather than being completely virtual creations. As the physical and the virtual worlds seem to come really close to each other, the distinction between the two begins to fade and the need arises for security systems capable of working in the contexts of interreality and augmented reality [37]. In his dissertation 'Architecture of a Cyber Culture' published in 2003, Van Kokswijk describes this phenomenon as "the hybrid and absolute experience of physical and virtual reality". Interreality is the creation of a hybrid total image of and in both the physical and virtual worlds. Unfortunately, currently available biometric systems are not designed to handle visual and behavioral variations observed in non-human agents and consequently perform extremely poorly if applied outside of their native domain.

The question of security and identification of avatars in this "interreality" consistently arises. Based on research and polls performed on Internet forums, people often complain about the insufficient security in Second Life, with almost 40% of the respondents asking for additional security [40]. More than half of the respondents admit they have been harassed (this includes imprisoning, stalking, gossiping and using inappropriate language) and 40% indicate that certain actions should not be permitted in Second Life. Thus, the definite need in increasing and enforcing security is apparent, which motivates emerging research on security in the increasingly complex and interrelated virtual worlds.

It is interesting to note that some biometric methods came very close to avatar development and intelligent robots/software authentication on a number of different instances. For example, in 1998 M.J. Lyons and his colleagues published a report: "*Avatar Creation using Automatic Face Recognition*", where authors discuss specific steps and processing techniques that need to be taken in order for avatar to be created almost automatically from the human face [36]. In fact, the process described in the above article is essentially the process of biometric synthesis, conceptualized and generalized in the book devoted specifically to this subject [69]. Users of virtual words have also noted that avatars very often resemble the characteristics of its creator, and not only facial characteristics, but also body shape, accessories and clothes.

But what about other less obvious resemblances such as manner of communication, various situation response, nature of work, style of house, leisure/recreational activities, time of appearing in virtual world etc.? All of the above encompasses behavioral characteristics that can be exploited by the fusion of biometric-based techniques, with methodology tailored to specifics of virtual world. Such behavioral characteristics, as authors of this article would postulate, are even less likely to change than the avatar's facial appearance and clothes during the virtual world sessions, as users typically invest a lot of time and money into creation of a consistent virtual image but would not so easily change their patterns of behavior.

The rest of this paper is organized as follows: a literature review is presented in Section 2, a comprehensive survey of non-biological entities (avatars) is given in Section 3, an overview of methodology under development, focusing on dataset creation, synthetic biometrics, visual, behavioral and multi-modal artimetrics constitutes Section 4, applications and implications of this emerging area are outlines in Section 5 and finally concluding summary is provided in Section 6.

2 Literature Review

To the best of our knowledge, no paper surveying automatic visual or behavioral authentication of software agents, virtual reality entities or hardware robots has been published to date. While no research has been reported in automatic robot authentication or behavior analysis some relevant research has been published on robot emotion recognition [13]. In addition to experiments on understanding of emotional states of robots, some work has been started on general analysis of *avatar behavior*. One of the projects is developed under the heading Avatar DNA. Together the segments define the makeup of an avatar. The genes of the avatar are unique and include user biometric data, public key information, personal information, authentication information, creation data, etc. Verification modules in the virtual world collect information directly from the avatar to establish the roles and rights that should be granted to this user [58].

In another experiment linking real world and the world of avatars, William Steptoe asked eleven volunteers some personal questions. During the interviews the volunteers wore eye-tracking devices. A second group of volunteers watched videos of avatars as they delivered first group's answers. Some avatars had eye movements that mirrored those of the original volunteers, while others did not. The volunteers had to determine if the avatar was lying to them. Eye-movement allowed increasing accurate detection of truthful statements from 70% to 88% and detection of lies from 39% to

48% clearly demonstrating importance of even subtle body language in virtual world communication and avatar behavioral analysis [12].

Multiplayer online computer games are quickly growing in popularity, with millions of players logging in every day. While most play in accordance with the rules set up by the game designers, some choose to utilize artificially intelligent assistant programs, a.k.a. bots, to gain an unfair advantage over other players. A recently published paper by one of the authors of this work demonstrated feasibility of applying strategy-based purely behavioral biometrics developed for recognition of human beings to the recognition of intelligent software agents [65]. The paper lays the theoretical groundwork for the research in authentication of non-biological entities. The possibility that behavior-based biometric systems can be spoofed in particular by artificially intelligent software agents [63] was also addressed, which lead to research on automatically telling bots and humans apart [66]. Authors of the paper demonstrate how an embedded non-interactive test can be used to prevent automatic artificially intelligent players from illegally participating in online game-play. Specifically, they demonstrated that behavioral biometrics is a great approach to intelligent software authentication.

3 Survey of Non-biological Entities

There are three main types of non-biological entities, that can be broadly classified as Virtual Beings (avatars), Intelligent Software Agents (bots), and Hardware Robots [21]. Virtual Being are at the focus of the following survey, while bots and robots, while equally interesting, are beyond the scope of the current paper.

According to a dictionary, the word "Avatar" means: "embodiment: a new personification of a familiar idea"; or the manifestation of a Hindu deity (especially Vishnu) in human or superhuman or animal form. In an on-line community, Avatar is a virtual representation of a player in an on-line world, a software creation that exists in virtual environment but is controlled by a human player from the physical world. A comprehensive summary of avatar types is given in an on-line book by John Suler, Department of Psychology Professor at Rider University [55]. The book itself is not your ordinary collection of printed articles – it exists only in the on-line form and evolves with time to reflect constant changes in virtual gaming communities. According to [55], the following types of avatars exist based on preferences and behavior of its human creator:

Odd/shocking avatars are unusual, strange, or bizarre; Abstract avatars may be represented by abstract art; Billboard avatars are announcements of some kind; Lifestyle avatars depict a significant aspect of a person's life; Matching avatars are designed to accompany each other; Clan avatars are worn by members of the same social group; Animated avatars contain motion; Animal avatars are typically associated with person's pets or self association with nature; Cartoon avatars are based on famous drawn characters; Celebrity avatars tend to follow trends in popular culture; Evil avatars are scary looking; Real Face avatars are uploaded pictures of the actual users; Idiosyncratic avatars are strongly associated with a specific user; Positional avatars are designed by the member to be placed into specific locations; Power avatars are symbols of omnipotence; Seductive avatars partially naked or scantily clothed figures.

Identification of such avatars can be carried out through analysis of their appearance, attributes, behavioral patterns, frequency and type of changes, using a combination of traditional image pattern recognition techniques and biometric behavioral identifiers. Classifying further the types of behaviors that avatars might exhibit can assist significantly in the task of avatar authentication. According to [55], such behavior can be expressed in Mischievous Pranks (such as smearing someone else's room, spoofing someone with "msay" command, or popping text balloon over someone's head), Flooding of the server by users who make rapid multiple changes of their avatars, Blocking (placing one's avatar on top or too close to another person's prop), Sleeping (by users who have walked away from their computer and their avatar fails to react), Eavesdropping (by reducing avatar to a single pixel and usernames to only one character, someone may become "invisible" and secretly listen in on conversations), Prop Dropping (placing an inappropriate or obscene prop in an empty room), Identity Disruption - people suffering from disturbances in their identity may act it out through frequently changing props they wear. Imposters - stealing someone's avatar, wearing it and also using that person's name (or a variation of it) – one of very serious crimes in cyberworld as it is essentially "stealing someone else's entire identity". Those behaviors resemble typical criminal behaviors of humans and so require high degree of attention from those in charge of security of the virtual communities.

Author of [55] describes one such act "Sometimes, it's hard even for sympathetic people to resist the antics and game-playing. One night, although trying to remain a neutral observer, I eventually found myself as an accomplice to another member in a prank where we set up an unmanned female prop in the spa pool. We used "msay" to talk THROUGH the prop while also talking to it as if it were another user. Essentially, it was a virtual ventriloquist act. Honey " (the prop) was rather seductive towards the guests, and the guests all thought it was a "real" person. It was quite funny, although perhaps a bit mean to the poor naive guests who were unaware of the msay command." The paragraph above is highly interesting as it describes the process of another virtual entity creation, or a "fake avatar", separate from legitimate avatars, that does not corresponds to a real person, but "appears" to be just like them and can sometimes fool even experienced users. Utilizing methods from biometric research as well as developing new approaches targeting specifically avatar authentication and behavior recognition can assist in identifying those "fake avatars" as well as classifying real ones.

4 Avatar Authentication

In this section, we first take a look at techniques for collecting and classifying databases of avatars and bots, moving on to propose a new way to synthesis the new images through application of biometric synthesis methods based on geometric processing and multi-resolution techniques. We then study the two main types of authentication in virtual world: visual and behavioral, and introduce the multi-resolution system for enhanced performance.

4.1 Datasets Generation

In the well-established fields such as biometrics, numerous standardized and publicly available datasets exist [49] making it possible to compare experimental results achieved by different algorithms and to test developed systems. Labeled public datasets of avatar faces, robot faces, or attributed conversations from artificially intelligent agents are currently unavailable. Techniques for creation of standardized and consistent with real world datasets can be learned from examining approaches to generation and evaluation of facial datasets [29, 15] utilized by biometric security systems or from chat mining research applied to gender attribution [9] and human versus bot classification [18].

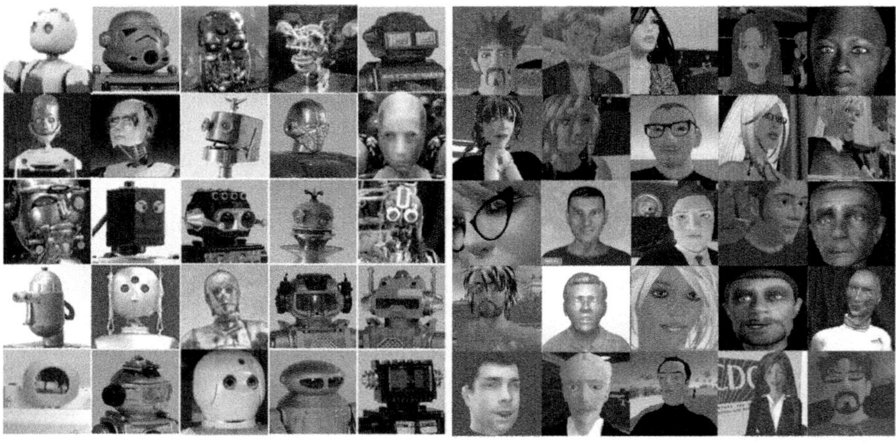

Fig. 2. *Left:* Sample images for a robot-face dataset, currently limited to manual collection; *Right:* Automatically generated random avatar-faces [43] [65]

The authors of this paper have begun work on generation of a publicly available avatar face dataset [43] by designed and implemented a scripting technique to automate the process of avatar face collection. Using the programming language AutoIt as well as a scripting language native to *Second Life*, better known as Linden Scripting Language (LSL), a successful generation of random avatars was achieved. The following is a walkthrough of this process for the creation of randomly generated dataset of avatar faces:

1) Using the scripting language AutoIt, it was possible to simulate key presses and mouse control in a Windows environment. During the first run of the AutoIt script, simulated keyboard commands are used to circle the *Second Life* camera around the avatar such that the front of the avatar's face is exposed.
2) The script is paused and requests the user to center the avatar's face with the horizon using the movement control. This is only needed on the first run and constitutes the last interaction with the user.
3) The AutoIt script then activates the LSL script by clicking on a button attached to the avatar's hub.

4) The LSL script locks the *Second Life* camera's position and rotation as well as controls from the game's automated functions (such as camera changing on clicking).
5) The AutoIt script then takes a screen shot of the avatar using the *Second Life* tool "screen shot". The script then labels avatar "Avatar 'x' face 'y'", where x corresponds to the number of avatar created (1 - N) and y corresponds to the screen shot for that avatar (1 – 10).
6) The script then zooms into the avatar's face before taking another screen shot and using the same labeling system as in step 5.
7) The AutoIt script then rotates the camera at eight specific angles (upper left, center left, lower left, upper center, lower center, upper right, center right, and lower right) taking screen shots at each.
8) The script then selects "edit", then "appearance", bringing up the avatar editing tool. From here the script randomizes a body for a new avatar. Body height, torso length, and leg height all must be set to 50% in order to preserve the camera angle, which is done automatically by the script.
9) The AutoIt script then clicks on the body parts sub menu items "skin", "hair", and "eyes" randomizing each of them as they are entered. The save "all button" is pressed, saving the avatar to begin the screen shot process again.
10) The script zooms away from the avatar before taking the new avatar's center body screen shot.

After step 10, the AutoIt script restarts at step 7 until all the images have been taken. A sample segment of Autoit source code responsible for GUI interaction is given below [43]:

```
Func snapshot ($picture)
dim $picture
mouseClick("Left", 440, 756, 1)   ;snapshot button
sleep(2000)
mouseClick("Left", 102, 296, 1)   ;save button
sleep(3000)
send("{DOWN}{ENTER}")
findname($picture)                EndFunc
```

The datasets generated by the scripted approach consists of ten pictures for each avatar taken from different angles. The images captured are in the Portable Network Graphics (PNG) format at a resolution of 1024 X 768 resulting in each image being between 110KB and 450KB in size. One upper body picture is taken as well as nine facial pictures, all differing in angles. These angles include the top, center, and bottom of the left, center and right side of each avatar's face. The images are named in a consistent format; stating the program, gender, avatar number, and angle. For example, the image "SecondLife Male Avatar 4 gesture 5.png" refers to the image of an avatar that looks like a male character, the fourth in the dataset, and the fifth picture taken in this avatar's set of 10. The gender of the avatar is dependent upon the user's selection at the beginning of the process.

In a separate project we are also working on collection of speech corpora from intelligent agents. We are assembling a a text corpus from intelligent agents who have performed extremely well in the recent Lobner.net prize in Artificial Intelligence competitions. With the assistants of the developed tools any researcher in the field can effortlessly generate virtually unlimited amount of data for visual and stylometric robot authentication experiments.

Currently, it is only possible to specify the desired amount of data and the gender of the avatars' faces and overall area of knowledge about which intelligent agents communicate. It is however already possible to generate multiple samples for each non-biological entity making it easy to perform training and testing on disjoint datasets. Additional work is still necessary to make it possible to generate data with specific characteristics, in which we propose to utilize some of the recently developed biometric synthesis processes as outlined below.

4.2 Synthetic Biometric and Artimetrics

A link between two areas - avatar generation and synthetic biometric generation, is very weak at the moment. One of the first examples can be accredited to 1998 report "*Avatar Creation using Automatic Face Recognition*", where authors discuss specific steps and processing techniques that need to be taken in order for avatar to be created almost automatically from a human face [36]. However, authors of this article postulate that the process of avatar creation and authentication can be further augmented by applying techniques from both biometric synthesis and biometric authentication.

Synthetic biometric is defined as "inverse problem of biometric" [70] and is intended to create artificial phenomenon that does not exist in physical reality, but resembles it. The extensive research on synthetic biometric has been conducted at the Biometric Technologies Laboratory, University of Calgary, and results has been recently reported in the World Scientific book "Image Pattern Recognition: Synthesis and Analysis in Biometrics" [69]. In that study, link between biometric synthesis and inverse logic has been established, as the same principles can be applied to solve inverse logical problems and generate new synthetic biometric data. There are numerous applications and high demand for new biometric databases to test new systems and study various phenomena, and many novel methods based on feature selection, pattern analysis, functional decomposition of spaces, signal processing, image decomposition and multi-resolution has been employed to generate new synthetic biometric data. However, looking at the problem from another point of view, synthetic biometric creations such as fingerprints, irises, faces, ears, hands, behavioral trends and virtual bodies are similar to avatars. They are created artificially, using computer means and sophisticated algorithms, to resemble human and human features. However, there are some substantial differences that make synthetic biometric be recognized in their own category. This is discussed next.

Synthetic biometric, at least up to day, is completely non-personalized. It usually does not correspond to a single human or function, but possesses characteristics of multiple biometrics that were used in the process of new biometric entity synthesis. However, exactly this property might prove most beneficial for new virtual dataset creation. In general, data synthesis refers to the creation of new data to meet some intended purposes, and includes areas such as texture synthesis, domain specific rendering and biometric synthesis. Due to logistical and privacy issues with collecting and organizing large amounts of biometric data, a new direction of biometric research concentrates on the synthesis of biometric information. One of the primary goals of the synthesis of biometric data is to provide databases for testing newly developed biometric algorithms [17]. For instance, author of this paper proposed an approach for facial synthesis and expression modeling based on the underlying mesh modification for both 2D and 3D face models. Selection of control points in this method is guided by the three-dimensional Voronoi diagram which represents the [72]. A general overview of related work on utilizing geometric algorithms in facial expression modeling can be found in [16].

However, not all synthetic biometric comes from multiple sources. While opportunities for customization and specific feature selection are endless when creating a new synthetic data, some data, such as synthetic face, might be created based on a single source – a single photograph or face drawing. The source in this case can be both real (actual photograph or a face scan) or artistically created (cartoon character, caricature etc.), but the resulted synthetic face can resemble the source and have its own personally customized features. For example, one can create a facial image in any pose, with chosen illumination, given color of the eyes, selected hair style and accessories (mustache, glasses, beard etc). Moreover, a concept of time can be introduced and the same synthetic face can be of a young person, middle-aged individual or an old human. The same is true for an avatar – it can resemble its creator, but is fully customizable in appearance, gender, age, voice or interaction with other characters. Thus, it is only natural to make a link between biometric synthesis studies and avatar creation and recognition domain, which is the scope of artimetrics.

4.3 Visual Recognition

We now would like to concentrate specifically on visual recognition of avatars problem. Face Recognition is the task naturally performed by humans, and it remains in the center of biometric research over the last few decades. Hundreds of papers have been published on the topic, with comprehensive surveys of facial biometric research found in [68, 73, 56] as well as in a recent book presenting state-of-the-art in the area [23]. Dozens of different approaches ranging in accuracy of face recognition from low 60% to 99% have been proposed [68]. Knowledge based methods, such as the multi-resolution based approach [67], capture the relationship between facial features. Feature invariant approaches look for structures consistency under a variety of poses and lighting conditions, examples include grouping of edges [71], space gray-level dependence matrix [11], and mixture of Gausians [38]. Template matching extracts standard patterns of the face which are later compared to regions being tested to determine the degree of correlation, classical examples include shape template [10] and Active Shape Model [34]. Finally, appearance-based methods such as Eigenvector decomposition [60], Support Vector Machines (SVM) [42], Hidden Markov Model [45], Naïve Bayes Classifier [50] and Neural Networks [48] learn facial templates from a set of training image.

It is interesting to note that the performance of such adapted technique on avatar face recognition would be superior to method performance among humans. Consider an actual scene in the natural world. The effects of air quality, lightning, reflections, person's posture, clothing, and possible movement, as well as the type of the physical medium used to capture the image (film, camera, cell phone) and the distance/positioning of this capturing device from the person make the problem of face recognition extremely difficult and not resolved up to date. However, in the virtual world, while some variability still exists, the nature of the avatar being a computer generated entity makes it much easier to extract the "ground truth" - the way avatar face was initially created, and thus to develop a standardized approach to avatar face recognition. An example of application of feature-based (geometry-based) method to avatar recognition is given in Figure 3 below.

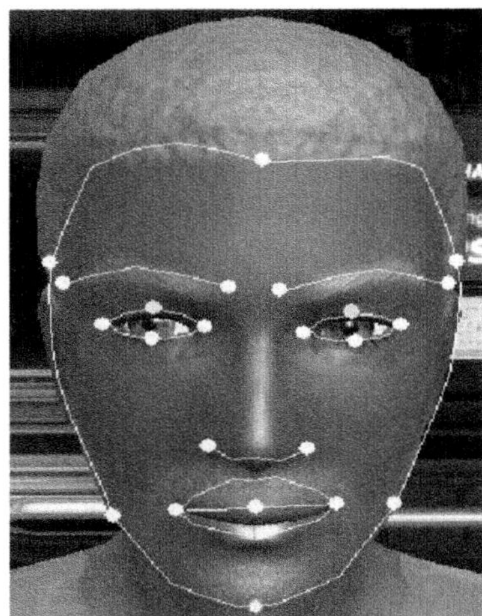

Fig. 3. Feature-based facial recognition applied to an avatar's face

Another important fact to consider is that, as mentioned in the introduction, some research confirms a strong resemblance of avatar to its human creator, which makes it possible to use the results of successful avatar recognition for human recognition, and vice versa. This will, in turn, open a new area of *virtual biometric*, or augmenting the actual biometric with results of recognition in virtual world.

4.4 Behavioral Authentication

As mentioned above, facial recognition, alone with expression analysis and face synthesis, are highly prominent and actively researched areas of biometric [17, 23]. Facial expression analysis has been an active research topic for behavioural scientists since the work of Darwin in 1872. Emotion recognition was studied in paralinguistic communication, clinical psychology, psychiatry, neurology, pain assessment, lie detection, intelligent environments, and multimodal human-computer interface (HCI). From comprehensive book chapters devoted to state of the art research on facial expression and modeling, to tutorials in Biometric conferences and Biometric conference themes devoted exclusively to face animation, morphing, expression analysis and 3D models - the area is receiving a spur of attention from biometric communities, consortiums and industries worldwide. One of the emerging recent trends is capturing subtle details such as wrinkles, creases and minor imperfections that are highly important for biometric modeling as well as matching.

A novel approach to the problem recently introduced to the scientific community takes into an account subtle expression changes and performs morphing expression images in 2D and 3D based on the powerful computational geometry methods [72]. This work makes a number of important contributions to the field of expression modeling and morphing: it is one of the first applications of the sketch-based approach to facial image generation and the first one that preserves and utilizes subtle expression lines. It provides a simple fully automated algorithm based on the distance transform that computed the mapping between pixels in a monochrome image through a clever process of sweeping the image and reusing the information obtained on the previous step. It also provides a combination of Sibson coordinates and Delaunay triangulation mesh to generate and morph 3D facial models. Because all the generated facial models have the same underlying structure, animation created by developed tools can be easily retargeted to various models. Thus, it allows generating facial models with different expressions suitable for further utilization in biometric testing and behavioral research.

Forensics and more specifically authorship recognition, sometimes called stylometry, is another area related to *behavior-based authentication* of identity. In particular, a lot of research has been done in vocabulary analysis and profiling of plain text [25, 32, 33], emails [54, 61] and source code [51, 19, 14]. Written text or spoken word, once transcribed, can be analyzed in terms of vocabulary and style to determine its authorship. In order to do so a linguistic profile needs to be established. Many linguistic features can be profiled such as: lexical patterns, syntax, semantics, pragmatics, information content or item distribution through a text [20]. Commonly utilized text descriptors include: word count, punctuation mark count, noun phrase count, word included in noun phrase count, prepositional phrase count, word included in prepositional phrase count and keyword count [52]. Once linguistic features have been established Support Vector Machines [24], Bayesian classifiers [28], multiple regression and discriminant analysis [53] algorithms (among others) have been applied to determine the authorship of the text. Applying the techniques above to pattern recognition and behavior authentication of avatars based on the way they present themselves, perform their tasks, and communicate in the virtual world is another emerging area of research.

4.5 Multi-modal Systems for Avatar Recognition

Biometric system based solely on a single biometric may not always identify the entity (human or avatar) in the most optimal or precise way. Thus, multibiometric system research is emerging as a trend which helps to overcome limitations of a single biometric solution [47]. This is especially useful in the presence of complex patterns, conflicting or misleading behavior, abnormal data samples, intended or accidental mischief etc. A reliable and successful multibiometric system normally utilizes an effective fusion scheme to combine the information presented by multiple matchers. Over the last decade, researchers tried different biometric traits with sensor, feature, decision, and match score level fusion approaches to enhance the security of a biometric system [47], thus enhancing security and performance of authentication system.

Multimodal biometric approaches improve overall system accuracy and address issues of non-universality, spoofing, noise, and fault tolerance. Multimodal biometrics can referrer to a number of different approaches such as [46]:

- Multi-Sensor – employ multiple sensors to capture a single biometric trait.
- Multi-Algorithm – utilize a number of feature extraction or matching algorithms on the same data.
- Multi-Instance – utilize data from multiple instances of the same trait such as multiple fingerprints Multi-Sample – collect multiple instances of the same trait via a single sensor.
- Multi-Modal – utilize multiple biometric traits (ex. face and fingerprint and voice).

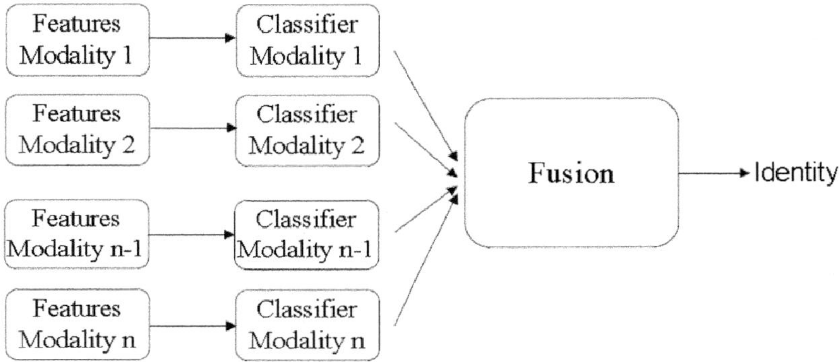

Fig. 4. An example of a multi-modal system architecture

An example of multi-modal biometric system architecture is shown in figure 4. Here, features and classifiers can be obtained from single biometric but acquired but multiple sensor devices, from multiple biometrics (i.e. physiological or behavioral), or from both single and multiple biometrics (with various acquisition, pre-processing and feature extraction techniques). The results of classification are then fused to determine the identity, and this type is generally referred to as post-classification fusion. An architecture will be different for fusions methods.

In a similar manner, together behavioral and physical artimetrics in the virtual worlds can be utilized as a part of a Physoemotional Artimetric system which is a multimodal system. In fact, this approach could be particularly beneficial for artificial entities recognition as there are more ways to "disguise" yourself in the virtual world than in the real world. For example, plastic surgery to change someone's appearance might be an expensive, time consuming and risky way to forge identity of an individual, but changing avatar appearance is much easier, cheaper and faster. Thus, behavioral traits in the virtual world start to play a much more important role. While appearance recognition value somewhat diminishes, the behavior pattern study, popular activities analysis, social surroundings, manner of speech, favorite places to visit, hobbies, skills, art and even wealth in virtual world can supply the crucial information for virtual entity authentication. Combining the visual and behavioral artimetrics using multi-modal biometric approach is emerging area of research that we introduce. In addition, we introduce another concept of Multi-Dimensional system, crossing over between virtual and real world. The concept proposes visual authentication of avatar through its creator authentication, and vice versa, in both virtual and real domains.

5 Applications

There are numerous applications for the methodology, some of which are listed below.

One is preventing malicious intelligent software from obtaining access to information or system resources and granting it to authorized agents. By doing so, one can improve security of virtual communities, social networks, and country's cyber-infrastructure especially vulnerable in the post 9/11 world.

With exponential growth in abilities of artificially intelligent agents (bots, software weapons, viruses, etc.) comes the pressing need to secure information and resources from access by the unauthorized agents, while at the same time allowing seamless access for the "goodware". Behavior based profiling of software agents provides an unobtrusive way of separating helpful bots from malware. Additional research in artimetrics is likely to produce novel behavior-profiling approaches specifically designed to take advantage of the unique "psychology" of artificially intelligent programs. Current explosive research on CAPTCHAs [66, 1, 3, 4, 39] demonstrates one promising direction of future research.

Finding out which agent has performed a given task in case a number of possible alternatives exist, either for demanding responsibility or assigning reward, is another application area. Behavioral profiling can be used to uniquely identify a specific type of avatar and potentially the avatar owner. Examples of such work can be found in Click Fraud and virus detection research. In both domains unique behavioral signatures can be obtained (sometimes indirectly) from the software agent and matched up with known behavioral signatures leading to attribution of the attack to a particular hacker or a mischievous group.

Securing interaction between different pieces of intelligent software or between a human being and an instance of intelligent software in a virual world is also an important domain [30, 26]. Botnents, groups of intelligent cooperating agent and mixed robot/human teams are quickly emerging in numerous domains. Being able to secure their communications is important for further progress in e-commerce, crowdsourcing, virtual community development, construction, military and any other industry with heavy reliance on team based efforts. In order to communicate securely identity of all parties wishing to exchange information needs to be determined with high degree of accuracy. Consequently, it is important to develop automatic algorithm which would give robots ability to recognize other robots and human beings they are working with. While numerous algorithms exist to authenticate identity of specific robots/computers based on digital signatures and cryptographic networking protocols in human dominated environments, robots have a better chance of "fitting-in" if they utilize humanlike biometric approach to identity management advocated in this paper.

Other applications include detecting cheating in games based on assistance from AI software, for example in chess, provide visual and behavioral search capabilities for the virtual worlds, such as Second Life, based on descriptions of individuals, and targeting merchandise marketing in virtual worlds only to agents matching a certain profile.

6 Conclusions and Future Work

This review paper describes a new subfield of security research which transforms and expands the domain of biometrics beyond biological entities to include virtual reality entities, such as avatars, which are rapidly becoming a part of society. Artimetrics research builds on and expands such diverse fields of science as forensics, robotics, stylometry, computer graphics, biometrics and security. The paper describes how verification and recognition of avatars can be carried out via visual properties and behavioral profiling. It also introduces a multimodal system, simultaneously profiling multiple independent physical and behavioral characteristic of en entity, and postulates the feasibility of creating a multimodal system capable of authenticating both biological (human being) and non-biological (avatars) entities.

Potential directions for future Artimetrics research include the investigation of other visual and behavioral approaches to avatar/robot security based on appearance of new characteristics and abilities in the avatars/robots of tomorrow. Even today it is possible to expand robotic biometrics beyond faces and vocabulary to intelligent software agents which mimic higher order human intelligence (such as composing inspiring music, drawing beautiful paintings, and writing poetry). As AI and virtual reality research progresses, it will in turn stimulate creation of new security solutions to identity management across both human and artificial entity worlds.

Acknowledgments. The authors would like to acknowledge the contributions of the members of Biometric Technologies Laboratory (BTLab) at the University of Calgary, as well as Prof. Alexei Sourin for his valuable help in manuscript preparation. The authors also would like to acknowledge the support of NSERC Funding Agency, Canada.

References

[1] L., Ahn, L.v., Blum, M., Hopper, N., Langford, J.: CAPTCHA: Using Hard AI Problems for Security, Eurocrypt (2003)

[2] Asoh, H., Hayamizu, S., Hara, I., Motomura, Y., Akaho, S., Matsui, T.: Socially embedded learning of the office-conversant mobile robot jijo-2. In: 15th International Joint Conference on Artificial Intelligence, IJCAI (1997)

[3] Baird, H.S., Bentley, J.L.: Implicit CAPTCHAs. In: Proceedings of the SPIE/IS&T Conference on Document Recognition and Retrieval XII (DR&R2005), San Jose, CA (January 2005)

[4] Bentley, J., Mallows, C.L.: CAPTCHA challenge strings: problems and improvements, Document Recognition & Retrieval, January 18-19 (2006)

[5] Boyd, R.S.: Feds thinking outside the box to plug intelligence gaps, http://www.mcclatchydc.com/2010/03/29/91280/feds-thinking-outside-the-box.html (retrieved April 10, 2010)

[6] Charles, J.S., Rosenberg, C., Thrun, S.: Spontaneous, Short-term Interaction with Mobile Robots. In: IEEE International Conference on Robotics and Automation, pp. 658–663 (1999)

[7] Chen, K.-J., Barthes, J.-P.: Giving an Office Assistant Agent a Memory Mechanism. In: 7th IEEE International Conference on Cognitive Informatics (ICCI), Compiegne, pp. 402–410 (2008)

[8] Cole, J.: Osama bin Laden's Second Life, Salon, (2008) `http://www.salon.com/opinion/feature/2008/02/25/avatars/` (retrieved June 7, 2009)

[9] Corney, M., Vel, O.d., Anderson, A., Mohay, G.: Gender-preferential text mining of e-mail discourse. In: 18th Annual Computer Security Applications Conference, Brisbane, Australia, pp. 282–289 (2002)

[10] Craw, I., Tock, D., Bennett, A.: Finding Face Features. In: Second European Conference on Computer Vision, Santa Margherita Ligure, Italy, pp. 92–96 (1992)

[11] Dai, Y., Nakano, Y.: Face-Texture Model Based on SGLD and Its Application in Face Detection in a Color Scene. Pattern Recognition 29(6), 1007–1017 (1996)

[12] Fisher, R.: Avatars can't hide your lying eyes, New Scientist, vol. (2755) (April 8, 2010), `http://www.newscientist.com/article/mg20627555.600-avatars-cant-hide-your-lying-eyes.html`

[13] Fong, T.W., Nourbakhsh, I., Dautenhahn, K.: A survey of socially interactive robots. Robotics and Autonomous Systems 42, 143–166 (2003)

[14] Frantzeskou, G., Gritzalis, S., MacDonell, S.: Source Code Authorship Analysis for Supporting the Cybercrime Investigation Process. In: 1st International Conference on eBusiness and Telecommunication Networks - Security and Reliability in Information Systems and Networks Track, pp. 85–92. Kluwer Academic Publishers, Setubal Portugal (August 2004)

[15] Gao, W., Cao, B., Shan, S., Chen, X., Zhou, D., Zhang, X., Zhao, D.: The CAS-PEAL Large-Scale Chinese Face Database and Baseline Evaluations. IEEE Transactions on Systems, Man and Cybernetics 38(1), 149–161 (2008)

[16] Gavrilova, M.L.: Algorithms in 3d real-time rendering and facial expression modeling, 3IA'2006 Plenary Lecture. Eurographics, 5–8 (May 2006)

[17] Gavrilova, M.L.: Computational geometry and image processing techniques in biometrics: on the path to convergence in Image Pattern Recognition. In: Synthesis and Analysis in Biometrics, World Scientific Publishers, Singapore (2007)

[18] Gianvecchio, S., Xie, M., Wu, Z., Wang, H.: Measurement and classification of humans and bots in internet chat. In: 17th Conference on Security Symposium, San Jose, CA, pp. 155–169 (2008)

[19] Gray, A., Sallis, P., MacDonell, S.: Software Forensics: Extending Authorship Analysis Techniques to Computer Programs. In: Proc. 3rd Biannual Conf. Int. Assoc. of Forensic Linguists, IAFL1997 (1997)

[20] van Halteren, H.: Linguistic profiling for author recognition and verification. In: Proceedings of ACL- (2004)

[21] Holz, T., Dragone, M., O'Hare, G.M.P.: Where Robots and Virtual Agents Meet. A Survey of Social Interaction Research across Milgram's Reality-Virtuality Continuum International Journal of Social Robotics 1(1) (January 2009)

[22] Ito, J.: Fashion robot to hit Japan catwalk, PHYSorg, `http://www.physorg.com/pdf156406932.pdf` (retrieved June 2009)

[23] Jain, A., Li, S.Z.: Handbook on Face Recognition. Springer, New York (July 2004)

[24] Joachim, D., Jorg, K., Edda, L., Paass, G.: Authorship Attribution with Support Vector Machines. Applied Intelligence, 109–123 (2003)

[25] Juola, P., Sofko, J.: Proving and Improving Authorship Attribution. In: Proceedings of CaSTA-04 the Face of Text (2004)

[26] Kanda, T., Ishiguro, H., Ono, T., Imai, M., Mase, K.: Multi-robot cooperation for human-robot communication. In: 11th IEEE International Workshop on Robot and Human Interactive Communication, pp. 271–276 (2002)

[27] Khurshid, J., Bing-rong, H.: Military robots - a glimpse from today and tomorrow. In: 8th Control, Automation, Robotics and Vision Conference (ICARCV), pp. 771–777 (2004)

[28] Kjell, B.: Authorship attribution of text samples using neural networks and Bayesian clas-sifiers. In: IEEE International Conference on Systems, Man, and Cybernetics. 'Humans, In-formation and Technology', San Antonio, TX, USA, pp. 1660–1664 (1994)

[29] Klimpak, B., Grgic, M., Delac, K.: Acquisition of a Face Database for Video Surveillance Research. In: 48th International Symposium focused on Multimedia Signal Processing and Communications, Zadar, pp. 111–114 (2006)

[30] Klingspor, V., Demiris, J., Kaiser, M.: Human-Robot-Communication and Machine Learning. Applied Artificial Intelligence 11, 719–746 (1997)

[31] Kobayashi, H., Hara, F.: Study on face robot for active human interface-mechanisms of facerobot and expression of 6 basic facial expressions. In: 2nd IEEE International Workshop on Robot and Human Communication, Tokyo, Japan, November 3-5, pp. 276–281 (1993)

[32] Koppel, M., Schler, J.: Authorship Verification as a One-Class Classification Problem. In: 21st International Conference on Machine Learning, Banff, Canada, pp. 489–495 (July 2004)

[33] Koppel, M., Schler, J., Mughaz, D.: Text Categorization for Authorship Verification. In: Eighth International Symposium on Artificial Intelligence and Mathematics, Fort Lauderdale, Florida (Januray 2004)

[34] Lanitis, A., Taylor, C.J., Cootes, T.F.: An Automatic Face Identification System Using Flexible Appearance Models. Image and Vision Computing 13(5), 393–401 (1995)

[35] Lim, H.-O., Takanishi, A.: Waseda biped humanoid robots realizing human-like motion. In: 6th International Workshop on Advanced Motion Control, Nagoya, Japan, pp. 525–530 (2000)

[36] Lyons, M., Plante, A., Jehan, S., Inoue, S., Akamatsu, S.: Avatar Creation using Automatic Face Recognition. In: ACM Multimedia 1998, Bristol, England, pp. 427–434 (September 1998)

[37] Lyu, M.R., King, I., Wong, T.T., Yau, E., Chan, P.W.: ARCADE: Augmented Reality Computing Arena for Digital Entertainment. In: IEEE Aerospace Conference, Big Sky, MT, March 5-12, pp. 1–9 (2005)

[38] McKenna, S., Gong, S., Raja, Y.: Modelling Facial Colour and Identity with Gaussian Mixtures. Pattern Recognition 31, 1883–1892 (1998)

[39] Misra, D., Gaj, K.: Face Recognition CAPTCHAs, International Conference on Telecommunications. In: Internet and Web Applications and Services (AICT-ICIW 2006), February 19-25, p. 122 (2006)

[40] Nood, D.d., Attema, J.: The Second Life of Virtual Reality., http://www.epn.net/interrealiteit/EPN-REPORT-The_Second_Life_of_VR.pdf (retrieved June 2009)

[41] Oh, J.-H., Hanson, D., Kim, W.-S., Han, I.Y., Han, Y., Park, I.-W.: In: International Conference on Intelligent Robots and Systems, Daejeon, pp. 1428–1433 (2006)

[42] Osuna, E., Freund, R., Girosi, F.: Training Support Vector Machines: An Application to Face Detection. In: IEEE Conference on Computer Vision and Pattern Recognition, pp. 130–136 (1997)
[43] : Parameterized Generation of Avatar Face Dataset. In: 14th International Conference on Computer Games: AI, Animation, Mobile, Interactive Multimedia, Educational & Serious Games, Louisville, KY (2009)
[44] Patel, P., Hexmoor, H.: Designing BOTs with BDI agents. In: International Symposium on Collaborative Technologies and Systems (CTS) Carbondale, USA, pp. 180–186 (2009)
[45] Rajagopalan, A., Kumar, K., Karlekar, J., Manivasakan, R., Patil, M., Desai, U., Poonacha, P., Chaudhuri, S.: Finding Faces in Photographs. In: 6th IEEE Intern. Conference on Computer Vision, pp. 640–645 (1998)
[46] Ross, A.: An Introduction to Multibiometrics. In: 15th European Signal Processing Conference (EUSIPCO), Poznan, Poland (September 2007)
[47] Ross, A., Jain, A.: Information fusion in biometrics. Pattern Recognition Letters 24, 2115–2125 (2003)
[48] Rowley, H., Baluja, S., Kanade, T.: Neural Network-Based Face Detection. IEEE Transactions on Pattern Analysis and Machine Intelligence 20(1), 23–38 (1998)
[49] Li, S., Jain, A. (eds.): Handbook of Face Recognition-Face Databases. Springer, New York (2005)
[50] Schneiderman, H., Kanade, T.: Probabilistic Modeling of Local Appearance and Spatial Relationships for Object Recognition. In: IEEE Conference on Computer Vision and Pattern Recognition, pp. 45–51 (1998)
[51] Spafford, E.H., Weeber, S.A.: Software Forensics: Can We Track Code to its Au-thors? In: 15th National Computer Security Conference, pp. 641–650 (October 1992)
[52] Stamatatos, E., Fakotakis, N., Kokkinakis, G.: Assoc. Computational Linguistics. In: Automatic authorship attribution, in Proc. nineth Conf. European, Bergen, Norway, pp. 158–164 (June 1999)
[53] Stamatatos, E., Fakotakis, N., Kokkinakis, G.: Computer-Based Authorship Attribution Without Lexical. Measures Computers and the Humanities 35(2), 193–214 (2001)
[54] Stolfo, S.J., Hershkop, S., Wang, K., Nimeskern, O., Hu, C.-W.: A Behavior-based Approach to Securing Email Systems. Mathematical Methods, Models and Architectures for Computer Networks Security 2776, 57–81 (2003)
[55] Suler, J.: The Psychology of Cyberspace, On-line book (2009), http://psycyber.blogspot.com
[56] Tan, X., Chen, S., Zhou, Z.-H., Zhang, F.: Face recognition from a single image per person: A survey. Pattern Recognition 39(9), 1725–1745 (2006)
[57] Tang, H., Fu, Y., Tu, J., Hasegawa-Johnson, M., Huang, T.S.: Humanoid Audio–Visual Avatar With Emotive Text-to-Speech Synthesis. IEEE Transactions on Multimedia 10, 969–981 (2008)
[58] Teijido, D.: Information assurance in a virtual world. In: Australasian Telecommunications Networks and Applications Conference (ATNAC 2009), Canberra, Australia, November 10-12 (2009)
[59] Thompson, B.G.: The State of Homeland Security, House.gov (2006), http://hsc-democrats.house.gov/SiteDocuments/ 20060814122421-06109.pdf (retrieved June 10, 2009)
[60] Turk, M., Pentland, A.: Eigenfaces for Recognition. Journal of Cognitive Neuroscience 3(1), 71–86 (1991)

[61] Vel, O.D., Anderson, A., Corney, M., Mohay, G.: Mining Email Content for Author Identification Forensics. ACM SIGMOD Record: Special Section on Data Mining for Intrusion Detection and Threat Analysis 30(4), 55–64 (2001)
[62] Yampolskiy, R.V.: Behavioral Biometrics for Verification and Recognition of AI Programs. In: 20th Annual Computer Science and Engineering Graduate Conference (GradConf), Buffalo, NY (2007)
[63] Yampolskiy, R.V.: Mimicry Attack on Strategy-Based Behavioral Biometric. In: 5th Interna-tional Conference on Information Technology: New Generations (ITNG 2008), Las Vegas, Nevada, April 7-9, pp. 916–921 (2008)
[64] Yampolskiy, R.V., Govindaraju, V.: Behavioral Biometrics for Recognition and Verification of Game Bots. In: The 8th Annual European Game-On Conference on Simulation and AI in Computer Games (GAMEON 2007), Bologna, Italy, November 20-22 (2007)
[65] Yampolskiy, R.V., Govindaraju, V.: Behavioral Biometrics for Verification and Recognition of Malicious Software Agents. In: SPIE Defense and Security Symposium, Orlando, March 16-20 (2008)
[66] Yampolskiy, R.V., Govindaraju, V.: Embedded Non-Interactive Continuous Bot Detection. ACM Computers in Entertainment 5(4), 1–11 (2007)
[67] Yang, G., Huang, T.S.: Human Face Detection in Complex Background. Pattern Recognition 27(1), 53–63 (1994)
[68] Yang, M.-H., Kriegman, D.J., Ahuja, N.: Detecting Faces in Images: A Survey. IEEE Transactions On Pattern Analysis and Machine Intelligence 24(1) (2002)
[69] Yanushkevich, S., Gavrilova, M., Wang, P., Srihari, S.: Image Pattern Recognition: Synthesis and Analysis in Biometrics. World Scientific Publishers, Singapore (2007)
[70] Yanushkevich, S., Stoica, A., Shmerko, V., Popel, D.: Inverse Problem of Biometric. CRC Press/Taylor&Francis, Boca Raton (2005)
[71] Yow, K.C., Cipolla, R.: Feature-Based Human Face Detection. Image and Vision Computing 15(9), 713–735 (1997)
[72] Yuan, L., Gavrilova, M., Wang, P.: Facial metamorphosis using geometrical methods for biometric applications. IJPRAI 22(3), 555–584 (2008)
[73] Zhao, W., Chellappa, R., Phillips, P.J., Rosenfeld, A.: Face recognition: A literature survey. ACM Computing Surveys 35(4), 399–458 (2003)

Range Based Cybernavigation in Natural Known Environments

Ray Jarvis and Nghia Ho

Monash University
ray.jarvis@eng.monash.edu.au

Abstract. This paper concerns the navigation of a physical robot in real natural environments which have been previously scanned in considerable (3D and colour image) detail so as to permit virtual exploration by cybernavigation prior to mission replication in the real world. An on-board high speed 3D laser scanner is used to localise the robot (determine its position and orientation) in its working environment by applying scan matching against the model data previously collected.

1 Introduction

This paper is about real robot navigation in real outdoor natural environment spaces. The intriguing title derives from the fact that the environment is first scanned off-line in considerable detail using a Riegl LMS-Z420i laser/camera scanner and this cyberspace database can be walked through and robotically navigated in virtual space. What is crucial, however, is that a real robot can be teleoperated or can autonomously navigate the real space represented in this virtual world of space and colour using an on-board fast laser range scanner, which instrument's data can, in real time, be matched with the model data to determine the location and orientation of the physical robot in real space. Optimal path planning in virtual (cyber) space can be executed simultaneously in the real and cyber worlds with high correspondence fidelity.

This approach is almost the opposite of the simultaneous localisation and modelling (SLAM) [15,7] methodology which has dominated the robotic navigation research field for one and a half decades. SLAM aims to use on-board sensors to simultaneously build a map of the working environment whilst moving through it, determining the location of the robot throughout. In such a manner entire maps can be generated incrementally with an accuracy dependent on the quality and richness of the sensor data and the algorithmic methodology applied. Many SLAM systems have been researched over time, using a variety of sensors and achieving a range of accuracies and computational efficiencies.

Whilst acknowledging the importance of the SLAM approach, the demonstrable success it has achieved and the generality of its application scope, there are still many circumstances where the availability of a map prior to navigation is of considerable advantage, particularly if it is already available (building plans, terrain maps etc.) or can be readily obtained and needs to be constructed just

M.L. Gavrilova et al. (Eds.): Trans. on Comput. Sci. XII, LNCS 6670, pp. 159–182, 2011.

once for an environment where ongoing robot navigation activity is to persist. Some arguments against the application of SLAM in certain circumstances and for particular purposes are listed below, not necessarily in order of importance:

(a) If one has a suitable map (eg. floor plan of a building, forestry terrain map, Google map etc.) why not use it?

(b) Optimal path planning can not really commence until a whole map is available, since access to certain spaces may not be known at the outset. Thus, for SLAM, considerable exploration tasks must be undertaken before goal oriented missions can proceed efficiently.

(c) If an instrument such as the Riegl LMS-Z420i laser/camera scanner (as is used in this paper) is available, it can be used to collect detailed spatial and visual models of a working environment and this needs to be done only once. If the environment is to be subject to ongoing robotic missions, this preparation step is easily justified as the map can be very accurate and can be annotated to identify important aspects such as shops, doors, stairs, fire extinguishers, heavy traffic regions, terrain roughness, water sources etc. for ongoing support of robot activity over a period of time. Both locations and visual cues can be listed.

(d) Detailed 'walk throughs' of the cyberspace which can be constructed from maps collected as in (c), above, can be used to plan robot missions and trial navigational paths can be generated in virtual space. Sensitive areas (eg. near stairs or water or where people often congregate) can be made 'no-go' or 'only if necessary' zones for path planning. Using 'walk throughs', good vantage points for security observation and disaster monitoring (eg. spotting victims) can be identified as potential goal points (perhaps way-points) for robot deployment. In fact, given the fidelity of such virtual spaces, aspects of navigation strategies which have a human behavioural dimension, perhaps a mix of curiosity and efficiency, risk avoidance or security can be learnt and applied to robot navigation in the real world.

(e) Using an off-line collected map such as discussed in (b) and (c), above, is now so easy to do and provides a much less complex strategy than SLAM to support ongoing navigation. Simplicity is often the way to reliability.

In any case, whether a good argument for or against SLAM can be made convincingly, this paper presents a clear alternative which is of value in its own right. In a short tutorial style, classical robot navigation relies on three essential sub-systems. Firstly, the location and orientation of the robot must be known in the context of its working environments. This is called localisation. Next the environment through which the robot will navigate needs to be known or constructed incrementally as the robot moves through it without collision. Finally a path planning strategy must be applied to determine the path of the robot. This path should preferably be optimal in terms of minimal length or time or include other factors such as visibility or tractability and certainly be collision-free. The plan can be updated when necessary, especially when new information comes to hand. We have, or course, assumed that the physical, sensor and communication capabilities of the robot are adequate for the required task fulfilment.

One may reasonably ask why, for outdoor localisation, does any complex method or expensive sensor need to be used when GPS (Global Positioning System) is almost universally available? The simple answer is that there are a number of situations where GPS data is not available or is unreliable (eg. in high rise building environments, deep valleys, in heavily wooded forests etc.) or where, whilst it is available, it is not sufficiently accurate (depending on the degree of clutter in the obstacle field).

To demonstrate and support the aforementioned arguments, we carried out outdoor experiments in two different environments surrounded by heavy vegetation. The first experimental site was on campus at Monash University and the second was at an outback property in Pomonal.

The paper continues with the following structure:

First the Riegl LMS-Z420i instrument, the data collection, the multiple scan consolidation process and cybernavigation will be described along with some parameters of the accuracy and quality of mapping which results. Aspects of a 'walk through' capability will also be touched upon. Then the experimental set up for collecting on-board real-time range scan data using a Velodyne HDL-54E S2 instrument and the type of data collected will be described. This is followed by a section describing the methodology of data correspondence used to match 'live' data with the cyberspace data to establish the location and orientation of the robot. Experimental results so far achieved, a section on path planning, discussion, future work and conclusions follow to complete the paper.

2 Constructing and Navigating the Cybermap

The Riegl LMS-Z420i is a laser time-of-flight range scanner with the option of collecting high resolution colour imagery (using a Nikon digital camera attached) which can be registered with the range data to allow a high fidelity 3D space/colour model of the scanned real world to be represented and explored in cyberspace. The instrument can collect 11,000 range samples/sec up to a range of 800m at a repeatability of 8 mm (and accuracy of 10 mm) with an angular field of view up to 80° (elevation) x 360° (azimuth). The instrument is splash and dust proof, can be powered by a 12V car battery, operates in the near infrared spectrum, and is class 1 eye safe. Its minimum range is 2m. The normal procedure for 3D mapping an environment is to set up the instrument at a number of locations such that the integration of the multiple scan data is able to cover most of the exposed surfaces in that environment. It is best that overlapped volumes be scanned to simplify and make more accurate the integration procedure. Depending upon the range scan density required, each collection phase can take minutes to hours. In our application approximately 20 minutes per location is sufficient to collect range/colour image data to suit our needs. This equates to about 5-6 million scanned points per location, which gives a practical coverage radius of 15-20 metres. Beyond that and the density of the points become too increasingly sparse to be usable. Since collecting the multiple scan data is carried out only once, this time imposition is not prohibitive if it is expected that the robot would operate in that domain for some considerable time and that domain would not change significantly during that period of operation.

2.1 Localisation Method

A total of 4 and 12 scans were collected at Monash and Pomonal (in Victoria), respectively. The location of each scan was chosen to provide a good balance between coverage and overlap. To consolidate all the scans to a global reference frame, an anchor scan was chosen as the reference and all subsequent scans were aligned to it. We use the Trimmed Iterative Closest Point (TrICP) algorithm [2] (a variant of the popular Iterative Closest Point (ICP) algorithm [1]) to register the scans. The principal of ICP is given as follows. Given two point clouds set, $\mathcal{P} = \{p_i\}_1^{N_p}$and $\mathcal{M} = \{m_i\}_1^{N_m}$, ICP will find try to find a R (rotation) and t (translation) matrix that will minimise the mean squared distances (MSE) between the corresponding points, mathematically expressed as

$$\arg\min \|m_i - (Rp_i + t)\|^2 \quad (1)$$

At each iterative step, ICP selects the closest pair-wise points between the two sets (using a nearest neighbour search) and calculates the (R, t) matrix to minimise the Euclidean distance between them. However, ICP assumes all points have a correspondence but this is not the case in practice. The percentage of correspondence between scans will depend on the overlap. A simple strategy to address this problem is to discard points that do not have a corresponding point within a certain distance. Even then, there will be still be a small percentage of incorrect matches. For this reason TrICP was chosen. TrICP uses least trimmed square (LTS) to increase robustness to bad matches by optimising only a percentage of the best matches. This parameter is chosen by the operator beforehand.

Our implementation of the TrICP allows the user to graphically select the corresponding points and control the iterative process via four parameters listed in Table 1. The MAX_POINTS parameter limits how many points to load into memory per scan. We found 1 million points to be suitable for a computer with 1GB of RAM; anymore and the operating system will start using hard disk swapping, which is very slow. The $OUTLIER$ parameter does the initial discarding of points that do not have a close correspondence, as mentioned earlier. This value was set to 2.0 metres. The LTS parameter is the percentage of points to optimise. This was set to 0.9 (90%). This means we assume there is about 10% of bad matches. The last two parameters, T and $MAX\_TERATIONS$, are the two terminating conditions for TrICP. The algorithm will terminate when either of these conditions are met. Parameter T was set to terminate the algorithm if the change in MSE falls below 0.01 (1% change). In practice, we found TrICP to typically converge in under 10 iterations. All the values were experimentally obtained.

Table 1. TrICP parameters used

Variable name	Description	Value
MAX_POINTS	Maximum number of points to process per scan	1,000,000
OUTLIER	Initial outlier distance after pre-registration (metres)	2.0
LTS	LTS value	0.9
T	Termination condition (change in MSE)	0.01
MAX_ITERATIONS	Maximum number of iterations	20

2.2 Localisation Results

Two sets of registered point clouds are shown in Figure 1. An aerial view is provided to give a better visual idea of the environment. The trees have a tendency to come out as a blue shade in the 3D data. We suspect this may be due to the blue sky saturating the camera.

Fig. 1. Registered point cloud and their corresponding aerial image. Top is a section of Monash campus. Bottom is a section of the property at Pomonal.

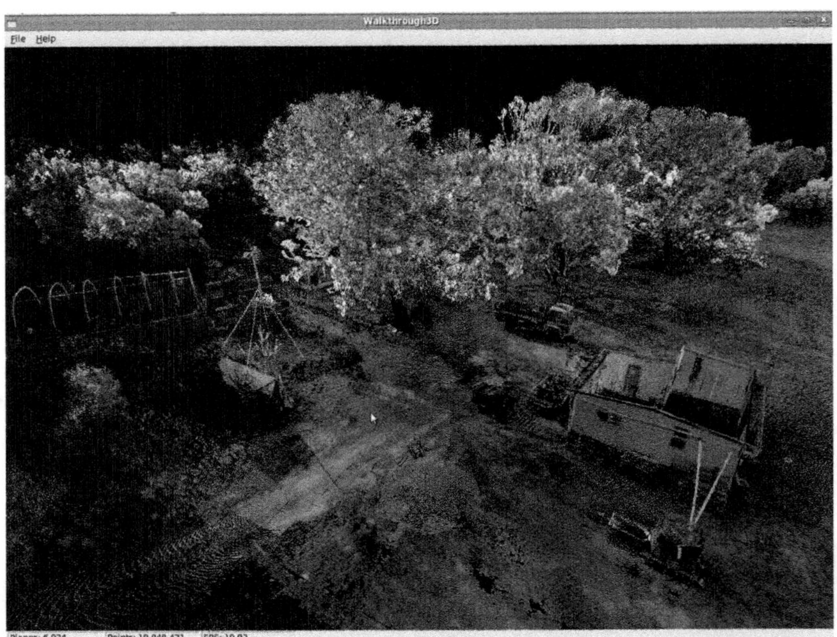

Fig. 2. 3D view of main operating area in Pomonal

A 3D view showing our main operating area at Pomonal is shown in Figure 2. The house can be seen in the bottom right aerial image in Figure 1, towards the bottom right hand corner.

3 Collecting 'Live' Range Data

The Velodyne HDL-64E S2 laser range scanner used in this work is designed for vehicle mounting and is weather and shock proof, as shown in Figure 3. It rotates at 5-15Hz and can return range readings up to 120 meters away (depending on surface albedo) with an accuracy of $<$ 2cm. at up to 1.8 million samples/sec. It has 64 laser emitters (4 groups of 16) and collects 64 line scans (360°) of range data as the head of the unit spins. In the elevation direction it collects data from +2° to –24.8° and thus is able to scan a volume suitable to determine where the vehicle can move without collision as well as the ground condition (roughness, holes, ditches etc.) The continuous stream of range data can be logged for replay. For our application, a data file is continuously replaced by the latest data. This data file is the source of near real-time data which is matched against the cyberspace model of the environment previously collected off-line by the Riegl laser scanner. A colour coded view of typical live data is shown in Figure 4. Intensity (reflective) data is also available for each range measurement but we do not use this for this project. It is eye-safe and uses 5 nano-second pulsed lasers at a wavelength of 905 nano-metres. It can operate in the complete

Fig. 3. Velodyne mounted on a vehicle

Fig. 4. A typical scan from the Velodyne. False colouring has been used to highlight the elevation information.

absence of ambient light. The unit communicates with its host computer via 100 Mbit/s UDP Ethernet packets and can be powered by a 12 volt car battery.

4 Localisation by Scan Matching

The 3D outputs of both the Riegl and the Velodyne laser range scanners are voluminous and matching the 3D raw data is its entirety much too time consuming on currently available standard lap-top computers to consider for near real-time localisation. On the other hand, matching a single 360 azimuth scan from each instrument would not provide reliable localisation, especially if some pitch/roll variations occur whilst collecting the Velodyne data from a moving vehicle, as would be expected. A common approach used by SLAM researchers is to match scans in 2D, either by using 2D laser scanners or projecting 3D scans onto a 2D plane. This assumes the robot traverses on a planar surface and can greatly simplify scan matching due to the reduction in dimension. Scan matching techniques can be categorised based on their data association method such as point to point, point to feature , or feature to feature. In feature to feature matching, features such as line segments [23] and corners [16] are extracted from the 3D range data. In point to feature matching, points can be matched to line features, as done by[3]. Some popular point matching algorithms are: ICP (mentioned earlier), iterative matching range point (IMRP) [18], iterative dual correspondence (IDC) [18], and polar scan matching (PSM) [6]. Point to point matching approaches do not assume the environment to be structured or contain pre-defined features. A simple, yet reliable compromise is to project both data sets on to the horizontal plane and detect distinct vertical structures in the environment (eg. walls, tanks, free trunks, fences) by thresholding. This procedure provides some immunity from pitch/roll variations, although more reliable and accurate results are possible if the pitch/roll changes are measured and used to correct the 3D data.

For localisation, a particle filter [5] or Kalman filter [15] is typically employed to localise and track a robot. These filters process sensor and odometry reading from the robot and output an estimate of the robot's pose. These type of filters fall under the general family of recursive Bayes filter, which, in its general form, can be expressed as

$$Bel\left(x_t\right) = \eta p\left(z_t \mid x_t\right) \int p\left(x_t \mid u_t, x_{t-1}\right) Bel\left(x_{t-1}\right) dx \qquad (2)$$

where $Bel\left(x_t\right)$ is the estimated state (pose) of the robot at time t, z_t are the sensor readings, u_t is the odometry information and η is a normalising constant. The particle filter represents this posterior belief by a sample of particles, where each particle is associated with a weighted hypothesis of the robot's pose. The particle filter is well suited for global localisation because it can localise a robot from an unknown starting pose by spreading particles all over the map. Usually, for such a project that utilises a global map and laser sensor, one would use a particle filter, but this is complicated by certain aspects of the 'live' range data collection process used for the experiments reported in this paper. Firstly, the vehicle does not come equipped with any sensors to measure odometry. Secondly, the scans from the vehicle were collected in a sparse manner (every 5 to 15 metres) for our preliminary experiments. This means we have to make some assumptions about how much the vehicle can move or turn, which ends up increasing the computational search space. In our experiments, we found that because we only collected a handful of scans (the most being 60 Velodyne scans in one experiment) spaced far apart, the particle filter does not have enough data to converge around the true pose. It should be noted that, in a fully operational mode, continual updates of the range data file would be made and localisation carried out as often as possible (processor limited). We subsequently localised continuously in real-time with the Velodyne mounted on a moving vehicle.

4.1 Localisation Method

For our preliminary experiments we opted for a simpler, though less sophisticated, method for tracking the robot. The tracking algorithm is given the initial starting position of the robot. When the first Velodyne scan is read, the algorithm performs a brute force search for the best possible match within a circular boundary centre on the initial starting position. The best match then becomes the new centre for the next Velodyne scan search. One unexpected issue we faced was the vehicle moving into incomplete sections of the map. By incomplete, we mean there is minimal scan coverage taken at that area. This would cause the tracking to fail. A simple strategy to deal with this is to augment the global map with the previous Velodyne scan (at $t-1$) when the probability of the best match is below a certain threshold, and then relocalise to find the best match. We found that we only need to augment the global map with the previous Velodyne scan temporarily, just for the duration required. This avoids cumulatively 'polluting' the global map with bad augmented data from the Velodyne scans. This simple tracking algorithm yields surprisingly good results and is summarised in Algorithm 1. The function $FindBestMatch$ finds the best pose in $GlobalMap$ by searching around the location at $CurrentPose$ constrained by $SearchConstraints$. For the $SearchConstraints$ parameter, we set the algorithm to search within a cone with a radius of $20m$ and $\pm 90°$ of the current pose. This constraint was chosen based on observation of the driver's behaviour in the vehicle. The parameter ε, the threshold to augment the global map with the previous Velodyne scan, was set to 0.6.

We use a point to point scan matching approach by creating a signature of the 3D data as follows:

1. Height threshold the 3D data and keep only points whose elevation falls between 0.5 and 1.0 metre.
2. Project the 3D point onto a 2D image, with each pixel equating to 0.1 metre.
3. Apply a 3x3 averaging mask to expand the point features slightly
4. From the centre of the scan (real world centre not 2D image centre), perform ray tracing at 2 degrees increment in a full circle. Ray tracing stops when

the ray hits an obstacle, exceeds 50 metres or exceeds the image's boundary. If the ray distance did exceed 50 metres or exceed the image's boundary, the distance is set to 0 (indicating an invalid value).
5. The signature is the 180 ray traced distances.

We found applying the averaging mask has a significant effect on localisation accuracy because it reduces the sensitivity of the rays (from ray tracing) to the presence of very small point features, that are easy to miss if the ray traced angle is off by even one degree. Originally, we tried projecting the 3D data into 2D histogram bins and experimented with thresholding the bin counts. The idea is to threshold for very strong vertical structures, such as tree trunks, but we found this had a tendency to cull out smaller vertical features, that are just as equally important.

Having a full 360 degree sampling of rays means we can effectively find the orientation of the robot by considering all directions at 2 degree increments, without having to perform ray tracing for each one. Given two signatures $sig1$ and $sig2$, the scoring function used is

$$s = \sum_i exp\left(-\left|sig1\left[i\right] - sig2\left[i\right]\right|^2 / 2\sigma^2\right) \tag{3}$$

$$x = \omega + \beta s \tag{4}$$

$$score = 1/\left(1 + exp\left(-x\right)\right) \tag{5}$$

Equation 3 calculates the 'raw' score using the sum of absolute differences between the signatures, weighted by an unnormalised Gaussian function. Equation 4,5 clamps the score to be between 0.0 and 1.0, using a sigmoid/logistic function. The weight ω and bias value β were found using logistic regression on a small set of training data. We skip over signature values that are either 0 (no valid reading) or those that have an absolute difference greater than 3 metres, we consider these values to be 'bad' readings. This approach is robust to noisy data and we found the scoring function can tolerate up to 65-69% of bad readings. To calculate the score for different rotations, $sig1$ is rotated by adding an offset to the starting index, and the score recalculated. The orientation returning the highest score is taken as the best match. A summary of the scoring function is presented in Algorithm 2. The value of the parameters used are: $\sigma = 1, \omega = -6.0, \beta = 0.1427$.

When the function *FindBestMatch* in Algorithm 1 is called, it performs a search for the best pose every metre, rather than every single grid position. This translates to skipping every 10 pixels for a grid quantisation of 0.1 metres per pixel. To speed up matching, the signatures for the global map are calculated offline. The cached signatures are not used when the global map is augmented with Velodyne data. In this case, the tracking algorithm has to perform the ray tracing on the spot. To improve localisation, the estimated pose is refined by performing a finer search in an area defined by a 1x1m (10x10 pixels) window around the estimated pose. In theory, we can expect the average localisation error to be around $\pm 0.1m$, assuming the errors from the laser range scanner are much smaller compared to the grid quantisation.

Algorithm 1. Tracking algorithm

```
CurrentPose ← InitialGuess
initialise SearchConstraints
initialise GlobalMap
for i ← 0 to |VelodyneScans| do
    [BestPose, Score] ← FindBestMatch(GlobalMap, CurrentPose, VelodyneScan_i)
    if Score < ε AND i > 0 then
        TmpMap ← Augment VelodynScan_{i-1} with GlobalMap
        [BestPose, Score] ← FindBestMatch(TmpMap, CurrentPose, VelodyneScan_i)
    end if
    VelodyneScan[i].pose ← BestPose
    CurrentPose ← BestPose
end for
```

Algorithm 2. Scoring function

```
procedure GetScore(sig1, sig2)
    HighestScore ← 0
    BestAngle ← 0
    for i ← 0 to |sig1| do
        s ← 0
        for j ← 0 to |sig2| do
            index ← modulus(j+1, |sig1|)
            if sig1[index] = 0 OR sig2[j] = 0 then
                skip
            end if
            d ← |sig1[i] − sig2[i]|
            if d > 3 metres then
                skip
            end if
            s ← s + exp(−d^2/2σ^2)
        end for
        if s > HighestScore then
            HighestScore ← s
            BestAngle ← (index × 2) // 2 degrees per index increment
        end if
    end for
    x ← ω + β×HighestScore
    score ← 1/(1 + exp(−x))
    return score, BestAngle
end procedure
```

4.2 Localisation Results

A total of 3 initial experiments were conducted, 1 at Monash campus and 2 at Pomonal. The estimated paths for all three experiments are shown in

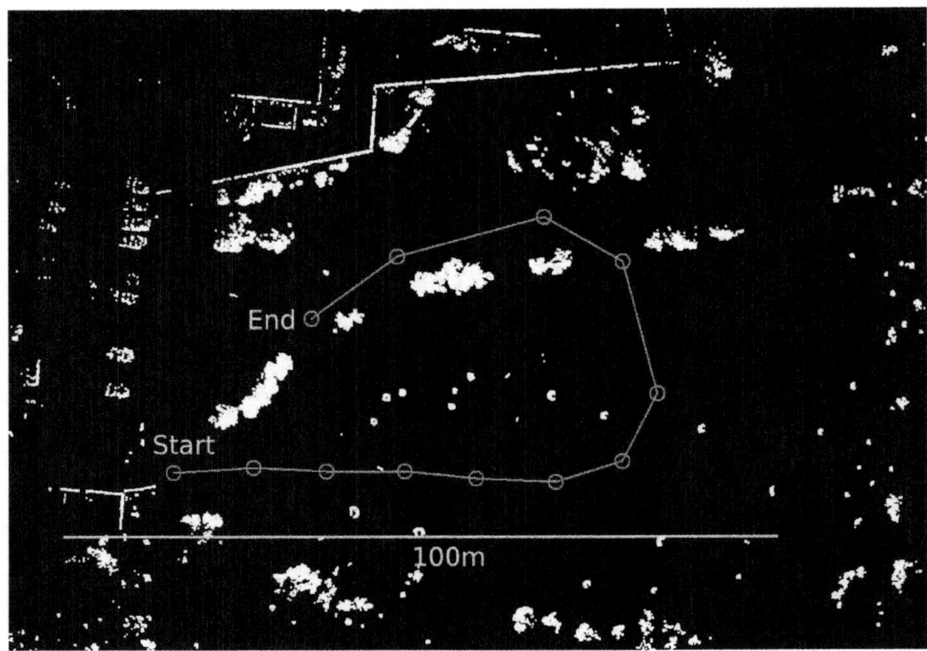

Fig. 5. Localisation results for Monash campus

Figure 5,6,7. The white pixels are the point features after thresholding and 2D projection, as outlined previously in Section 4.1. The green line indicates the path taken and the green circles were where the vehicle stopped to record a Velodyne scan. Ground truth was obtained by manually registering the Velodyne scan to the global map using TrICP. The average localisation errors are presented in Table 2. The localisation error for translation is about 15 cm and 1 degree for rotation. This error is certainly good enough for navigation related applications. A summary of the computational costs required to match a single scan, using an Intel i7 2.67GHz with 6GB of RAM, is presented in Table 3. When using the cache ray traced values the system is able to process just under 3 scans per second. When the map is augmented with a Velodyne scan, the rays have to be traced and takes around 2.5 seconds. At the moment, our implementation only uses one of the four available cores. We can expect some linear improvement if we parallel the matching process across the four cores, especially since it is a naturally parallelable task.

Some more results showing the localisation with the predicted view in the 3D world are shown in Figure 8. In these cases, localisation was carried out as continuously as possible with the Velodyne mounted on a moving vehicle. These experiments confirm the practicality of the method for realistic situations.

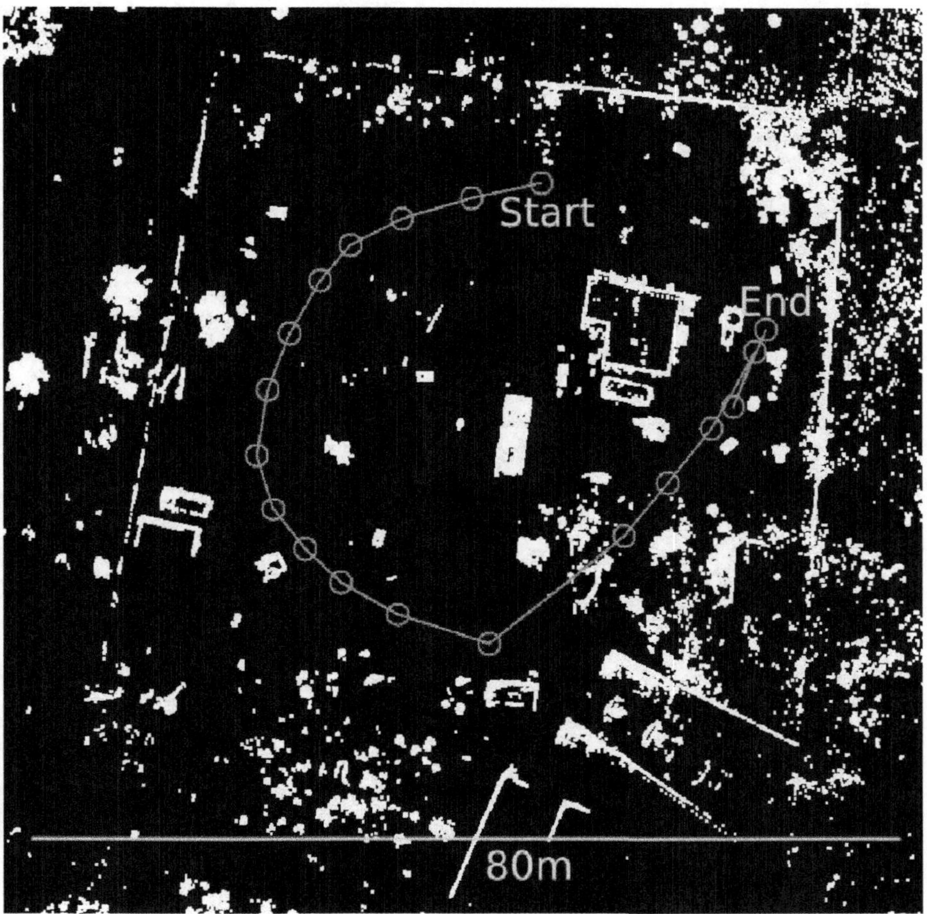

Fig. 6. Localisation results for Pomonal #1

Table 2. Average localisation errors

	mean	stdev
x (metres)	0.143	0.128
y (metres)	0.147	0.143
θ (degrees)	0.903	1.226

Table 3. Computational costs to match a single Velodyne scan

	mean	stdev
Matched using cache ray traced values (ms)	352	100
Matched using augmented map (ms)	2403	200

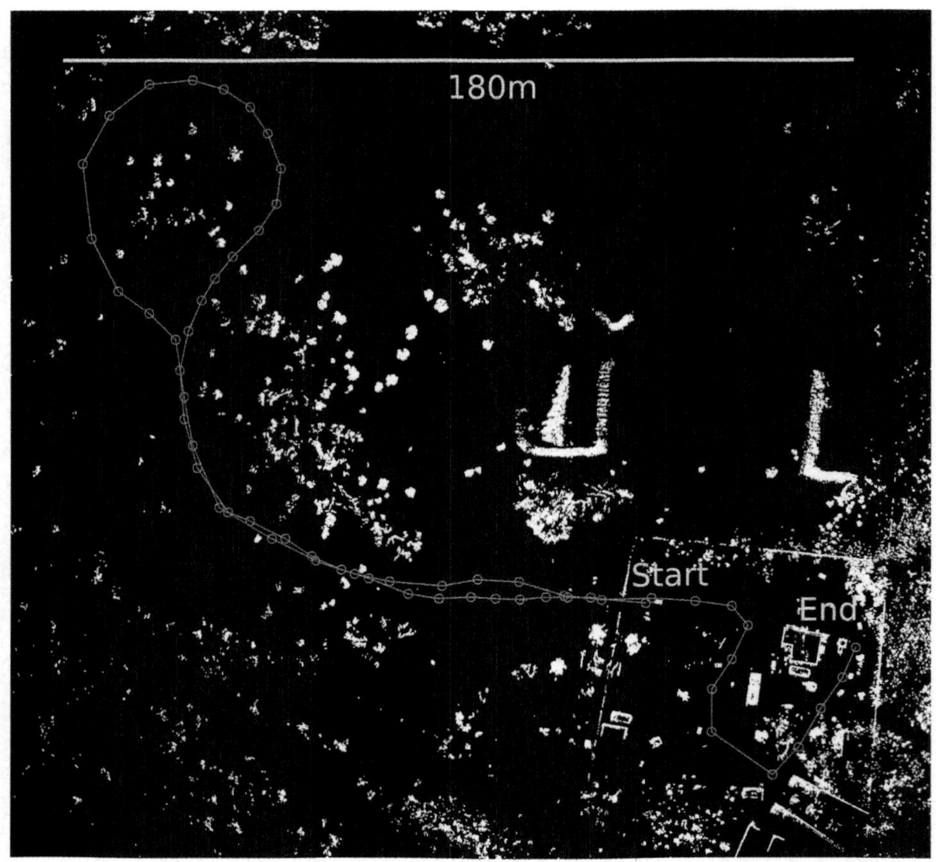

Fig. 7. Localisation results for Pomonal #2

5 Path Planning

A robot navigation system is not complete without path planning, since, even once a map of the environment is made available and the current location of the vehicle is continuously determined (in this case by matching on-board range scan data against a database of range scans collected off-line previously), the manner in which the vehicle is to proceed, without collision and with some degree of efficiency, towards its nominated goal from its starting point needs to be determined and this plan executed. There have been many methodologies proposed regarding path planning, each with their strengths and weaknesses in regard to computer complexity, global scope, flexibility, dimensional extensions and capability of coping with initially unknown and/or changing environments and terrain variability. Only three will be touched on here, but these suggest a wider scope of variations and give an overall flavour of the rich research field of path planning.

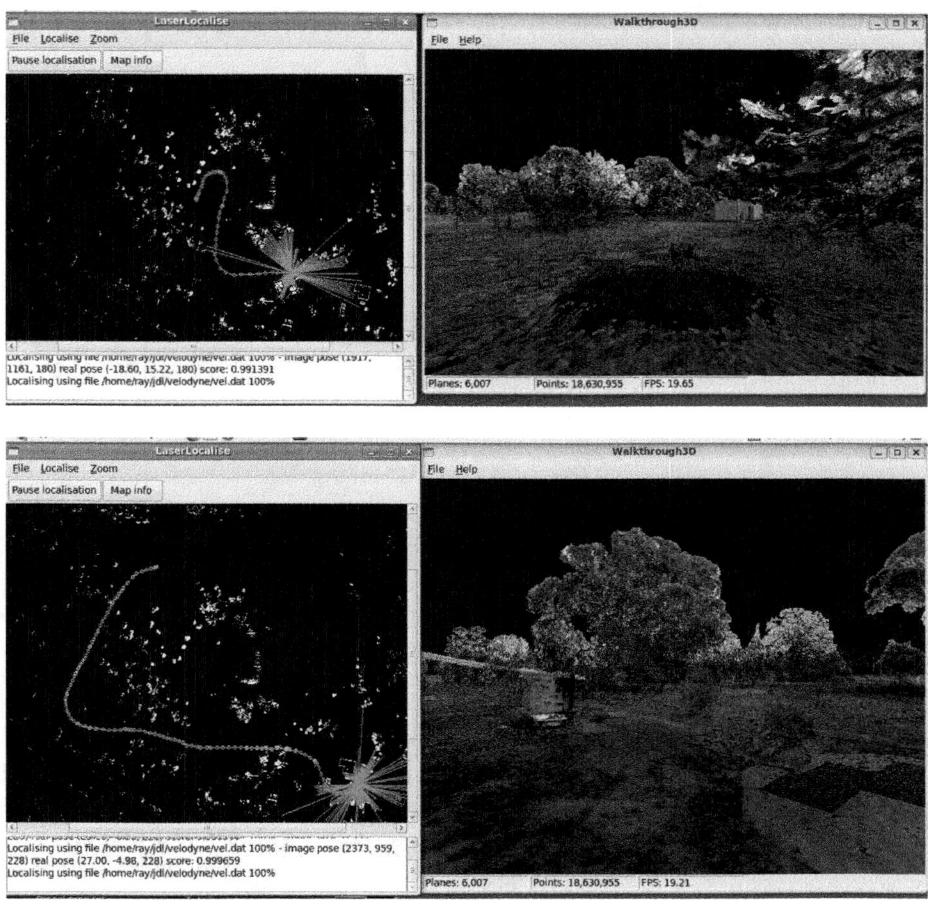

Fig. 8. Localisation results with predicted view in 3D world

A path planning methodology is usually based on a number of assumptions regarding what is known about the environment and whether it is subject to change and/or contains dynamic obstacles; it also depends on whether optimal or merely feasible paths are required to be discovered. Most path planners are also classed as either operating in continuous Euclidean spaces or discreet (grid tessellated) spaces. A brief tutorial on path planning and comparisons of several methods [12] provides further details.

One of the earlier, now regarded as a classic, methods of path planning applies the famous A* tree search optimisation strategy used in Artificial Intelligence [21] to find the shortest collision-free amongst convex polygon obstacles in a Euclidean 2D continuous space from a start point to goal point [17].

Non-convex obstacles can, of course, be segmented into convex components before applying the method. It is assumed that the environment is entirely known and unchanging. This method exploits the knowledge that the shortest path can

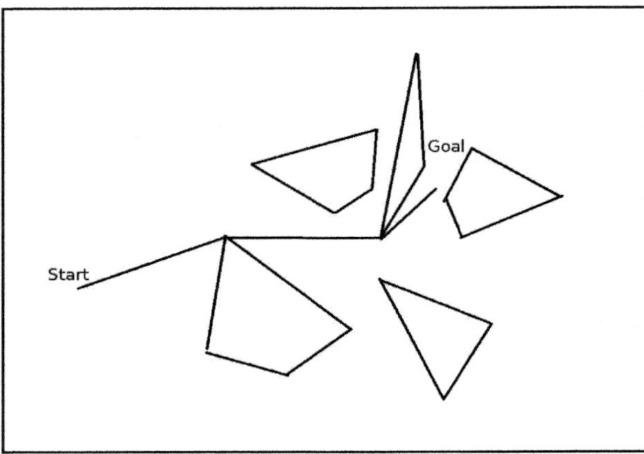

Fig. 9. A* Path Planning

only be made up of straight line components between vertices and including the start and goal point, there being a finite number of such possibilities. Unfortunately, the method's computational complexity grows very rapidly with the number of vertices [10] and does not extend gracefully to three or more dimensions. Should the obstacles be very complex in shape and not easily segmented into convex polygonal parts (as is likely in natural outdoor environments) then the method is not convenient to use.

Whilst the method clearly has shortcomings for application to complex realistic obstacle fields, it does have the virtues of elegance, clarity, optimality and theoretically provable robustness. A simple example is shown in Figure 9.

A much more recent methodology is based on growing random tree structures in Euclidean continuous space (of any dimension) such that, eventually, the tree includes both the start and goal point, no branch penetrating an obstacle, so that branch to branch linkages define a feasible, but not necessarily optimal, path. The simplest version of this algorithm [14], known as the Rapidly Exploring Random Tree (RRT) method, is as follows:

1. Let the start point be the root of the tree.
2. Choose a random point in free-space (not occupied by an obstacle). Points that do not meet this requirement are simply discarded.
3. Extend the tree to include a node along the line between the existing node and the random point such that its distance is some specified length from the existing node, provided that this point is in free-space. Discard the node if not so.
4. Repeat 2 and 3 , above, except for the nearest point of the existing tree being the candidate for expansion toward the new random point until some node of the tree is close enough to the goal point for the last link to be made without penetrating an obstacle.

5. The linking of branches from the root node to the goal node constitutes the sought after path.

The use of 'rapidly' in the title refers to the property of the method such that large, open as yet unexplored areas have a high probability of hosting a randomly selected point simply because the selection is uniform over the search space. In this way the tree extends rapidly.

The typically jagged non-optimal paths can be smoothed somewhat by removing nodes if the path skipping those nodes does not penetrate an obstacle [4].

Extensions of the basic method include extending each link by increments until just short of collision, growing trees from both start and goal points until a sufficiently shot link can connect them and weighting the random point selection probabilistically in some way which seems to have an advantage.

A typical example is shown in Figure 10 which shows both the original path and a smoothed alternative following the original.

The method is very fast due to its simplicity but does not guarantee optimality(in fact , more often than not, is clearly not so), but merely provides feasible paths. It can be applied efficiently and simply to very high dimensional search spaces without varying the basic approach.

Unfortunately, the RRT method does quite badly when narrow passages have to be negotiated since the development process takes no account of the

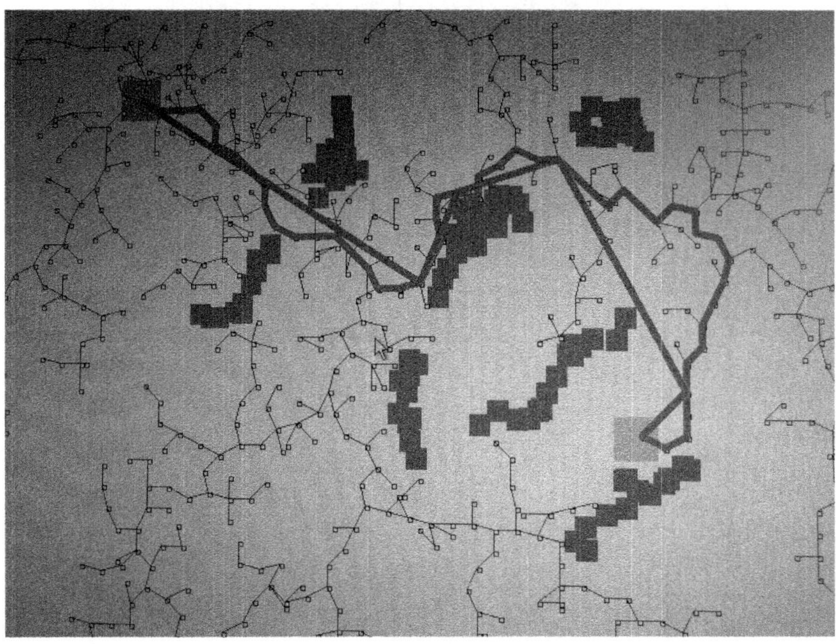

Fig. 10. RRT path planning in sparse obstacle space

Fig. 11. RRT path planning in obstacle spaces with narrow passages

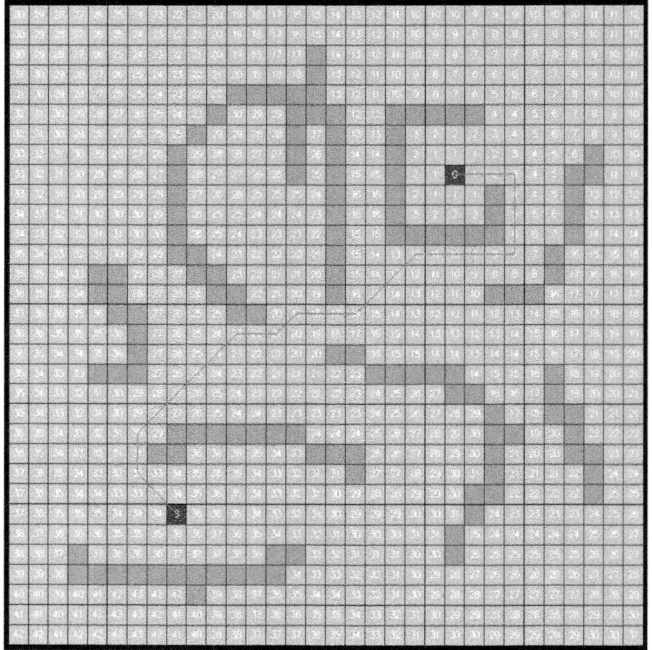

Fig. 12. Basic DT

topological structure of the obstacle field nor of any shape attributes. Figure 11 demonstrates this problem unambiguously.

The path planning method used for our project is based on Distance Transforms (DT), first introduced [22] to study shape properties of binary microscopic cell images digitised into a discreet rectangularly tessellated grid space. The DT was grown from the outside edges of blobs in towards their cores and the distances in from their edges used to derive various shape properties. One of the current authors [11] discovered that by propagating distance out from goal(s) around the blobs (now regarded as obstacles) though out all reachable free-space, using a multi-scan iterative algorithm, the resulting DT could be used as the basis of an optimal robot path planning method.

Initial costs are distributed through all cells in a rectangularly tessellated space representing the work environment of the robot with the obstacle cells set at computer infinity (or some other very large number) indicating inpenitrability, free-space cells as large numbers which get replaced by integer distance values and goal cells as zeros, indicating that no cost is incurred in going from a goal cell to itself. Goal cells clearly must be free-space cells. In a systematic way (full details found in [12]) distances are propagated out from the goals, flowing around obstacles until all reachable free-space cells are covered, every such cell eventually (perhaps after a number of passes) containing the value indicating exactly how

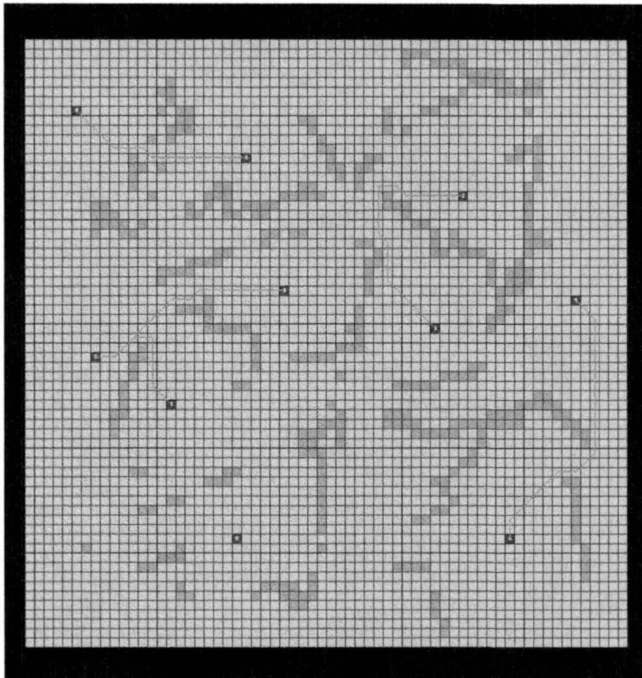

Fig. 13. DT with multiple start and goal points

many steps are required to reach the nearest goal in the most efficient way. Note that any number of goal cells can be specified.

If instead of allocating computer infinity to obstacle cells one treats them as pseudo goals by giving them the value of zero and temporarily removing the real goals from the field, the resulting DT can be thresholded by a value approximating the radius of the robot vehicle, thus expanding the obstacles to account for the physical extent of the vehicle, which can now be regarded as a single cell, making the following computations easier.

Note that the location of a single or a number of start points (perhaps for multiple robots) do not enter into the DT calculation, rendering the result starting point free. That is, shortest paths to nearest goals are always along the steepest descent trajectories from any free-space cell. Thus, should the robot stray from a planned path due to some control error or uncertainty the new steepest descent trajectory is the correct one to follow, thus rendering the method self correcting in this respect.

The simplest version applies a uniform travel cost to moving from any free-space cell to an immediately neighbouring free-space cell (up to eight such cells), including both diagonal and non-diagonal neighbours. If compensation is preferred to recognise that a diagonal move is longer than a non-diagonal one, the relative weighting should ideally be $\sqrt{2}$:1. If integer relative weights are preferred, 4:3 or 17:12 can be used. However, cell weightings can include more than

Fig. 14. DT with unexplored territory

just incremental distance costs. Tractability of terrain can also be added, as can one-way restrictions, time of day slow downs (e.g. expected traffic congestion on certain roads) etc. These can easily be accommodated as long as all costs are non-negative. Also, visibility constraints which may be associated with covert behaviour in security related operations [19] can easily be accommodated by the method.

As with the RRT approach, any number of dimensions can be coped with easily, but with the DT memory requirement growth may become unacceptable if the spatial resolution is fine. With the DT, a temporal dimension can be added as long as the irreversibility of time is recognised, unlike that of paths in real space [12].

All in all, the DT is a very versatile path planning method and is an ideal methodology for our project in natural outdoor environments where the unevenness of the terrain [13] can be built into the path plan.

Moving obstacles can be coped with, either by updating the DT continuously as moving obstacles are detected or by explicitly estimating the movement speeds and directions of obstacles in the spatio-temporal DT version.

Uncertainty can be modelled by applying Gaussian probability density functions to future estimated location of obstacles with the standard deviation increasing into the future as uncertainty of location grows. One can even use the DT to compute a new type of Gaussian distribution probability density spread

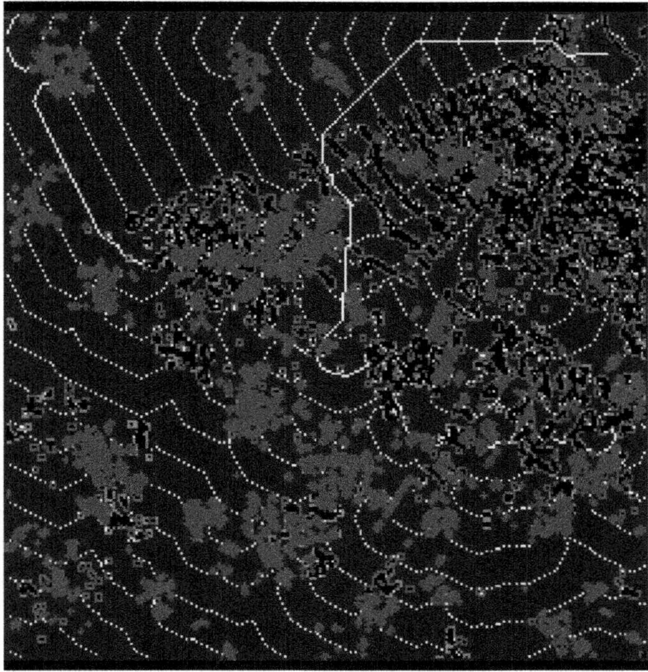

Fig. 15. DT utilising terrain roughness

function by replacing the Euclidean square distance exponent by the square of the DT value so that the function spreads around fixed obstacles [20], which clearly can not be penetrated by moving obstacles of uncertain location.

Figure 12 show a simple DT result with the DT values of all free-space cells marked. Figure 13 shows a more complex example with multiple start and goal points with an obstacle field which has been grown two cells from the original field (obstacles now smoother). Figure 14 illustrates a planning situation where an on-board vision sensor discovers viewable obstacle as navigation toward a known goal proceeds, not all the environment having to be explored to complete the task in a piece-wise optimal fashion (unexplored territory shown as darker).

Finally, Figure 15 shows a path planning result where the roughness of the terrain (as extracted from dense 3D range data) is factored into the DT algorithm.

6 Discussion

Whilst it may be thought that this matching procedure is grossly wasteful of the quality and density of the 3D range data, it should be remembered that the full richness of the sensory data captured by the Riegl scanner can be exploited during 'walk-through' and mission planning, where individual objects in the data can be chosen as navigation targets or locations to be avoided due to danger or fragility. Also, Velodyne data can be used for dynamic obstacle avoidance and accurate close up localisation for delicate tasks like sampling some details (branches, leaves, fruit etc.) in the natural environment, should such be required.

In some earlier work we carried out [8,9], we were able to localise using the Riegl data and an on-board panoramic camera alone. Real-time images were matched with those extracted from the Riegl database from hypothesised positions and orientations, with a particle filter to resolve the selections. In principle, range matching should be more accurate and reliable, as we have confirmed. However, there is nothing to stop one using both approaches simultaneously either independently or as a fusion process. Of course, the capabilities of the range only method being able to operate in the absence of ambient lighting cannot be shared by the vision only approach.

Furthermore, experiments are planned to test the effectiveness of the Velodyne in smokey and possibly raining situations where it is hoped functionality would not be seriously impaired as is likely for the pure vision approach. One further advantage of having live 3D Velodyne data is the possibility, in a civil disaster scenario, to detect gross building damage by comparison with pre-disaster scans and to adaptively adjust navigation missions to suit whilst at simultaneously negotiating rubble strewn terrain.

7 Conclusion and Future Works

We have presented a preliminary laser localisation system using a rich 3D map collected offline using the Riegl and online data collected from a Velodyne. We have tested the system in an outdoor environment surrounded by heavy vegetation and have successfully localised the vehicle, despite having no explicit odometry feedback, to an accuracy of around ± 14 cm and $\pm 1°$. Although the tracking algorithm used in this project does work, it does not have the ability to recover from failure or maintain multiple hypotheses, like the particle filter. This is something that could certainly be improved upon.

References

1. Besl, P.J., McKay, N.D.: A method for registration of 3-d shapes. IEEE Trans. Pattern Anal. Mach. Intell. 14(2), 239–256 (1992)
2. Chetverikov, D., Svirko, D., Stepanov, D., Krsek, P.: The trimmed iterative closest point algorithm. In: Proc. International Conf. on Pattern Recognition, Quebec, Canada. IEEE Comp. Soc., Los Alamitos (2002)
3. Cox, I.J.: Blanche-an experiment in guidance and navigation of an autonomous robot vehicle. IEEE Transactions on Robotics and Automation 7(2), 193–204 (1991)
4. Deak, Z., Jarvis, R.A.: Robotic path planning using rapidly exploring random trees. In: Australian Conference on Robotics and Automation (Decemeber 2003)
5. Dellaert, F., Fox, D., Burgard, W., Thrun, S.: Monte carlo localization for mobile robots. In: IEEE International Conference on Robotics and Automation (ICRA 1999), vol. 2, pp. 1322–1328 (1999)
6. Diosi, A., Kleeman, L.: Laser scan matching in polar coordinates with application to slam. In: Intelligent Robots and Systems, pp. 3317–3322 (August 2005)
7. Dissanayake, M.W.M.G., Newman, P., Clark, S., Durrant-Whyte, H.F., Csorba, M.: A solution to the simultaneous localization and map building (slam) problem. IEEE Transactions on Robotics and Automation 17(3), 229–241 (2001)
8. Ho, N., Jarvis, R.: Global localisation in real and cyberworlds using vision. In: Australasian Conference on Robotics and Automation (ACRA), Brisbane, Australia (December 2007)
9. Ho, N., Jarvis, R.: Towards a platform independent real-time panoramic vision based localisation system. In: Australasian Conference on Robotics and Automation (ACRA), Canberra, Australia (December 2008)
10. Jarvis, R.A.: Growing polygedral obstacles for planning collision-free paths. The Australian Computer Journal 15(3), 103–111 (1983)
11. Jarvis, R.A.: Collision-free trajectory planning using distance transforms. In: National Conference and Exhibition on Robotics, Melbourne, Australia, August 20-24 (1984)
12. Jarvis, R.A.: On distance transform based collision-free path planning for robot navigation in known, unknown and time-varying environments. In: Zang, P. Y.F. (ed.) Advanced Mobile Robots. World Scientific Publishing Co. Pty. Ltd, Singapore (1994)
13. Jarvis, R.A.: Terrain-aware path guided robot teleoperation in virtual and real space. In: ACHI, February 10-14, St. Maartins (2010)
14. LaValle, S.: Rapidly-exploring random trees: A new tool for path planning. In: Tech report. Dept. Of Computer Science, Iowa State University (1998)
15. Leonard, J.J., Durrant-Whyte, H.F.: Directed Sonar Sensing for Mobile Robot Navigation. Kluwer Academic Publishers, Norwell (1992)

16. Lingemann, K., Surmann, H., Nuchter, A., Hertzberg, J.: Indoor and outdoor localization for fast mobile robots. In: Intelligent Robots and Systems, vol. 3, pp. 2185–2190 (2004)
17. Lozano-Perez, L., Wesley, M.A.: An algorithm for planning collision-free paths among polyhedral obstacles. Communications of the ACM 22, 560–570 (1979)
18. Lu, F., Milios, E.: Robot pose estimation in unknown environments by matching 2d range scans. J. Intell. Robotics Syst. 18(3), 249–275 (1997)
19. Marzouqi, M., Jarvis, R.: Covert robotics: Covert path planning in unknown environments. In: Australasian Conference on Robotics and Automation, Canberra, Australia, December 6-8, p. 8 (2004)
20. Marzouqi, M., Jarvis, R.A.: Distance transform based gaussian distribution for probabilistic target tracking. In: International Conference on Intelligent Robots and System, Beijing, China, October 9-15 (2006)
21. Nilsson, N.J.: Problem-Solving Methods in Artificial Intelligence. McGraw-Hill, New York (1979)
22. Rosenfeld, A., Pfaltz, J.L.: Sequential operations in digital image processing. J. ACM. 13(4), 471–494 (1966)
23. Yaqub, T., Tordon, M.J., Katupitiya, J.: Line segment based scan matching for concurrent mapping and localization of a mobile robot. In: Control, Automation, Robotics and Vision, pp. 1–6 (December 2006)

Generating Situation Awareness for Time Critical Decision Making

Shang-Ping Ting[1], Suiping Zhou[2], and Nan Hu[1]

[1] Nanyang Technological University
ting0021@e.ntu.edu.sg, huna0002@e.ntu.edu.sg
[2] Teesside University
S.Zhou@tees.ac.uk

Abstract. The quality of situation awareness directly affects the decision making process for human soldiers in Military Operations on Urban Terrain (MOUT). It is important to accurately model situation awareness to generate realistic tactical behaviors for the non-player characters (also known as bots) in MOUT simulations. This is a very challenging problem due to the time constraints and the heterogeneous cue types in MOUT. Although there are some theoretical models on situation awareness, they generally do not provide computational mechanisms suitable for MOUT simulations. In this paper, we propose a computational model of situation awareness for the bots in MOUT simulations. The model forms up situation awareness quickly with key cues. It is also designed to work with some novel features. They include *case-based reasoning*, *qualitative spatial representation* and *expectations*. The effectiveness of the computational model is assessed with *Twilight City*, a virtual environment that we have built for MOUT simulations.

Keywords: Time Critical Decision Making; MOUT Simulations; Situation Awareness.

1 Introduction

Coup d'oeil - the rapid discovery of a truth which to the ordinary mind is either not visible at all or only becomes so after long examination and reflection.

Carl von Clausewitz used the term *coup d'oeil*, or "glance" to describe the importance of situation awareness when operating under the time constraints [1]. Clausewitz found that soldiers do not rely much on comprehensive rational approaches to assess the battlefield situations during war. Instead, they tend to rely more on experience and intuition to form up situation awareness. Clausewitz's observation still applies for modern soldiers despite the advancement of military technologies. This is especially true for Military Operations on Urban Terrain (MOUT).

Going into MOUT without proper preparation can be very dangerous [2], [3]. To survive, soldiers rely heavily on individual situational awareness. They often need to assess the current situation with incomplete information. They also

M.L. Gavrilova et al. (Eds.): Trans. on Comput. Sci. XII, LNCS 6670, pp. 183–205, 2011.

have to make rapid decisions under time pressure, uncertainty, high stakes and changing conditions [4], [5]. Therefore, it is crucial to produce MOUT simulations for the soldiers to achieve clear understanding of their own tactics and the adversarial threats [6], [7] before the actual combat.

In MOUT simulations, the virtual urban environments are populated with various characters. While some of these characters are controlled by human players (i.e., the trainees), most of them are non-player characters (or bots) which are usually represented by AI-driven agents. For an MOUT simulation to be effective, it is important for these bots to demonstrate some human-like tactical behaviors. Although different approaches may be used to this end, we believe that human-like behaviors should be generated by human-like cognitive processes.

After studying experts operating under time constraints, Gary Klein discovered that the "intuition" possessed by experts are a result of the expert's brain storing up cases from direct experience and the experience of others acquired through learning [8]. He pointed out that humans depend more on past experiences rather than deliberate rational analysis of possible alternatives during time-critical decision making.

In our previous work, we have proposed a time critical decision-making framework called *SNAP* that aims to imitate human decision-making processes in MOUT situations [9], [10]. Soldiers fighting in MOUT often need to pick up some key situation cues and retrieve past experiences to make quick decisions. Therefore, in *SNAP*, we have introduced the case-based reasoning (CBR) approach for decision making. CBR is the process of solving new problems based on the solution of similar past problems. It has been argued that CBR is not only a powerful method for machine reasoning, but also a pervasive way for humans to make decisions in everyday situations [11].

Although much work had been done on the *CBR process* in *SNAP*, the *Situation Awareness* component of *SNAP* is still rather primitive. A realistic situation awareness component has yet to be developed to provide useful information for the decision making process, which motivates us to work on a computational model of situation awareness. Although high quality decisions are not guaranteed with a high level of situation awareness, good situation awareness is expected to increase the probability of good decisions. The computational model should work together with the major features that we have built into *SNAP* such as *CBR process*, *qualitative spatial representation* and *expectations*.

Now let us use an example of a soldier bot on a patrol mission to illustrate how situation awareness is formed up in our model in dynamic MOUT situations. First, the bot perceives the current situation through the cues that fall within its hearing and vision range. The bot comprehends the current situation by comparing the perceived cues with its experience cases. The experience case with the highest similarity value to the current situation will be selected. The corresponding action of this case will be chosen to deal with the current situation.

Being on a patrol mission, the bot has certain *expectations* (i.e., projection) on how the situation may evolve. The perceived cues are also used to compare with

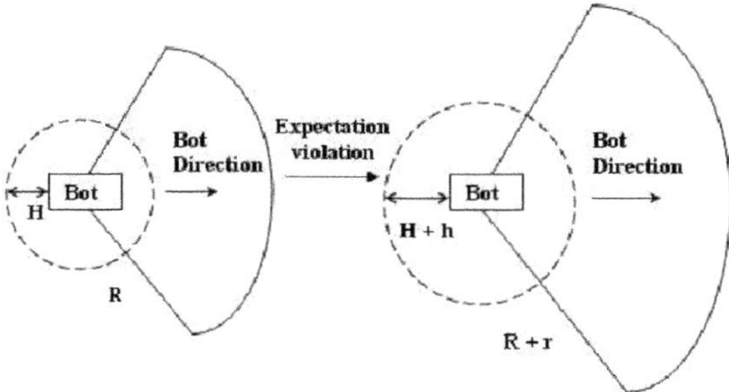

Fig. 1. Bots increasing their awareness on anticipation of impending threat

the bot's *expectations*. For example, the bot may expect that the street should be populated with at least fifty civilians since the street is normally crowded. If the street is too quiet, this could mean that local people already anticipated an impending ambush. Figure 1 shows the initial hearing and sight radius of the bot which are represented as H and R, respectively.

When the *expectation* on the size of the crowd is not met, the bot suspects an impending ambush. Thus, the hearing and sight radius are expanded (by h and r, respectively) which will allow the bot to pick up more cues in anticipation of a possible attack. This is consistent with the fact that a soldier tends to be more alert when he senses danger.

The remainder of this paper is organized as follows. The next section discusses the related work. The section, *Computation Model of Situational Awareness*, describes the details of our design. *Experiments and Analysis* discusses the results and impact of the computational model. The last section concludes our current work and provides some insights into our future direction.

2 Related Work

Due to the increasing importance of MOUT combat, there had been some work done in generating realistic human behaviors during MOUT simulations. Wray and Wood used the SOAR approach to produce intelligent bot behaviors in MAVEN-SA and MOUTbots respectively [12],[13]. Wood aims to improve situation awareness for company-level commanders while we seek to simulate situation awareness for squad-level soldiers. Wray designed a rule-based SOAR system that relies on adjusting statistical values to generate variable behaviors [14]. However, such rule based systems perform increasingly poorly as the possible problem states scale up. Our work aims to overcome this issue through case-based reasoning.

The recognition-primed decision (RPD) framework [8] proposed by Gary Klein emphasizes the role of experiences in human's decision-making process during time critical situations. Klein suggests that when people are making decisions in time-critical situations, they spend most of their time in making sense of the situation rather than in comparing and evaluating different alternative solutions. Similar ideas were also advocated in [15], [16]. However, these works in general still remain at high-level, which leave much room for a modeler while developing a specific computational model. It is critical to have a robust computational model for situation awareness for bots to perform proper matching of past experiences to current cues in a case-base reasoning approach.

Endsley's framework of situation awareness is a widely accepted abstract framework of situation awareness [17]. According to Endsley, there are three main stages for the formation of situation awareness which include *perception*, *comprehension* and *projection*. In the *perception* stage, key cues from the environment are picked up by a person's sensory and attention system. These cues will then be used for the person to understand the current situation in the *comprehension* stage. A person also has the capability to predict how situation may evolve (i.e., projection into the future). It should be noted that Endsley's framework describe situation awareness at high and abstract level, thus it leaves much room for a modeler when developing computational models of situation awareness for specific applications.

Like in the *perception* stage of the Endsley's model, Hill attempts to model the perceptual attention in virtual humans [18]. Similarly in [19], Herrero et al. introduced human-like hearing perception for intelligent virtual agents. With similar goals in mind, we adapted the sensory functions of the *Unreal* game engine to create the sight and hearing senses for our bots [20].

The *comprehension* stage of the Endsley's framework requires the transformation of the perceived information into relevant values for situation understanding. McCarley et al. [21] and Warwick et al. [22] separately developed computational models to support situation awareness. With their models, they seek to achieve greater simulation realism. However, their models do not cater to the specific needs of MOUT simulations. For instance, McCarley's model does not work with the CBR process, and Warwick's model can not handle heterogeneous cue types which are typical for MOUT.

For evaluation purpose, Endsley designed the Situation Awareness Global Assessment Technique (SAGAT) [23]. SAGAT provides an objective measure of situation awareness based on queries made in simulation freeze. During the freeze, the simulation is halted and evaluation of the situation awareness will be made on the test subjects. We will apply SAGAT methods to evaluate the effectiveness and realism of our proposed model.

3 Computational Model of Situation Awareness

In this section, we will describe the design of our computational model of situation awareness for MOUT simulations. We will also describe how the proposed model works together with other major features of *SNAP* which is a time critical decision-making framework that we had previously proposed for MOUT simulations. To this end, we will first give an overview of *SNAP*.

The key to producing realistic behaviours lies largely in the decision making process of the bots. Currently, conventional game AI techniques had been used to design bots in MOUT simulations. Although game AI is able to provide challenging opponents for gamers, they may not be realistic for simulation training. Building realistic behaviours for simulations require special considerations for the situations involved. An important factor that differentiates the behaviours of human in MOUT situations and normal life is the time constraints. As human are likely to make different decisions given different time frames. An optimal decision can be made with a long time frame and a satisficing decision is produced under time constraints. Although there is already a lot of work done in producing optimal decisions in computer simulations, the research on producing satisficing behaviours in MOUT simulations is still lacking. Producing optimal behaviours under time constraints is unrealistic and serves little operational benefits to the soldiers. Our work seeks to bridge this gap by producing a time critical decision making framework, *SNAP*, for bots in MOUT simulations.

In the proposed framework, rapid cognition is used to model human's ability to quickly form up situation awareness under uncertain and complex situations using key cues from partial information. In his book, Blink [24], Gladwell discussed how humans are able to achieve rapid situation awareness using only a narrow slice of information. Using this limited information, humans can focus on the most significant information (or key cues) about a situation, and make judgments quickly. We adopt a similar approach to imitate human's rapid situation recognition capability in complex and uncertain situations. Figure 2. shows a soldier bot and a militant bot in a close combat situation.

The soldier bot, *Ted*, needs to make a rapid decision to fight with the militant. Using the CBR process, *Ted* will retrieve past experiences that are similar to the current situation and reuse the previous solutions. If needed (i.e., the current situation is new), the proposed solution can be revised to adapt to the current situation. Once the revised solution had been successfully adapted for the new

Fig. 2. Close Combat

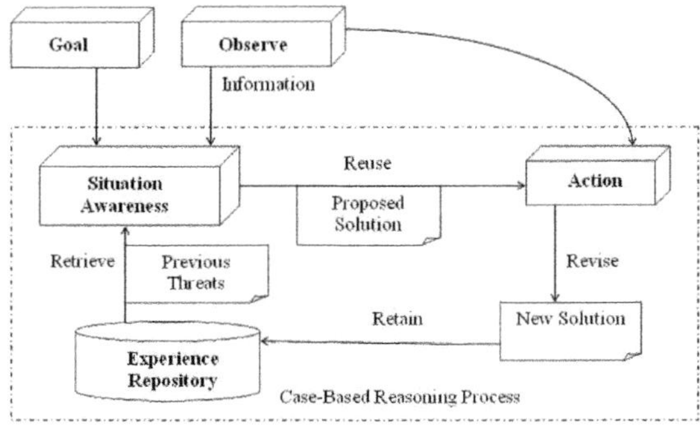

Fig. 3. SNAP: Time Critical Decision-Making Framework

situation, it will be retained in the memory of the *Ted* for future engagement. The matching of the current situation with past experiences is done by picking up key cues from the partial information gathered within the limited time, i.e., by rapid cognition. These cues may include the location of the militant, whether the militant is in open area or hiding behind a wall, the weapon being used and strength of the fire, etc.

As shown in Figure 3, the *SNAP* framework consists of five main components: *Goal*, *Observe*, *Situation Awareness*, *Experience Repository* and *Action*.

The *Goal* component defines the goals of the bots. The *Observe* component collects key cues of the current situation and sends the information to the *Situation Awareness* component. The components inside the dash line box form the CBR process. By matching the observed cues with the *experience cases* in the *Experience Repository*, a solution will be selected for execution by the *Action* component.

The CBR process consists of four steps: retrieve, reuse, revise, and retain. A bot first needs to retrieve past experiences from its *Experience Repository*. In our implementation, *experience cases* are represented by $\langle threat, solution \rangle$ sets as shown in Figure 4.

The threat in an experience case holds pre-condition cues which act like a pattern (or schema) for the bot to recognize the threat. In *SNAP*, these pre-condition cues mainly consist of some descriptions about the situation. The cue values can be nominal, quantitative or qualitative. Similarity matching is then conducted by matching the evidence cues of the current situation with the pre-condition cues. The key cues picked up by the bots are used to form an *evidence set*. When the evidence cues match the precondition cues of an *experience case*, the corresponding solution of the case will be used. The solution generally activates a new mission.

Each mission is organized into several phases to represent the temporal relationships between different parts of the mission. In each phase, there are some *expectations* providing constraints to the mission. The *expectations* are monitored while the mission is being executed. If the *expectations* are not met, the next phase may not be executed and the current mission could be under threat.

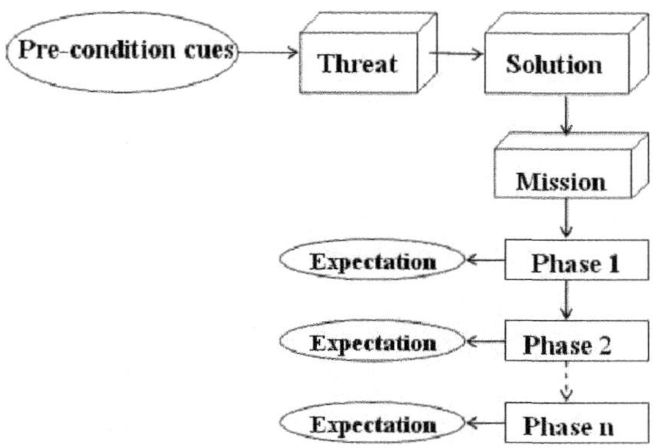

Fig. 4. Experience Case

We believe that expectation plays an important role in this regard. In MOUT warfare, by matching the current situation cues with his/her experience cases, a human soldier may have a quick assessment of the current situation and find a solution to deal with the current situation. In the meantime, the soldier will also form up expectations on the likely future events. As the situation evolves, if the expectations are not fulfilled or violated, the soldier may infer that his/her initial solution may be invalid due to dynamic situational changes. Then the soldier may abort the previous solution and find a new solution. Without a mechanism to assess their solutions during execution, the bots can only know the validity of the solution after the execution of the complete solution. This could result in the bots blindly executing the initial solutions even though they are no longer effective.

Therefore, we propose to handle the dynamic situational changes with expectations. In this work, we consider expectations as some events that are likely to occur in the near future. Bots continuously observe the ongoing situation after a solution is selected. If the observations are consistent with expectations, then the solution will be enhanced. Observations that do not match expectations may lead to the invalidation of the solution. In this case, the bots re-assess the situation to find new solutions and form up new expectations.

Now let us use a simple example to illustrate how these features of the *SNAP* framework could help a bot to make decisions in time-critical and dynamic MOUT situations. Figure 5 shows a sniper bot aiming at a soldier bot.

The soldier bot needs to make a rapid decision to counter the sniper. Using the *CBR process*, it will retrieve past experiences that are similar to the current situation and reuse the previous solutions. If needed (i.e., the current situation is new), the proposed solution can be revised to adapt to the current situation. Once the revised solution had been successfully adapted to the new situation, it will be retained in the memory for future engagement. The matching of the current situation with past experiences is done by picking up key cues from the partial information gathered within the limited time. These cues may include

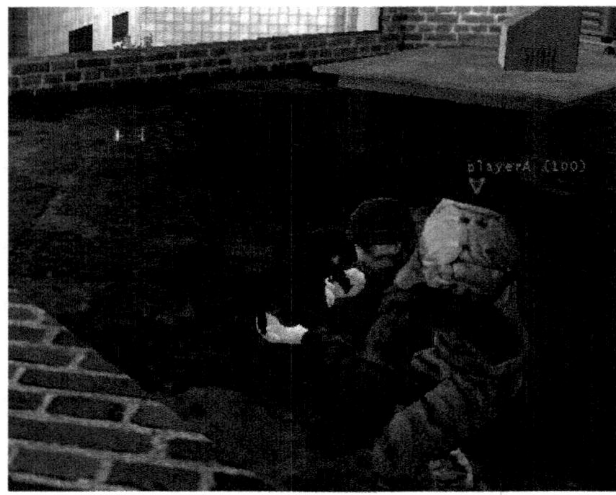

Fig. 5. Enemy sniper bot taking aim at soldier bots

the location of the sniper, whether the sniper is in open area or hiding behind a wall, the weapon being used and strength of the fire, etc.

Figure 6 shows the simulation time line of a group of soldier bots being attacked by a sniper and their subsequent actions. It is likely that the enemy sniper is attacking the soldier from a building. According to doctrine, a human soldier will attempt to take cover and form up with his squad before performing a counter sniper operation to neutralize the enemy threat. During the counter sniper operation, the soldier is expected to overpower the enemy through speed and superior fire power. If there are more enemies or stronger firepower than the soldier expects, then the initial solution to perform a counter sniper operation may become invalid. The soldier bots select the solution to perform a counter sniper operation at T_1. With this in mind, they hold a set of expectations on the enemy behaviour, e.g., the enemies are adopting a hit-and-run tactics and will not be waiting at the sniper location. They do not expect to face heavy enemy fires during the counter sniper operation. At T_2, an unexpected threat of *Close Combat Fire* is observed. This new threat violates the current expectation and therefore invalidates the counter sniper solution. The violation of the expectation implies that there might be an ambush waiting for the counter sniper squad and a new solution will need to be generated.

Our current work seeks to improve the *Situation Awareness* component in *SNAP*. Figure 7 shows how our model generates situation awareness in *SNAP*.

We consider 3 levels of situation awareness in our model. During the *Perception* stage, the *Sensory System* will pick up the external information from the virtual environment and produce observation cues. These observation cues will be used during the *Comprehension* stage for evaluation with the *expectations* of the bots. The mission can continue if cues do not violate the *expectations* of the bots. Violations will be used to identify the probable threat. When a new threat is identified, a new mission will be activated to neutralize the threat.

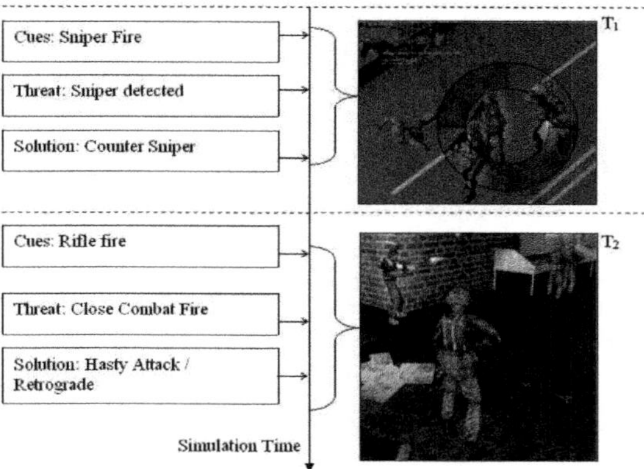

Fig. 6. Dynamic changes during MOUT simulations

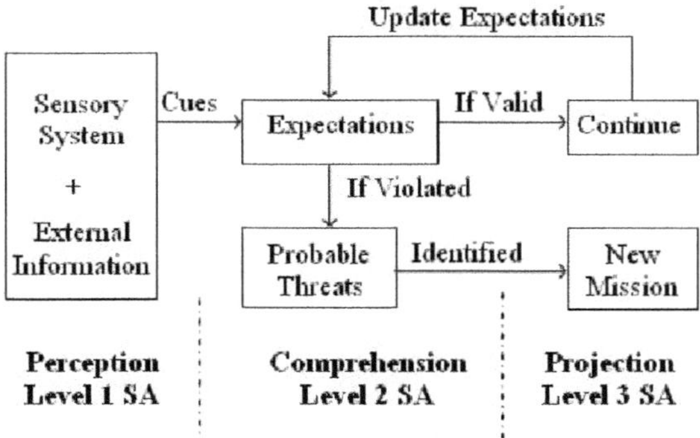

Fig. 7. Maintaining Situation Awareness in *SNAP*

3.1 Perception Stage during Situation Awareness

Figure 8 illustrates the sensory system that our bots employ to perceive the environment in MOUT. The solid sector in Figure 8 represents the field of view of a bot. This sector is characterized by the *field of view angle* and the *SightRadius* to simulate the *Sight* function of the bot. The dotted circle surrounding the bot shows the coverage of the *Hearing* function of the bot. The *Hearing Threshold* is used to control the hearing radius of the bot. The coverages of both the field of view and hearing radius are three dimensional and influenced by the head movement of the bot when scanning an area.

Although there may be many cues occurring within the MOUT environment, only cues that fall within the coverage of a bot's *Sight* or *Hearing* function are picked up. As shown in Figure 8, only Cue A and Cue C are picked up by the *Hearing* and *Sight* functions of the bot respectively. Cue B will be missed out by the bot. Such cues may have different types such as weapon, weapon fire or enemies. Studies had shown that human soldiers tend to generate

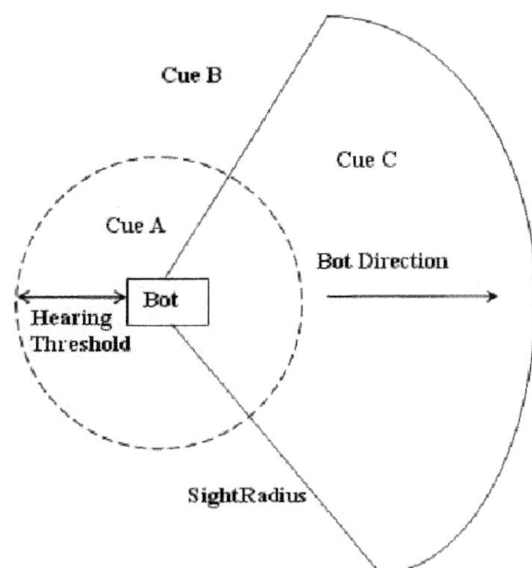

Fig. 8. Sensory System of A Bot

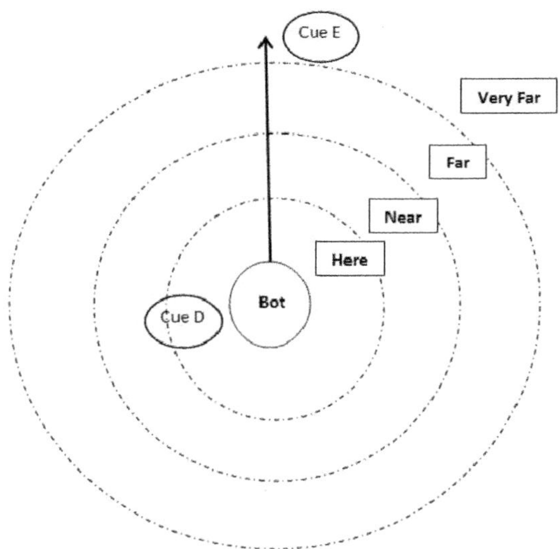

Fig. 9. Descriptive distance with the MOUT environment

numerical distance estimates that scale linearly with increasing distance within the ranges typically encountered in MOUT [25]. Thus, we apply a qualitative spatial reasoning method to set the descriptive distance of various cues as shown in Figure 9.

A descriptive distance, as seen in Figure 9, is how distant something is relative to the current visual horizon. Their ranges includes "here" (within 1/8 the distance to the horizon), "near" (between 1/8 and 1/4 the distance to the horizon), "far" (between 1/4 and 1/2 the distance to the horizon), and "very far"(beyond 1/2 the distance to the horizon). In the current context, the smallest range, "here", is used to describe cues within the immediate proximity of the bot. The second two distances, "near" and "far", can be used to divide the space used for ranges within a building such as rooms and doorways. Finally, "very far" allows for description of distant objects and architecture. For example, Cue E can be a visual cue of an enemy sniper at a "very far" distance (at least 800 metres) and Cue D is an audio cue of the enemy sniper round hitting the immediate proximity, "Here" of the bot.

Cues are gathered and organized into *observation* sets for further evaluation during the *Comprehension* stage. An example observation set is shown as follows:

$$[\text{Bot(Militant, Rifle)} \wedge \text{Gun\_Fire(Rifle)} \wedge \text{Grp(Civilian, 50),...}]$$

which essentially means that the bot has observed 1) "a militant bot with a rifle"; 2) "rifle fire"; 3) "50 civilians"; and "...".

3.2 Comprehension Stage during Situation Awareness

In this stage, the *Situation Awareness* component receives the *observation sets* from the *Observe* component. With the cues from these *observation sets*, a bot determines if its current mission is under threat. If so, it will generate an *evidence set* and match with a list of probable threats. Once the threat is identified, the corresponding *experience case* will be retrieved from the *Experience Repository* and the corresponding solution will be used.

Each bot also has a set of expectation cues. It will monitor the expectation cues with the *Sensory* system as discussed in the *Perception Stage*. The expectation cues are further classified as *invariant* or *variant*. Violation of the *invariant expectations* means that the mission is under threat, therefore it should be terminated and a new mission may be started. In contrast, violation of *variant expectations* may raise certain concern of the bot (for instance, the bot may become more alert) but are insufficient to show that the current mission is under threat.

Cue Types. In *SNAP*, cues observed from the environment can be qualitative, nominal or quantitative. In our implementation, quantitative cues (for instance, number of civilians) are represented using integer values. A nominal cue is a discrete cue whose values are not necessarily in any order (e.g, weapon type). The qualitative cues are used for qualitative spatial reasoning.

Similarity measures. Similarity measures are used to assess the degree of resemblance between the current situation and past experiences. We assume that the cues in current situation and the cues in the past experiences have the same set of features.

Suppose that the *Evidence Set*, E_c, of the current situation and the past *Experience Case*, E'_c, are represented as:

Evidence Set, $E_c = (C_1, C_2, ..., C_i, ..., C_n)$

Experience Case, $E'_c = (C'_1, C'_2, ..., C'_i, ..., C'_n)$

where C_i and C'_i can be nominal, quantitative or qualitative values. To assess the the similarity between E_c and E'_c, a global similarity value is computed by aggregating the local similarity between cue pairs C_i and C'_i. The following heterogeneous similarity computation method is proposed to calculate the local similarity between C_i and C'_i:

$$S_L(C_i, C'_i) = \begin{cases} 0 & \text{if } C_i \text{ or } C'_i \text{ is unknown} \\ \text{N\_Match}(C_i,\ C'_i) & \text{if } C_i \text{ is nominal} \\ \text{Quant\_Similar}(C_i,\ C'_i) & \text{if } C_i \text{ is quantitative} \\ \text{Quali\_Spatial}(C_i,\ C'_i) & \text{if } C_i \text{ is qualitative} \end{cases}$$

If either C_i or C'_i is unknown, the cue similarity between them is set to 0. For the nominal values, the similarity is either 0 or 1 as determined in the following way:

$$\text{N\_Match}(C_i,\ C'_i) = \begin{cases} 1 & \text{if } C_i = C'_i \\ 0 & \text{otherwise} \end{cases}$$

Assuming the bot picks up the cue, C_1 = Weapon_Fire(Rifle), and compares it with C'_1 = Weapon_Fire(Any). N_Match(C_1, C'_1) will give a result of 1. This is because C'_1 includes all types of weapon fire.

The similarity value between a pair of quantitative cues are determined as follows:

$$\text{Quant\_Similar}(C_i,\ C'_i) = 1 - \frac{|C_i - C'_i|}{\triangle_i}$$

where $\triangle_i = \text{Max}_i - \text{Min}_i$ is used to normalize the cue difference, and Max_i and Min_i are the maximum and minimum values for cue i, respectively. When the *experience case* contains C'_2 = Group(Civilian, 50) and the bot observes C_2 = Group(Civilian, 12), Quant_Similar(C_2, C'_2) = 1 - $\frac{|12-50|}{50-0}$ = 0.24. In this case, the low similarity value between C_2 and C'_2 implies that the current cue does not match closely to the *experience case*.

It has been suggested that humans rely more on the qualitative description than the quantitative description of their environment [26]. In our previous work, we had organized the virtual environment into hierarchical zones and added qualitative descriptions [27] to facilitate efficient path planning in dynamic environments. Using the qualitative spatial information, Quali_Spatial(C_i, C'_i) computes the similarity values for qualitative spatial cues. A typical qualitative cue is the combat zone where the conflict is taking place.

The model activates the recursive algorithm below to check if the soldier bot and the enemy have the same parent zone:

```
Query_Zone(zone a, zone b){
  //Let parent_a & parent_b be parents of a & b
if(parent_a = parent_b)
  return parent_a;
  return Query_Zone(parent_a, parent_b);
}
```

The Query_Zone algorithm terminates when a common parent zone is found. This parent zone is the lowest level in the spatial hierarchy that contains both the soldier and its enemy. Thus, the parent zone is considered as the combat zone.

Quali_Spatial(C_i, C'_i) will compare the qualitative spatial relationship in the following way:

$$\text{Quali\_Spatial}(C_i, C'_i) = \begin{cases} 1 & \text{if } C_i = C'_i \\ 0 & \text{otherwise} \end{cases}$$

For example, the *Close Combat Fire experience case* has C'_3 = Combat_Zone(Room) stating that the threat can be considered *Close Combat Fire* only if the combat takes place within a room. However, the observed cue, C_3 = Combat_Zone(StreetA), indicates that the soldier is attacked on *Street A*. Thus, C'_3 and C_3 does not match and Quali_Spatial(C_3, C'_3) = 0. This result implies that the soldier is not engaging his enemy within the close proximity of a room.

After obtaining the similarity values of all cue value pairs, the next step is to integrate them into a global similarity value, $S_G(E_c, E'_c)$, using the following formula:

$$S_G(E_c,E'_c)= \begin{cases} 0 & \text{if } w_i = 1,\ S_L(C_i,C'_i) \leq \beta \\ S_p + \frac{\sum_{i=1}^{n} w_i S_L(C_i,C'_i)}{\sum_{i=1}^{n} w_i} & \text{otherwise} \end{cases}$$

where $\beta \in [0,1]$ is a threshold parameter determined by the user. It is used to determine whether C_i and C'_i is close enough. $w_i \in [0,1]$ is the weight of the cue i, which represents the relative significance of the cue i and is also assigned by the user. Often, there are some important cues or cues that are required to be satisfied before the past experience can be reused. That is, if the value of a required cue in the current situation is not close to the value of the same cue in a past experience, this *experience case* cannot be used to solve the current problem no matter how similar the other cues are. For example, spotting sniper fire is a required cue to assess for the *Sniper Assault* threat. For required cues, we will set their weight to be 1. If a required cue in the *evidence set* is not close enough to the corresponding value in the *experience case*, we shall assign 0 to $S_G(E_c, E'_c)$.

In our implementation, each mission will have a pre-determined list of possible threats. Thus, the global similarity values are computed for the threats in the pre-determined list and the threat with the greatest similarity will be identified.

Certain threats are more common in some missions. S_p modifies the similarity value according to the current mission. For example, patrolling soldiers are common targets of enemy snipers. When violation of *expectations* occur during a patrol, it is likely that a sniper assault had taken place. When performing matching with the *Sniper Assault experience case*, the S_p adds x to $S_G(E_c, E'_c)$. The addition of x enhances the likelihood of a sniper assault in the patrol mission. With S_p, the decision bias of the bots towards certain threats during specific missions can be modeled. The values of the parameters such is x in S_p obtained from interviews with human experts.

An example of experience matching with the partial information about the environment is given here. It is achieved by the *Goals*, *Observe* and *Situation Awareness* components of the *SNAP* framework. First, the *Goal* component determines the constraint set of a bot according to the current scenario, mission and roles of the bot. Secondly, the *Observe* component monitors the virtual environment and forms the observation set. Finally, *Situation Awareness* component uses the observation set to detect any violation of the constraint set. The observations that violate the constraints will be used to form the evidence.

For example, consider a soldier bot *Ted* in a patrolling mission. The constraint set of *Ted* can be expressed as:

$$[\neg \text{ Gun\_Fire(All)} \wedge \neg \text{ Vehicle(Militant)} \wedge \neg \text{ Bots(Militant)}]$$

This constraint set means that *Ted*, as a soldier patrolling on the street, will not allow any type of weapon fire, enemy vehicles and militants to appear. These constraints are expressed with some qualitative cues, which will be used to compare with the observation cues. The *Observe* component monitors the virtual environment. Suppose the observation set of *Ted* about the current situation is:

$$[\text{Bot(A, Militant, Male, Sniper gun)} \wedge \text{Gun\_Fire(Sniper gun)} \wedge \text{Vehicle(Friendly)} \wedge \text{Grp(Civilian, 50)} \wedge \ldots]$$

It means that *Ted* has seen that "bot A who is a male militant with a sniper gun, the militant is shooting with his gun, there is a friendly vehicle coming by, and a group of civilians are around, and ..." This set of observation cues is used to compare with *Ted*'s constraint set. The observations $Bot(A, Militant, Male, Sniper\ gun)$ and $Weapon\_Fire(Sniper\ gun)$ violate constraints $\neg Weapon\_Fire(All)$ and $\neg Bots(Militant)$ respectively. Thus, the following evidence set is formed:

$$[\text{Bot(A, Militant, Male, Sniper gun), Gun\_Fire(Sniper gun)}]$$

This evidence set is then used by the *Situation Awareness* component to match with the precondition cues of *Ted*'s experiences.

3.3 Projection Stage during Situation Awareness

Expectations help the bots to anticipate certain future events and pick up key cues rapidly. They are used during the *projection* stage to monitor the status of the mission and detect key cues.

We have introduced two types of *expectations*: *invariant expectations* and *variant expectations*. *Invariant expectations* must be fulfilled at all times or the current mission will be considered to be under threat. Unlike the *invariant expectations*, the *variant expectations* need not be fulfilled at all times. The violation of the *variant expectations* merely alerts the bots that certain threats may be forming.

Now let us use a simple example to further illustrate how *expectations* work. Suppose that *Ted*, a soldier bot, is patrolling a street. His current *invariant expectations* are:

[¬Gun_Fire(Any) ∧ ¬Vehicle(Militant) ∧ ¬Bot(Militant)]

This *expectation* means that *Ted* does not expect to see any type of weapon fire, enemy vehicles and militants during his movement. Suppose that the *observation set* about the current situation is:

[Bot(A, Militant, Rifle) ∧ Gun_Fire(Rifle), ...]

It means that *Ted* has been engaged by "bot A who is a militant with a rifle, and gun fire is observed, and ...". This set of observation cues is used to compare with *Ted*'s *invariant expectations.* The comparison reveals that the observations Bot(A, Militant, Rifle) and Gun_Fire(Rifle) violate *expectations* of ¬Bot(Militant) and ¬Gun_Fire(Any) respectively. This tells *Ted* that its *invariant expectations* is not fulfilled and the mission is under threat. The cues violating the bot's *expectations* are used to form the *evidence set.* This *evidence set* is then used to generate similarity values with the *experience cases.* In this case, the *evidence set* should state that *Ted* had spotted a militant bot and also observed sniper fire. The current *evidence set* is:

[Bot(A, Militant, Sniper gun) ∧ Gun_Fire(Sniper gun)]

A *variant expectation* for the patrol mission could be [Grp(Civilian $\geq$ 10), ...]. This *expectation* states that the soldier bot expects to see at least ten civilian on the streets. It is not normal for the streets to be deserted. In real life situation, this could mean that the locals sense an impending attack and are staying away from the danger zone. The violation of this *expectation* alone is not sufficient to determine any threat. Yet, this cue is significant enough to be included in our *evidence set.*

When *variant expectations* are violated, the perception senses could be sharpened by increasing the *SightRadius* and *Hearing Threshold* of the bots. This reflects that the bots are expecting danger and increasing their alertness.

4 Experiments and Analysis

In this section, we first give a brief overview of *Twilight City*, a virtual environment built for MOUT simulations. Subsequently, we shall present our experiment results and analysis.

4.1 Overview of *Twilight City*

Twilight City aims to provide a high fidelity simulation platform for MOUT simulations. It is built on top of the *Unreal* engine [20] with various modifications [28] [29]. *Twilight City* simulates urban warfare in an area of approximately 10 km by 10 km. There are more than fifteen buildings in the virtual environment. Figure 10 shows some screen shots of *Twilight City*.

Fig. 10. *Twilight City*

The common cues in MOUT are simulated in *Twilight City*. They include weapon, weapon fire, bots, obstacles, bombs,vehicles. The properties of these cues are type, strength, distance, location and speed.

4.2 Experiment Results

In this section, we summarize the major results of these experiments. These experiments were conducted on computers with Intel T2500@2GHz processor and 2GHz RAM. To evaluate the performance of our situation awareness model, we have conducted some experiments. In particular, we wish to verify its realism and efficiency during time-critical situations.

In our work, we focus on achieving both procedural and end-result realism. In fact, the procedural accuracy to real-life human situation awareness is more important than the end result. To verify our procedural realism, the *SAGAT* approach is adopted. A group of 7 infantry section commanders are put through the experiments individually. The test subjects are instructed to recognize threats in the virtual environment. Sudden freezes will be activated during runtime and the subjects will answer a set of questions. Although the subjects were informed that freezes would occur, they were not informed of the timing. The questions are structured such that a proper understanding of the perception, comprehension and projection stages during situation awareness can be judged. Subsequently, the results of the human subjects are compared with the results of the bots simulation.

A total of three test scenarios are conducted. All three scenarios start with a patrol mission. In Scenario A, a single enemy bot will be used to engage the subject at a close range. A group of enemy bots will ambush the subject with heavy fire in Scenario B. Finally, in Scenario C, an enemy sniper will attack the subject from a distance. The questions for the human subjects are listed below:

1. What is the current mission?
2. What are the possible threats?
3. What is the current threat identified?
4. List the cues that identify the current threat
5. What is the solution?

For each question, a similarity measure, S_t, is computed to judge the closeness of the answers with the actual situation.

$$S_t = \frac{\text{Number of matches}}{\text{Total number of cases}}$$

For example, if there are a total of 5 threats in *Question 2* and the subject identify 3 of them accurately, a score of $\frac{3}{5} = 0.60$ is given for answers to this question. The average scores of all the subjects are then taken for comparison with the scores of the bots as shown in Figure 11.

As all 3 scenarios starts off with a *Patrol* mission, the results for *Current Mission* and *Possible Threats* are similar in all 3 scenarios.

We can observe that, in Scenario A, the scores for spotted cues differ by only 0.12 between the human subjects and the bots. All the subjects identified the *Close Combat Fire* threat accurately.

	Scenario A - Close Combat Fire				
	Current Mission	Possible Threats	Identified Threat	Cues spotted	Solution
Humans	1	0.89	1	0.77	1
Bots	1	1	1	0.89	1

	Scenario B - Ambush				
	Current Mission	Possible Threats	Identified Threat	Cues spotted	Solution
Humans	1	0.89	0.87	0.67	1
Bots	1	1	1	0.91	1

	Scenario C - Sniper				
	Current Mission	Possible Threats	Identified Threat	Cues spotted	Solution
Humans	1	0.89	0.55	0.4	1
Bots	1	1	0.8	0.86	1

Fig. 11. Results of Situation Awareness

In Scenario B, the difference on spotted cues is increased to 0.24. However, both the humans and the bots are able to identify the threat accurately. Upon further investigation, we found that the drop in the human subjects' score on cue spotting is due to the fact that most human subjects pick up the key cues of "heavy fire" and "a group of enemies". They make their judgment of an ambush threat and stop picking up the rest of the cues. This explains why threat identification remains relatively accurate even though the number of observed cues dropped.

In Scenario C, the human subjects' score for the observed cues and threat identification dropped to 0.4 and 0.55 respectively. The main reason for the drop is that the human subjects are not familiar with the visual and audio effects of the sniper fire in the virtual environment. Upon clarification that they had witnessed sniper fire, all the human subjects could identify the threat accurately.

As the score of the solutions are only considered if the subjects could identify the correct threat, all solution generation scores achieved a maximum of 1.0. We found that the solution generation is very accurate once the threat is identified correctly. This is in fact consistent with Klein's observation that experts can generate rapid solutions through proper situation awareness and relevant experience cases.

In all three experiments, the cues picked up by the bots are more than the humans. This is because the bots are able to pick up all cues in their field of view. The humans can miss out certain cues in their field of view if they are overly excited with certain prominent cues. Moreover, the human subjects are not familiar to the virtual environment yet. This is evident in their failure to recognize the gunshot from a sniper rifle in Scenario C. More exposure to the virtual environment can help the human subjects to improve their recognition of cues.

In another experiment to study the effect of situation awareness on the bots behaviours, a set of bots were tested with simulated scenarios. In *Twilight City*, there are mainly three types of bots: soldier bots, militant bots, and civilian bots. Typical scenarios in *Twilight City* include street combat between soldiers and militant groups, and the fighting between soldiers and militant sniper, etc.

As shown in Figure 12, seven experience cases for soldier bots have been implemented. These cases reflect the typical situations a soldier may face in urban warfare. The cues and solutions of these experience cases are extracted from various sources including military doctrines and interviews with experts. For instance, the *Close Combat Fire* situation refers to soldier bots being attacked at close range by enemy bots. This situation requires two experience cases. When the enemy bot is in open area, the *Hasty Attack* experience case will be retrieved so that the soldier bot will return fire at the enemy. However, when the enemy is firing under concealment, the soldier bot should retrieve the *Retrograde* experience so that it will quickly move out of the killing zone.

The testing scenarios are performed within *Twilight City*. They include engaging a set of 20 soldier bots with various threats such as *Sniper Assaults*, *Close Combat Fires*, *Bomb Assaults*, and *Air Strikes*. The objectives of the experiments

MOUT Situation	Experience Case
Sniper Assault	Counter Sniper
Air Strike	Counter Air Strike
Close Combat Fire	Hasty Attack, Retrograde
Ambush	Counter Ambush
Bomb	Counter Bomb
Night Assault	Night Mode

Fig. 12. Experience Cases and Scenarios

are to test the behaviors of the soldier bots under different situations and the impact of different experience cases on the soldier's behaviours. In particular, the behaviors of soldier bots during the Sniper Assault threat and the *Close Combat Fire* threat are compared. Three sets of 20 bots were used for this experiment. They are S(A) bots which only have *Counter Sniper* experience, S(B) bots which only have *Hasty Attack* and *Retrograde* experience, and UT bots which have no experiences but are equipped with the default tactics in UT. The average results from 10 runs are shown in Figure 13. The results in Figure 13 show how a type of soldier bots reacted to in different tactical situations. For example, the data in the first column show that for the 20 soldier bots of type S(A), under *Close Combat Fire* situation, 2.1 of them just stood around, none of them made smoke screen, 15.7 of them returned fire immediately, and 2.2 of them took cover first. In the *Close Combat Fire* situation, we let the militant bot to shoot at the soldier bot from behind the walls. For experienced human soldiers, the natural reaction is to take cover first. In this situation, returning fire at the militant will only expose the soldiers to unnecessary danger as there is no clear line of fire at the militant. It can be observed from the experimental results that the 15.7 of the S(A) bots and 16.9 of the UT bots started returning fire to the militant immediately. 14.0 of the S(B) bots took cover. UT bots returned fire as they are designed to retaliate upon being attacked. The S(A) bots did not have the relevant experience case to handle the *Close Combat Fire* situation, so they had to rely on the basic reaction of retaliation with fire. However, these behaviors are unrealistic in this situation. The S(B) bots, on the other hand, behaved more like real soldiers. They were able to retrieve the *Retrograde* experience when there is no clear line of fire to the militant bot.

During the *Sniper Assault* threat, 12.1 of the S(A) bots created a smoke screen to before moving to a cover, whereas 15.1 of the S(B) bots simply moving to a cover without creating a smoke screen. This is because the S(A) bots are equipped with the Counter Sniper experience, as a result, they were able to act with the correct counter sniper tactics. For the 20 UT bots, 17.9 of them just stayed in the open area and searched for enemy, since they could not differentiate between a close combat fire and a sniper assault. This behavior is extremely

	Close Combat Fire			Sniper Assault		
	S (A)	S(B)	UT	S(A)	S(B)	UT
Stand Around	2.1	2.1	3.1	1.3	3.9	**17.9**
Smoke Screen	0	0	0	**12.1**	0	0
Return Fire	**15.7**	3.9	**16.9**	0	1.0	0
Take Cover	2.2	**14.0**	0	6.6	**15.1**	2.1

S(A) – Bots with Counter Sniper Experience Only

S(B) – Bots with Hasty Attack and Retrograde Experience Only

UT – Unreal Tournament bots

Fig. 13. Behavior of Soldier Bots with Different Experiences

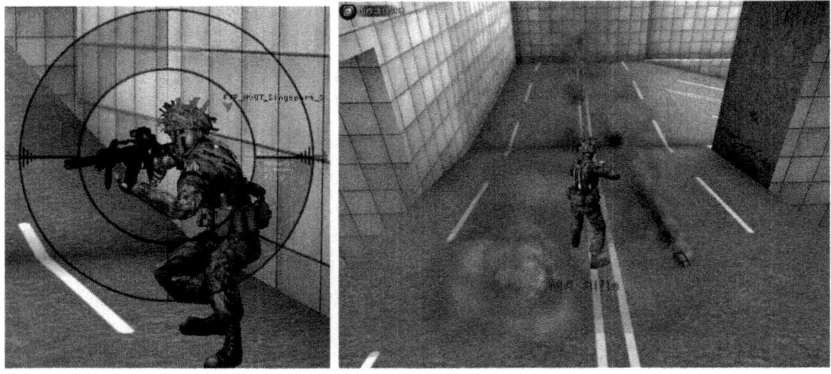

Fig. 14. Sniper Assault Scenario in *Twilight City*

unrealistic as the bots remain exposed to sniper fire[30]. Figure 14 shows a soldier bot being attacked by a sniper. Figure 14 also shows the behavior of a soldier bot of type S(A) upon being attacked by the sniper - the soldier bot created a smoke screen to prevent him from being hit easily by the sniper.

To further evaluate the effectiveness of our model in supporting real-time simulations, we have also conducted a frame rate analysis using the same scenarios. Although the frame rate can be affected by a number of factors, the decision making process (which includes the situation awareness model) may incur significant computation overhead thus may impair the frame rate greatly.

We have conducted a series of experiments. Two different types of bots are used in these experiments for comparison. As shown in Figure 15, the *Normal* bots are the default bots provided by the *Unreal* engine which have very primitive cognitive capability and are supposed to consume little computation resource. The *SNAP* bots are equipped with our decision making and situation awareness model. The results shown in Figure 15 are the average of 10 runs.

It can be seen that the frame rates for both implementations decrease as the number of bots increases. The results show that the performance of our *SNAP* bots are comparable to the Unreal bots in terms of simulation frame

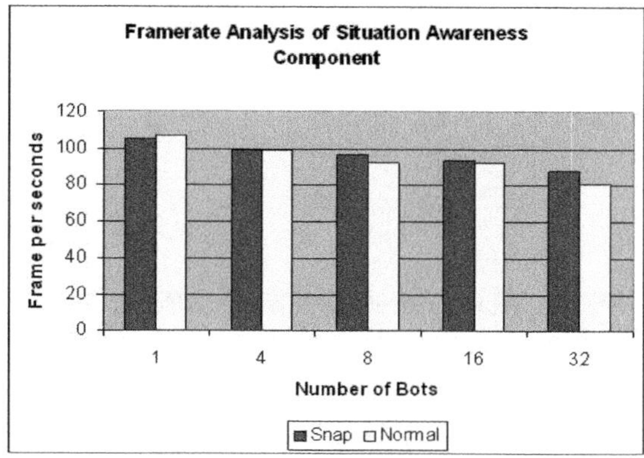

Fig. 15. Frame rate analysis with Situation Awareness Model

rate. This suggests that the decision making process (including the situation awareness model) in *SNAP* does not incur significant computation overhead to the simulation.

5 Conclusion and Future Work

Recent findings in decision making processes show that human experts often make decisions by comparing the current situation with their past experiences. This is very different from the goal-oriented or rational reasoning of conventional models.

The *SNAP* decision-making framework that we discussed in this paper aims to imitate how human make decisions in time-critical tactical situations for MOUT simulations. It uses the CBR cycle to enable the bots to make decisions in such situations with past experience cases. To enhance the effectiveness of *SNAP*, we produced a computational model of situation awareness. Our experimental results demonstrate that the computational model can help to generate realistic behaviors for bots under time pressure.

The computational model follows the three stages of the Endsley's model. In the perception stage, the bot senses are limited and can pick up heterogeneous cues. During the comprehension stage, similarity measures were computed and experience cases are retrieved. Finally, the novel feature of *expectations* is applied during the projection stage.

Our experimental results also shows that the computational model closely models the situation awareness of human soldiers. Further frame rate analysis also shows that the computational cost is not high.

We will continue to work on the proposed computational model for more complex situations. More cues will be investigated and added into the environment. The feasibility of building a group situation awareness model will also be investigated.

Acknowledgment

This work is supported in part by the Singapore National Research Foundation under Grant NRF2007IDM-IDM002-052.

References

1. Clausewitz, C.V.: On War. Princeton University Press, New Jersey (1976)
2. Stacey, C.P.: Official History of the Canadian Army in the Second World War Volume III. Queen's Printer, Ottawa (1960)
3. Ballard, J.R.: Fighting For Fallujah - A New Dawn for Iraq. Greenwood Publishing Group, Connecticut (2006)
4. Barrett, L.F.: Science of Emotion: What people believe, What evidence shows and Where to go from here - Human Behavior in Military Contexts. The National Academies Press, Washington (2007)
5. McDermott, P., Battaglia, D.A., Phillips, J., Thordsen, M.: Military Operations in Urban Terrain (MOUT): Decision Making in Action. US Army Research Institute. Fort Benning (2001)
6. Military Operations on Urbanized Terrain FM 90-10. Department of the Army, Washington (1979)
7. Urban Operations, FM 3-06. Department of the Army, Washington (2003)
8. Klein, G.: Sources of Power: How People Make Decisions. MIT Press, Massachusetts (1998)
9. Ting, S.P., Zhou, S.P.: Snap: A Time Critical Decision-Making Framework for MOUT Simulations. Computer Animation and Virtual Worlds 1(3-4), 505–514 (2008)
10. Ting, S.P., Zhou, S.P.: Dealing with dynamic changes in time critical decision-making for MOUT simulations. Computer Animation and Virtual Worlds 20(2-3), 427–436 (2009)
11. Lopez, R., et al.: Retrieval, Reuse, Revision and Retention in Case-Based Reasoning. The Knowledge Engineering Review 20, 215–240 (2005)
12. Wood, S.D., Zaientz, J.D., Holt, L.S., Amant, R.S., Healey, C., Endsley, M., Strater, L.: MAVEN-SA: Model-Based Automated Visualization for Enhanced Situation Awareness. United States Army Research Institute for the Behavioral and Social Sciences (2006)
13. Wray, R.E., Laird, J.E., et al.: Synthetic Adversaries for Urban Combat Training. In: 2004 Innovative Applications of Artificial Intelligence Conference, San Jose, pp. 82–92 (2004)
14. Wray, R.E., Laird, J.E.: Variability in Human Behavior Modeling for Military Simulations. In: Behavior Representation in Modeling and Simulation Conference, Scottsdale (2003)
15. Azuma, R., Daily, M., Furmanski, C.: A Review of Time Critical Decision Making Models and Human Cognitive Processes. In: 2006 IEEE Aerospace Conference, MT (2006)
16. Osinga, F.: Science, Strategy and War: The Strategic Theory of John Boyd. Routledge, Abingdon (2007)
17. Endsley, M.R.: Toward a theory of situation awareness in dynamic systems. Human Factors 37(1), 32–64 (1995)

18. Hill, R.W.: Modeling perceptual attention in virtual humans. In: Conference on Computer Generated Forces and Behavioral Representation, Orlando (1999)
19. Herrero, P., Antonio, A.: Introducing human-like hearing perception in intelligent virtual agents. In: Second International Joint Conference on Autonomous Agents and Multiagent Systems, Melbourne, pp. 733–740 (2003)
20. Epic Games: Unreal Engine (1998), http://unrealtechnology.com
21. McCarley, J.S., Christopher, D.W., Goh, J., Horrey, J.W.: A computational model of attention/situation awareness. In: Conference on Computer Generated Forces and Behavioral Representation, Santa Monica (2002)
22. W. Warwick, Hutton, R.: Developing computational models of recognition-primed decision making. 10th Conference on Computer Generated Forces. Norfolk (2001) 232-331
23. Endsley, M.R.: Situation awareness global assessment technique (SAGAT). In: IEEE Aerospace and Electronics Conference, pp. 789–795 (1988)
24. Gladwell, M.: Blink, The Power of Think without Thinking. Little, Brown and Co., Boston (2005)
25. Loomis, J.M., Knapp, J.M.: Visual perception of egocentric distance in real and virtual environments. In: Virtual and Adaptive Environments. Lawrence Erlbaum, Mahwah (2003)
26. Forbus, K., Mahoney, J.V., Dill, K.: How qualitative spatial reasoning can improve strategy game AIs. In: 15th International Workshop on Qualitative Reasoning, San Antonio (2001)
27. Ting, S.P., Zhou, S.P.: Quartz: an autonomous navigation system for MOUT simulations. Computer Animation and Virtual Worlds 18(4-5), 383–394 (2007)
28. Zhou, S.P., Ting, S.P., Shen, Z.Q., Luo, L.B.: Twilight City - A virtual environment for MOUT. International Journal of Computer and Applications 30(2) (2008)
29. Ting, S.P., Zhou, S.P.: Qualitative Physics for MOUT. In: 39th Annual Simulation Symposium, Huntsville (2006)
30. Karagosian, J.W.: Streetfighting: The Rifle Platoon in MOUT. Infantry Magazine (September 2000)

HumDPM: A Decision Process Model for Modeling Human-Like Behaviors in Time-Critical and Uncertain Situations

Linbo Luo, Suiping Zhou, Wentong Cai, Michael Lees, Malcolm Yoke Hean Low, and Kabilen Sornum

Parallel and Distributed Computing Centre,
School of Computer Engineering,
Nanyang Technological University, Singapore 639798
{lbluo,asspzhou,aswtcai,mhlees,yhlow,sornum}@ntu.edu.sg

Abstract. Generating human-like behaviors for virtual agents has become increasingly important in many applications, such as crowd simulation, virtual training, digital entertainment, and safety planning. One of challenging issues in behavior modeling is how virtual agents make decisions given some time-critical and uncertain situations. In this paper, we present HumDPM, a decision process model for virtual agents, which incorporates two important factors of human decision making in time-critical situations: experience and emotion. In HumDPM, rather than relying on deliberate rational analysis, an agent makes its decisions by matching past experience cases to the current situation. We propose the detailed representation of experience case and investigate the mechanisms of situation assessment, experience matching and experience execution. To incorporate emotion into HumDPM, we introduce an emotion appraisal process in situation assessment for emotion elicitation. In HumDPM, the decision making process of an agent may be affected by its emotional states when: 1) deciding whether it is necessary to do a re-match of experience cases; 2) determining the situational context; and 3) selecting experience cases. We illustrate the effectiveness of HumDPM in crowd simulation. A case study of two typical crowd scenarios is conducted, which shows how a varied crowd composition leads to different individual behaviors, due to the retrieval of different experiences and the variation of agents' emotional states.

Keywords: intelligent virtual humans, time-critical decision making, crowd simulation.

1 Introduction

Human behavior modeling for virtual agents has gained tremendous momentum in recent years. It is well recognized as an essential component in many applications, such as crowd simulation, military training, digital entertainment, and safety planning. One of the key issues in human behavior modeling is how to

M.L. Gavrilova et al. (Eds.): Trans. on Comput. Sci. XII, LNCS 6670, pp. 206–230, 2011.

realistically model the human decision making process. Compared to traditional decision analysis, which aims at searching for optimal solutions given a set of alternatives, the decision making for human behavior modeling concerns how virtual agents can make decisions in a similar manner as real humans.

As human decision making is complex and dynamic in nature, it is challenging to develop a decision model that can effectively produce human-like decision outputs under different circumstances and social context. Many existing decision models ([7], [9], and [21]) apply some decision analysis mechanisms (e.g., decision networks, utility function) to human decision problems. For example, Yu and Terzopoulos [21] use the decision networks to represent the casual relationships between the actions that an agent can take and the human factors that may influence the action selection. In these models, the decision making process is based on the applied mechanism and the human factors (e.g., personality, social preference) are incorporated as parameters or decision nodes that may influence the action selection. Although these decision models are effective to produce realistic decision outputs, the process of decision making in these models is not modeled as the way how a real person actually makes decisions.

In our work, we emphasize the procedural realism of the decision model. Our objective is not to develop a model that can generate realistic human behaviors by itself. Rather, we aim to provide the general mechanism (i.e., a process model) to define the major processes, representations and relationships of some key factors (e.g., experiences and emotions) involved in real human decision making. Based on our process model, model developers may design their own systems and define realistic human responses in different scenarios according to the relevant domain knowledge. One important consideration of our design is to make sure that the decision process being modeled can be mapped easily to the common way people usually follow. The purpose is to make it easier and more intuitive for model developers to follow such decision process to design their own virtual agents with human-like behaviors. Furthermore, the design of our decision process model is based on a review of the cognitive studies on human behaviors in real-life situations. Some key assumptions on human behaviors are made based on the literature in social psychology and the model is designed to reflect the major cognitive and physical processes based on these assumptions.

Specifically, we focus on the modeling of the human decision making process in time-critical situations, such as emergency evacuations, riots and disasters. In this paper, we present HumDPM, a decision process model for virtual agents, which incorporates two important factors of time-critical decision making: experience and emotion. In cognitive research, the key role of experience in time-critical decisions is supported by Klein's Recognition-Primed Decision (RPD) model [11]. The RPD model offers a high-level cognitive framework that describes the general process on how human makes decisions based on the recognition of situation and the retrieval of past experiences. Although it appears to be a valid model to simulate human-like fast decision making process, the RPD model only describes the process at highly abstract level. In HumDPM, we propose the detailed representation of experience cases and investigate the mechanisms of *situation assessment*, *experience matching* and *experience execution*.

While experience clearly plays an important role in time-critical decision making, another important factor is the influence of emotions. In HumDPM, we also investigate how to incorporate emotion into RPD-based decision making process. To this end, we incorporate the emotion elicitation through an appraisal process according to Appraisal Theory [12]. The appraisal process is performed in the *situation assessment* step of decision making process and the updates on the emotional states affect how an agent matches experiences with the current situation. In addition, as we are particularly interested in crowd simulation, how a person's emotion is affected by others in a crowd is also taken into account.

In general, our objective for human behavior modeling is not to develop some specific behavior rules for the agents in some typical applications. Instead, we aim to develop a generic behavior modeling framework that models the cognitive and physical processes involved in human's daily-life behaviors. In terms of realism of behavior model, our philosophy is that a human behavior model should not only be able to generate seemingly realistic behaviors in some given situations, but also *work like* a human brain in the sense that the decision-making process of an agent should be similar to that of a human-being.

The remainder of this paper is organized as follows. In section 2, we introduce some basic assumptions on human behaviors and related cognitive theories. In section 3, we present the overview of decision making process design in HumDPM. In section 4, we describe the situation assessment process in HumDPM. In section 5, we provide a detailed description on the experience representation and the action selection procedure. In section 6, we explain how emotion is incorporated into the decision making process. In section 7, we show our case study in two crowd simulation scenarios. In section 8, we discuss the related work. In the last section, we finish with some conclusions.

2 Assumptions on Human Behaviors

The research on human behavior modeling is different from the traditional AI, in the way that the traditional AI research focuses on the machine intelligence that generates "smart" agents capable of producing the best or optimal decisions and solutions for real-life and complex situations, whereas human behavior modeling has more emphases on designing the agents with "believable" behaviors similar to real human beings. For human behavior modeling, it is necessary to understand how human makes decisions and how human brains work in real-life situations. The research in this area is still pre-mature. In the existing work, most research results suggest the best way to do something based on some formal theory of reasoning (e.g., Game Theory [4], utility optimization [15], and Bayesian network [17]). However, they do not investigate how brains make decisions (i.e., humans are modeled as rational utility optimizers). These studies are insufficient to address issues like time pressure, incomplete information, uncertainty, emotional and culture effects involved in real human's decision making.

In our approach, to understand how a real person actually makes decisions, we first investigate some existing psychological and cognitive theories/observations. However, in the literature, there are too many theories/observations on human behaviors and decision making. Different theories may be only applicable to specific situations and contexts, and there are even exception and conflicts among these theories. To provide a reasonable model based on the psychological and cognitive study, we have to limit our scope by clearly identifying the context of our modeling. In this research, we mainly focus on how human makes decisions in small to medium size crowd in daily-life and time-critical situations. Under this limited scope, we aim to design a decision process model for modeling human-like behaviors. Our design of the decision process model starts with some fundamental assumptions on human behaviors that are relevant to the limited scope. These assumptions are based on social psychology theories and observations in the literature.

Assumption 1: *A person's behavior is largely determined by her/his experiences rather than by some complex decision rules.*

It is fascinating to observe that humans are able to handle various everyday tasks with ease, although many of these tasks seem to be quite challenging from a computational intelligence point of view. We have heard frequently of something like "*I act in this way simply because I used to do so*" as the reply when we ask a person the reason that she/he reacts to certain events. It suggests that people rely on their experiences, rather than deliberative analysis of all alternatives, to make decisions in daily-life situations. For such situations, we believe that the *Naturalistic Decision Making* (e.g., Recognition-Primed Decision [11]) is a good model of human's decision-making process, which emphasizes the role of a person's experiences in determining her/his behavior. As opposed to traditional theories of decision-making, the naturalistic decision-making theories emphasize situation assessment over the comparison of alternatives. Therefore, considerable effort should be put on representing an individual's experience and the recognition process.

Assumption 2: *In many daily-life situations, people spend more time on situation assessment than on working a solution when making decisions.*

As suggested by Recognition-Primed Decision (RPD) [11] theory, situation assessment plays an essential role in human's daily-life decision making. To make a decision, people usually spend more time on situation assessment than working out a solution. In most cases, a suitable course of action will emerge based on a person's experiences, once the situation is understood. How a person understands the situation directly affects what action(s) she/he will perform. Therefore, it is important to understand what and how the understandings about the situation are acquired through situation assessment. According to Schema Theory [1], cognitive representations of social objects (refereed as schemas) are exist in human's memory, and these schemas influence how a person understands the situation and makes inference. The common way that people assess a situation is through pattern matching. When a person faces a new situation, she/he matches the patterns in the new situation with the similar situations (schemas) people have experienced before. Different people may match with different schemas in the memory. Therefore, different understandings about a given situation may be acquired. To model the process of situation assessment, the social schemas and the salient features (or cues) of the given scenario need to be investigated.

Assumption 3: *Emotions, as intrinsic states of a person, greatly influence her/his behaviors in different situations.*

In real-life, people are often unknowingly affected by different kinds of emotions inside themselves. In some extreme cases, such influence becomes explicit. For example, people exhibit some irrational behaviors in face of danger. When a bomb explosion occurs, some people may become panicked due to fear and anxiety, and start to run aimlessly. To model the cause and effect of human's emotions, Appraisal Theory [12] can serve as a good cognitive account of how emotions arise. The theory describes emotion as a two-stage cognitive process: appraisal and coping. Appraisal refers to the process of assessing person's relationship with the environment, while coping provides some general strategies to maintain or alter this relationship [6]. To model the real-life situations, especially emergency situation, how emotions can be incorporated into the decision making process needs to be studied.

3 Overview of Decision Making Process

Generally, the modeling of human behavior can be regarded as a process of mapping a human's perceptional, mental, and physical functions into a computationally tractable approximation at certain degree of accuracy. Therefore, it mainly involves the modeling of a person's perception, decision making and action systems and the cognitive functions and processes within these systems. In our previous work [13], based on the assumptions on human behaviors, we have proposed a generic framework for human behavior modeling, which follows the *perceive-decide-act* paradigm of human behaviors and models the naturalistic patterns and processes in the paradigm. In this work, we focus more on the decision making process in the behavior modeling and present a detailed decision process model (named HumDPM), as shown in Fig. 1. In HumDPM, the decision making process of virtual agents is modeled as a continuous process, which consists of three iterative steps: *situation assessment*, *experience matching* and *experience execution*. In *situation assessment*, an agent assesses the situation according to the sensory inputs, updates the emotional states based on the assessed situation and determines the general context of the situation. The updated information about the situation and the agent's states are used for experience matching. In *experience matching*, the agent compares the situation with the experiences in one of experience sets. The experience set is determined based on the context of the situation. From the chosen experience set, the agent selects the experience, which matches with the situation with the highest similarity. In *experience execution*, the agent performs the actions according to the selected

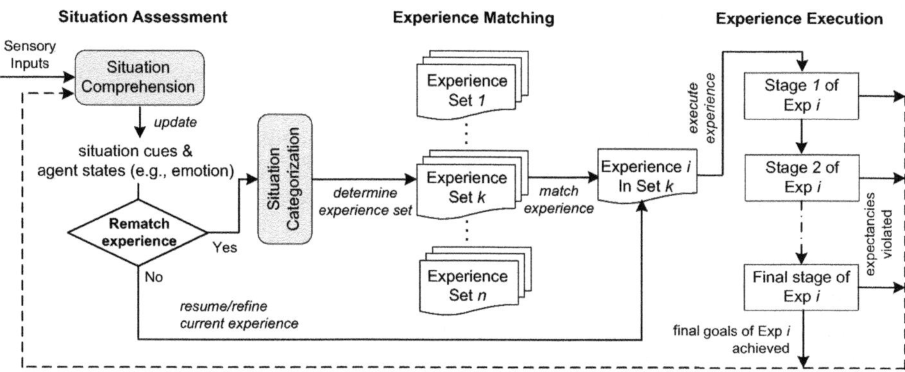

Fig. 1. Overview of decision making process in HumDPM

experience. The remainder of this section provides a brief overview of each of the three steps. The subsequent sections will discuss the steps in more details.

Situation Assessment: The *situation assessment* step in HumDPM is further divided into two phases: *situation comprehension* and *situation categorization.* In the *situation comprehension* phase, an agent synthesizes the sensory data to gain a high-level understanding of the situation and evaluates the situation cues. The agent also updates its internal states, such as the emotional states, based on the assessed situation. In HumDPM, an agent's emotional states are predicted from the evaluation of some appraisal variables in an emotion appraisal process. In the *situation categorization* phase, the agent further assesses the situation by categorizing the current situation into certain context. The contexts in HumDPM are defined as the set of frames of all possible situations to be modeled. A context of the situation could be an environmental context, such as "subway station" or "football stadium" context, which is determined by the physical factors (e.g., location) of the environment. A context of the situation could also be a situational context, which is determined by both the environmental data and the agent's internal states. For example, a situation may be categorized into the situational context of "normal", "emergency" or "high danger" based on the detection of danger events and the fear level of the agent.

One important consideration in our design is that an agent may need to decide whether it needs to rematch experience while assessing the situation. This decision is made in the *situation comprehension* phase and is based on the assessment of the situation with the reference of the currently matched experience and the updates of the agent's internal states. The *situation categorization* phase of *situation assessment* is required only if the agent needs to rematch experience. If the agent decides not to do a re-match, the comparison of the situation with experiences is not needed, which can make the decision process more efficient. In such a case, the agent will either refine (e.g., change the goal's location in navigation) or simply continue to execute the currently matched experience.

Experience Matching: In the *experience matching* step, an agent retrieves the experiences from one experience set. The experience set is determined based on the context of the situation obtained in *situation categorization*. The experiences from the chosen experience set are compared with the current situation. The experience, which is the most similar to the current situation, is then selected for execution. The categorization of the situation into the general context gives two advantages to our model. First, it helps the model developers to systematically organize the experience sets based on different categories. Second, the categorization also helps to reduce the number of comparisons required for experience matching, as only the experiences in the chosen experience set are retrieved for comparison. This may improve the execution efficiency of the system, when the number of experience cases is large.

Experience Execution: In the *experience execution* step, an agent executes the experience by performing the actions at different stages of the experience. In HumDPM, an experience of an agent may consist of single or multiple stages. Each stage has a list of goals to achieve. The agent advances to the next stage of the experience if the goals of the current stage are met. While executing the experience, the agent constantly monitors the virtual environment. The agent may go back to the *situation assessment* step under two conditions: the final goal defined in the final stage of the experience is achieved, or the expectancies of the experience are violated. In the case that the final goal of experience has been achieved, the agent always does a re-match of experiences. However, when violation of expectancies occur, the agent needs to decide whether it needs to rematch to another experience by reassessing the violated situation in the *situation assessment* step as explained above.

4 Situation Assessment in HumDPM

As suggested by Recognition-Primed Decision (RPD) theory [11], how a person makes a decision is largely determined by how she/he understands the situation. In HumDPM, the situation assessment is modeled as a part of decision making process. In the step of the situation assessment, the situational cues are abstracted or synthesized from sensory inputs whose value are obtained from agent's perception. In HumDPM, we also refer to the sensory data obtained through the agent's perception as *observed cues* and the situation cues obtained through situation assessment as *assessed cues*. Having *observed cues* as inputs and *assessed cues* (also called situation cues) as outputs, the situation assessment process is depicted in Fig. 2.

As shown in Fig. 2, the situation assessment process contains a set of situation assessment functions (i.e., SAFunctions), which are activated by SA Trigger. In HumDPM, SA Triggers are referred as to some situational conditions which can cause an agent to invoke the situation assessment process and the relevant situation assessment functions. In other words, the detection of SA Trigger is the necessary condition for an agent to start the situation assessment

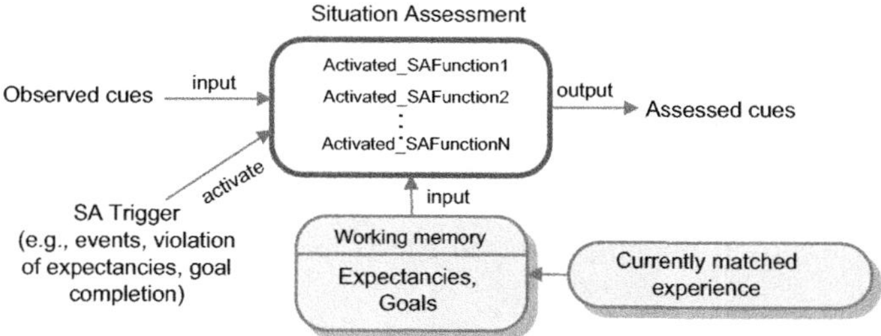

Fig. 2. Situation assessment process

process. Examples of SA Trigger include the external events (e.g., emergency alarm, bomb detonation), the violation of expectancies (e.g., group member missing), and the completion of the current goal (e.g., reaching the destination). When an SA Trigger is detected, an agent assesses the situation with some activated situation assessment functions. For example, if a bomb detonation (i.e., an external event) is detected, the activated situation assessment functions may include *check_crowd_density()*, *get_panic_level_of_nearby_agents()*, *observe_conditions_of_exits()*, and *determine_unexpectedness_level()* in a crowd simulation scenario. The SA functions may take the observed cues (e.g., number of nearby agents) obtained from the perception system and/or the expectancies and goals from the currently matched experience as the inputs. The SA functions determine the values of the assessed cues (e.g., crowd density) based on these inputs.

The set of SA functions activated by an SA Trigger and the actual implementations of SA functions are scenario and context-dependent. To organize SA functions in a clear manner, separate SA modules are implemented in our system for different scenarios (e.g., a subway station scenario and a food distribution scenario). In each of these SA modules, the scenario is further classified into different contexts or situational categories (e.g., normal situation, emergency

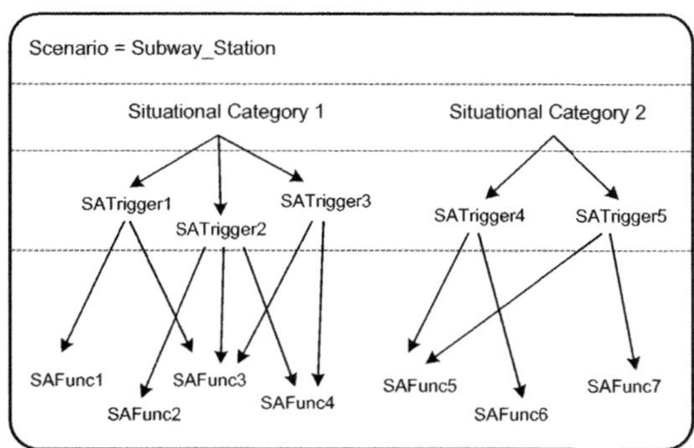

Fig. 3. Hierarchical organization of situation assessment module

situation). Under different contexts, an agent monitors the statuses of SA Triggers associated with the context. If the agent detects a particular SA Trigger, the corresponding SA functions will then be activated and the assessed cues will be generated. The hierarchical organization of a SA module in HumDPM is illustrated in Fig. 3. Note that it is possible that a single SA function is activated by different SA Triggers. All the SA modules for different scenarios follow the same the structure as demonstrated in Fig. 3.

5 Experience Representation and Action Selection in HumDPM

In this section, we give descriptions on the representation of the experience cases and the action selection mechanism in HumDPM. In general, our design of experience has two main characteristics as follows.

- An experience consists of single or multiple stages (S).
- Each stage of an experience contains a set of feature cues (C), a set of goals (G), a set of expectancies (E) and a set of actions (A).

In HumDPM, we consider the organization of experience by representing an experience with single or multiple stages. We define the stages of an experience as a sequence of intermediate steps of the experience, which have the temporal relationships among each other. In most cases, the stages of an experience are in a sequential order. An agent advances to the next sequential stage if the goals specified in the current stage are met. Apart from the sequential relations, our model also allows other relations between the stages (see Fig. 4).

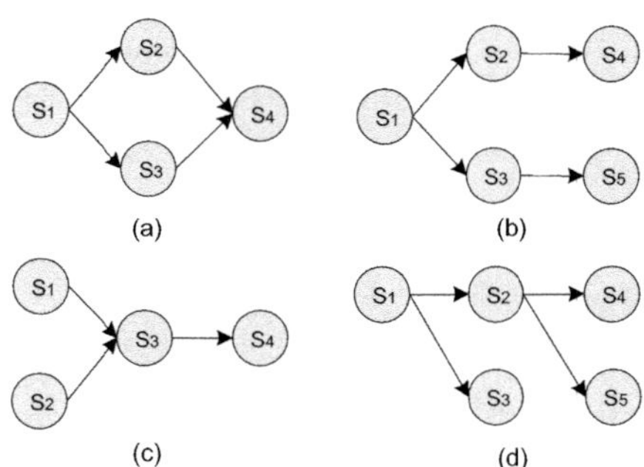

Fig. 4. Variations of stage structure

As shown in Fig. 4, there may exist some parallel stages in an experience, such as S_2 & S_3 in (a), (b) and (d), S_1 & S_2 in (c), and S_4 & S_5 in (d). Here, the term 'parallel' does not mean that these stages are executed in parallel. It only refers to the stages that are parallel structurally. When an agent reaches to the parallel stages, it needs to select one of the parallel stages to advance to. To define the relationship between stages, we use $S_i \rightarrow S_j$ to denote two stages that are in sequential order and use $S_i \parallel S_j$ to denote two stages that are in parallel. The stage structures of the experiences in Fig. 4 can then be represented as: $S_1 \rightarrow (S_2 \parallel S_3) \rightarrow S_4$ in (a), $S_1 \rightarrow ((S_2 \rightarrow S_4) \parallel (S_3 \rightarrow S_5))$ in (b), $(S_1 \parallel S_2) \rightarrow S_3 \rightarrow S_4$ in (c), and $S_1 \rightarrow (S_3 \parallel (S_2 \rightarrow (S_4 \parallel S_5)))$ in (d).

According to the RPD framework, when a person tries to recognize the situation for decision making, four by-products may be produced: *relevant cues* (what the person is aware of about the situation), *plausible goals* (what is to be achieved); *expectancies* (what is expected to happen as situation evolves) and *actions* (what are the workable actions likely to succeed). In HumDPM, these four by-produces naturally form the "content" of each stage of an experience. A stage S_i of an experience can be represented as a 4-tuple:

$$S_i = (C_i, G_i, E_i, A_i),$$

where the C_i, G_i, E_i, A_i represent the sets of feature cues, goals, expectancies and actions in the stage S_i respectively.

The feature cues defined in different stages of an experience are used to compare with the situation cues C_{sit} obtained from situation assessment. The comparison is needed in two cases when: 1) selecting an experience from the experience set, and 2) selecting a stage from parallel stages. For selecting experience, the set of feature cues C_1 in the first stage S_1 of each experience is used to compare with C_{sit}. For selecting stage, the set of feature cues C_i in each parallel stage S_i is used to compare with C_{sit}. Note that $C_i = \emptyset$, if S_i is not parallel with other stages. This means there is no comparison required for a stage in a sequential order (i.e., a non-parallel stage). The agent automatically advances to the next stage once the goals of a sequential stage are met.

To compare a set of feature cues C with the situation cues C_{sit}, each feature cue c_j in C is compared with a corresponding situation cue c'_j in C_{sit} with the same type. A cue similarity value $sim_c(c_j, c'_j)$ is calculated by comparing the value difference between c_j and c'_j. It is possible that the situation cue c'_j corresponding to c_j cannot be found. In such case, $sim_c(c_j, c'_j)$ is set to zero. Here, we do not specify how the $sim_c(c_j, c'_j)$ is formulated, as the formulation varies according to the types of cues and the scenarios being modeled. To obtain an overall similarity between the situation cues and the set of feature cues being compared, individual cue comparisons are aggregated to yield a weighted average similarity value. Given that each $sim_c(c_j, c'_j)$ is determined, the weighted average similarity value sim_a can be calculated as follows:

$$sim_a(C, C_{sit}) = \frac{\sum w_j sim_c(c_j, c'_j)}{\sum w_j}, \tag{1}$$

where w_j is the weight for c_j, which indicates the relative importance of a feature cue to an experience. For selecting an experience, the agent chooses the experience with the highest sim_a value. Similarly, for selecting a parallel stage, the agent chooses the parallel stage with the highest sim_a value. For experience matching, it is possible that some of agents (e.g., foreigners in the subway station scenario) cannot match any experience from their experience sets in certain situations. In this case, the experience selected by the majority of other agents will be used to control the agent's behaviors. This implies that an inexperienced agent will imitate the other's behaviors when no suitable experience is matched.

Once an agent has matched a certain experience, it goes to the *experience execution* step of decision making and executes the matched experience stage by stage. When the agent is in a certain stage of the experience, it needs to perform three tasks in every time step of simulation: *checking goals*, *checking expectancies*, and *performing actions*. First, the agent checks the set of goals G_i defined in the current stage. If all the goals in the current stage are met, the agent will advance to the next stage or branch to one of the parallel stages subsequently. Second, the agent checks if there is any violation of expectancies. The set of expectancies E_i in the stage defines an agent's expectations on what are likely to happen or not to happen, given the current situation. The violation of expectancies will cause the agent to go back to the *situation assessment* step of decision making and reassess the situation. However, it may or may not directly cause the re-match of experiences. The agent needs to make a decision on whether to rematch experience in *situation assessment* as we explained previously. Third, when the goals of the current stage are not met yet and there is no violation of the expectancies, the agent simply continues performing all the actions A_i specified in the current stage. The actions specified in the stage refer to some high-level action strategies, which are used to guide the agent's physical motions. In our crowd simulation, examples of such actions include seeking to a location, moving as a group, and searching along a direction.

6 Incorporating Emotion into HumDPM

In real-life, the influence of emotion plays an essential role in human decision making. In time-critical situations, emotions such as fear and anger can be easily elicited and affect a person's behaviors. For example, when a bomb explosion occurs, some people may become panicked due to fear and start to run aimlessly. In HumDPM, we focus on the modeling of the process of human decision making. As the emotion has a pervasive influence on the entire human decision making process [2], it is necessary to model how the emotion elicits, affects and integrates within the agent's decision making process. Our modeling of the emotion elicitation process is based on Appraisal Theory [12], a leading cognitive theory of emotion that explains how emotions arise through a person's subjective assessment of her/his relationship to the environment. According to Appraisal Theory, the person-environment relationship is assessed along several dimensions, called appraisal variables or dimensions, such as goal congruence, casual attribution, urgency and coping potential. The emotion of a person is elicited through the combination of these appraisal assessments. In HumDPM, the elicited emotions affect an agent's decisions on how to match experiences with the situation. The remainder of this section describes the design of appraisal variables assessment, emotion prediction and emotion effects on agent's decisions in HumDPM.

6.1 Appraisal Variables Assessment

In HumDPM, we consider the emotion elicitation as a part of decision making process. According to Appraisal Theory, we introduce an *emotion appraisal process* for emotion elicitation in the *situation comprehension* phase of the *situation assessment* step of decision making. In *emotion appraisal process*, the emotion is elicited through *appraisal variable assessment* and *emotion prediction*. In *appraisal variable assessment*, an agent needs to assess the situation through the evaluation of appraisal variables. Appraisal variables refer to some essential evaluation criteria, which are identified by a number of appraisal theories, for the prediction of emotions. A list of key appraisal variables from appraisal theories has been summarized in [6]. In our model, the values of appraisal variables may be evaluated based on both the sensed environmental data as well as the knowledge of currently matched experience. For example, one of appraisal variables, *goal_congruence*, is used in our model to indicate the extent to which the situation facilitates the achievement of agent's goals. To update *goal_congruence* value in *emotion appraisal process*, the agent needs to evaluate the situation against the goals specified in the currently matched experience.

In HumDPM, the selection of appraisal variables and the assessment mechanisms are considered to be context-dependent and flexible. A model developer may choose sets of suitable appraisal variables for different scenarios and specify the way how various situational inputs in these scenarios can affect the states of the selected appraisal variables. In our crowd simulation scenarios, we have used a subset of appraisal variables to predict the emotional states of agents (e.g., fear and anger). The selected appraisal variables and the semantic meaning of these variables are summarized in Table 1.

Table 1. Selected appraisal variables

Appraisal variables	**Semantic meaning**
goal_congruence	Extent to which the situation facilitates agent's goals
unexpectedness	Degree of how the situation violates agent's expectancies
coping_potential	Agent's ability to control, change, or adapt to the situation
predicability	Extent whether the outcome is predictable
intrinsic pleasantness	Whether a stimulus is likely to result in pleasure or pain

In HumDPM, the values of the selected appraisal variables are evaluated based on both the sensed environmental data as well as the knowledge of the currently matched experience. For example, to check *goal_congruence*, an agent needs to evaluate the situation against the goals defined in the currently matched experience. For each goal defined in agent's experience, the model developer may associate a set of feature cues in a specific scenario, which are considered to be relevant to the goal. For example, if the goal is to reach certain destination, the set of goal-relevant cues may include the cues like estimated distance to the destination and crowd density around the destination area. By assessing the sets of goal-relevant cues, the *goal_congruence* value can be derived. The evaluation of the appraisal variables is also influenced by the agent's characteristic attributes (e.g., threat vulnerability), which are used to model the individual traits of the agents and reflect the personalities of different types of agents.

6.2 Emotion Prediction

In *emotion prediction*, the emotions are predicted based on the values of the appraisal variables. A theoretical prediction of some major modal emotions from appraisal variables has been given in [18]. In this theoretical prediction, the values of the appraisal variables are specified as some fuzzy values (e.g., low, medium and high). As the values of these appraisal variables are finite discrete numbers, the combinations of different values of the appraisal variables form a number of so-called appraisal patterns. A prediction table is constructed by mapping emotions to these appraisal patterns. In our model, to computationally model emotions, we consider both the appraisal variables and the emotional states as linguistic variables. Each of these linguistic variables, such as *goal congruence* and *fear level*, takes some linguistic values (e.g., "low", "medium", "high"). The membership functions are defined to map the numeric values of the variables into a degree of membership in the linguistic value.

By referring to the predication table in [18], we may derive some fuzzy rules for the predication of emotion from the appraisal variables. To predicate the *fear_level* of agent using *goal_congruence*, *unexpectedness* and *coping_potential*, we may define the traditional fuzzy rules such as the two examples below:

- If *goal_congruence* is *low* and *unexpectedness* is *high* and *coping_potential* is *low*, then *fear_level* is *high*.
- If *goal_congruence* is *high* and *unexpectedness* is *medium* and *coping_potential* is *low*, then *fear_level* is *medium*.

However, given that we have only three appraisal variables and each of these appraisal variables takes three linguistic values (i.e., "low", "medium", "high"), it requires total of $3 \times 3 \times 3 = 27$ fuzzy rules for predicating *fear_level*. If we use more appraisal variables in the emotion predication, the number of fuzzy rules will increase exponentially. To reduce the number of the fuzzy rules, we adopt the Combs Method [3] by defining the fuzzy rules to individually specify each appraisal variable's relationship to the emotional state. Three examples of fuzzy rules adopting the Combs Method are as follows:

- If *goal_congruence* is *high*, then *fear_level* is *low*.
- If *unexpectedness* is *high*, then *fear_level* is *high*.
- If *coping_potential* is *medium*, then *fear_level* is *medium*.

Given that each appraisal variable has three linguistic values, it only requires 3 fuzzy rules for each appraisal variable by adopting Combs Method. Therefore, there are only $3 \times 3 = 9$ fuzzy rules in total, if we use 3 appraisal variables in the prediction. Using the membership functions defined for each appraisal variables, the degree of membership (DOM) can be derived from the numeric value of the appraisal variable. Then, in each fuzzy rule as in the above example, the DOM of emotional state (i.e., output of a fuzzy rule) can be easily inferred from the DOM of appraisal variable. For all fuzzy rules defined, fuzzy OR operator is used to combine the output from each fuzzy rule. The final result from the fuzzy inference produces a logic sum of DOMs for each linguistic value of emotional state. For instance, the *fear_level* may be predicated as *high*: 83%; *medium*: 53%; and *low*: 0% and the agent's fear state will then be set to the corresponding high value.

6.3 Emotion Effects on Agent's Decisions

The updates on the emotional states may affect an agent's decision on how to match experiences with the situation in three cases. First, in the *situation comprehension* phase of the *situation assessment* step of decision making, the updates of emotional states may affect an agent's decision on whether it is necessary to do a re-match of experiences. For example, when an agent detects an emergency alarm in a subway station scenario, the agent's decision on whether to rematch the experience case for escaping is affected by its fear level. Some agents may assess the alarm as a faulty alarm and their fear levels do not increase greatly in such a situation. These agents then choose not to rematch experience and continue their currently matched experiences (e.g., go into the station) after assessing the situation. Second, in the case that an agent has decided to rematch experience, the emotional states of the agent may be used to determine the situational context in the *situation categorization* phase of *situation assessment*.

Here, we illustrate how the *fear_level* of agent may affect the agent's decisions under an emergency situation. Suppose an unexpected event (e.g., emergency alarm) is detected by the agent, while the agent is performing the actions according to the currently matched experience (e.g., go to train). The unexpected event causes the violation of expectancies, which forces the agent to go into *situation assessment* step of decision making. In *situation comprehension* phase of *situation assessment* step, the *fear_level* of agent is updated through the *emotion appraisal process*. Depending on the updated *fear_level* and currently matched experience, the agent may decide whether to rematch experience using some simple rules as follows:

- If *current_exp=go_to_train* and *fear_level* is *high* or *medium*, then rematch.
- If *current_exp=go_to_train* and *fear_level* is *low*, then do not rematch.

If the agent chooses to rematch experience, the *fear_level* of agent is used to determine the situation category. The agent may recognize the situation as "highly dangerous" or "emergency" based on its *fear_level*. Once the situation category is determined, the agent compares the situation with the experiences from the experience set selected. In some cases, the emotion may also play a role in the comparison process. For example, in "highly dangerous" situation, the agent with missing group members may match either the experience for "searching group members" or "escaping alone" depending on its *fear_level*.

7 Case Study

To demonstrate the effectiveness of HumDPM, we apply HumDPM to model the decision making of virtual agents in two typical scenarios in crowd simulation. The first scenario targets on the daily-life urban environment, where we simulate the crowd behaviors in a subway station under both normal and emergency situations. The second scenario is oriented towards the military operation scenario, where a simulation of food distribution process in a post-disaster area is realized. In general, we aim to use these simulated scenarios as the test cases to testify the capabilities of HumDPM decision model to support the modeling of human's decision making in different kinds of situations and scenarios. Beyond that, we also intend to explore how the individual behaviors and emotional changes modeled by HumDPM can affect the global patterns and variations in a crowd.

7.1 Scenario Design

(a) Scenario I: urban subway station scenario

In urban underground station scenario, we created the testing environment by reconstructing a MRT station in Singapore. The MRT station is a three-storey underground transportation connected by escalators and lifts. Fig. 5 (a) shows the 2D top-down layout of the first storey in the reconstructed MRT station. Fig. 5 (b) shows the corresponding environment in 3D.

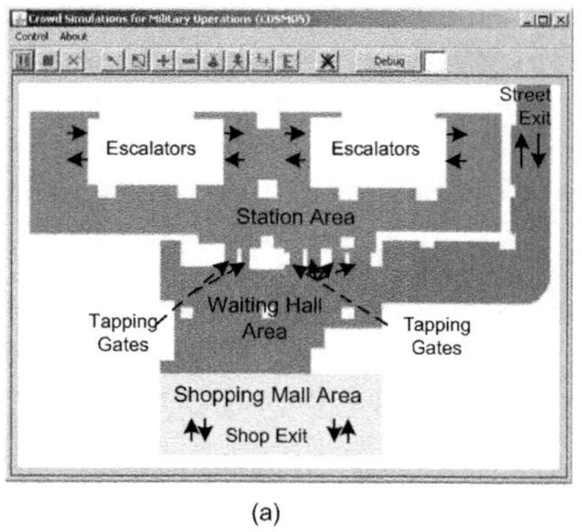

(a)

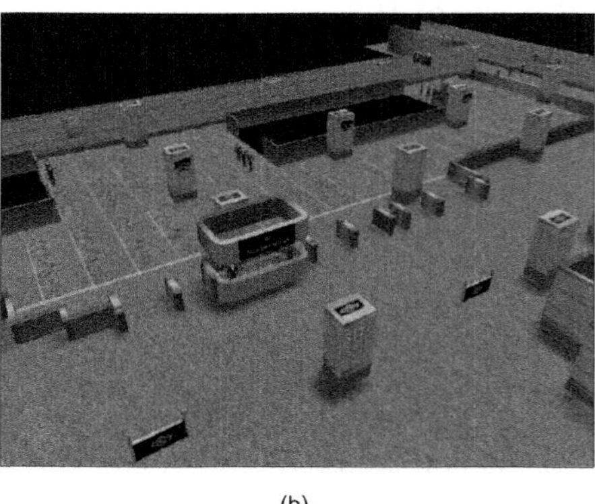

(b)

Fig. 5. Virtual environment construction in Scenario I: (a) Top-down 2D layout, (b) 3D environment

As shown in Fig. 5 (a), the first storey of the station is divided into three areas: station area, waiting hall area, and shopping mall area. In this scenario, agents are created from escalators in the station area, exit to shops and exit to street. For the agents created from escalators, their goals could be either to go to the shops or to go to the street. For the agents created at the exit to shops or exit to street, their goals are to get into the station and reach one of the escalators. To achieve their goals, they first have to get out or into the station by passing a tapping gate. After that, some of them may stop in the waiting hall area to wait for other group members. Other people simply move towards the directions to their goals. People in the crowd could be either individuals or in some groups. We model how people behave in normal situation as well as in some emergency events, such emergency alarm and bomb detonation.

(b) Scenario II: food distribution scenario

In the food distribution scenario, we create the testing environment, which is a virtual village in a post-disaster area with hundreds of local residences. Due to the disaster (e.g., e.g., earthquake, tsunami, or flood), the village is short of water and food supplies and civilians are waiting for the humanitarian aid from the international support team. Fig. 6 (a) shows the 2D top-down layout of the simulated virtual village. Fig. 6 (b) shows the corresponding environment in 3D. In the village, we define two major areas: food distribution area and village area. The food distribution area is a square in the village, which will be used for placing the food distribution points and other humanitarian relief facilities. The village area is the residential area of the villagers. When the food distribution starts, the villagers are generated from the buildings in the village area.

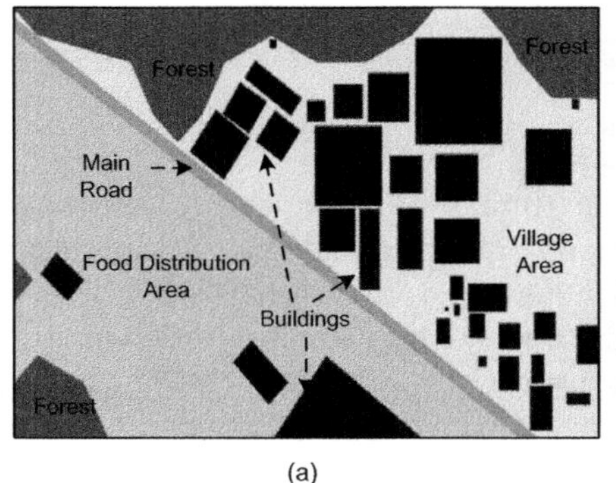

(a)

(b)

Fig. 6. Virtual environment construction in Scenario II: (a) Top-down 2D layout, (b) 3D environment

The majority of agents in this scenario are the local civilians, whose goal is to get the food packets from the distribution point. Among the civilians, there are some instigators and criminals, who try to instigate the crowd and steal food from the normal civilians. The soldiers, who are carrying out the food distribution mission, interacts with the crowd and try to maintain the order in the distribution process.

7.2 Simulation Results and Analysis

(a) Scenario I: urban subway station scenario

Our crowd simulation results are visualized in both a Java 2D display and a 3D game engine. To have a global view of our simulated crowd scene, Fig. 7 captures the 2D and 3D simulation result in the top-down view of the station in urban subway station scenario. We use colors to indicate the agents belonging to different types of relationships. For example, in 2D display, the color representation is as follows: black - individual agents, blue - strong-tie group agents, and red - normal-tie group agents.

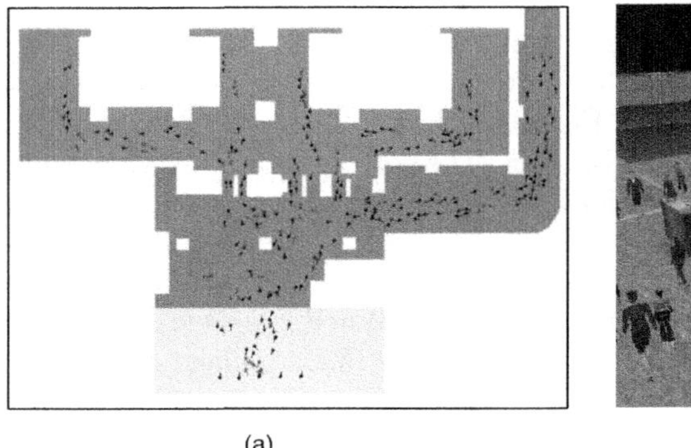

(a)

(b)

Fig. 7. Visualization results in urban subway station scenario: (a) 2D result, (b) 3D result

Following the representation defined in HumDPM, experience cases are implemented based on the observation and study of typical behaviors of real people in normal and emergency situations under subway station scenario. For different types of the agents (e.g., individual agents and group agents), different experience cases are implemented. Fig. 8 captures the screen shots from our 3D visualization to show how an individual agent (the boxed one) performs its behaviors according to a simple experience case for escaping. The agent matches to this experience case when it detects a bomb detonation. The agent tries to escape from the station by following the four sequential stages defined in the experience case. In the first stage, the agent tries to get out of the station area by searching and moving towards the tapping gates. Once the agent detects the tapping gates, it advances to the second stage with the goal of selecting one of the tapping gates with fewer people. The agent passes through the selected tapping gate in the third stage. In the last stage, the agent's goal is set to reach to the nearest exit, which connects to the area outside the station.

The emotional states of an agent are dynamically updated when it assesses the situation. In the appraisal process, the values of the appraisal variables are updated based on not only the detected events and objects, but also the states

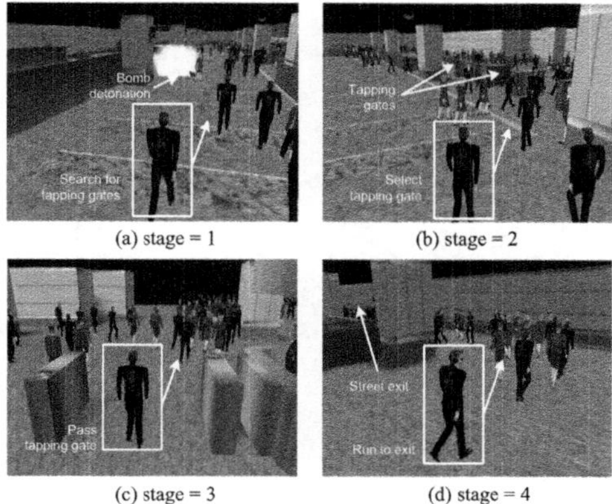

(a) stage = 1 (b) stage = 2

(c) stage = 3 (d) stage = 4

Fig. 8. Behaviors of an individual agent in emergency situation

of the surrounding agents. For example, when an agent observes that many surrounding agents' fear levels are high, it affects the evaluation of some appraisal variables (e.g., unexpectedness), which in turn causes the agent's own fear level to increase. As more and more agents increase their fear levels, the emotion propagates among the crowd. Fig. 9 shows such emotion propagation effect from our 2D visualization. Fig. 9 (a) captures the screen shot at the time that a bomb is just detonated. As the bomb's influence area (i.e., the area where the bomb either can cause physical damage or can be explicitly detectable) is set to be within a certain range (e.g., the circle), not all the agents may detect the bomb detonation. Therefore, it can be observed that there are some non-escaping agents (unfilled dots) in the area far away from the bomb location. Fig. 9 (b) captures another screen shot after the bomb has been detonated for a period of time. It can be observed that many non-escaping agents have turned into escaping agents (filled dots) as they find that the surrounding agents' fear levels are increasing.

In urban subway station scenario, two experiments have also been conducted to evaluate whether the decision making modeling using HumDPM could effectively reflect the differences of the evacuation behaviors among different types of agents. In the first experiment, the crowd population are initialized with two types of compositions: composition 1 - 70% of "neurotic" agents and 30% of "normal" agents, and composition 2 - 30% of "neurotic" agents and 70% of "normal" agents. By "neurotic" and "normal" agents, we refer to the agents with different personalities. Compared to a "normal" agent, a "neurotic" agent is set with lower values of some personality-related attributes, such as threat vulnerability and in-danger time susceptibility. The values of these attributes affect how an agent assesses the situation cues and updates the emotional states. Fig. 10 shows the percentage of the agents who have decided to escape (referred to as escaping agents) after the bomb is detonated in the station. It can be seen that the number of escaping agents increases much faster when there are more "neurotic" agents in the crowd composition. This is consistent with real-life situations, as

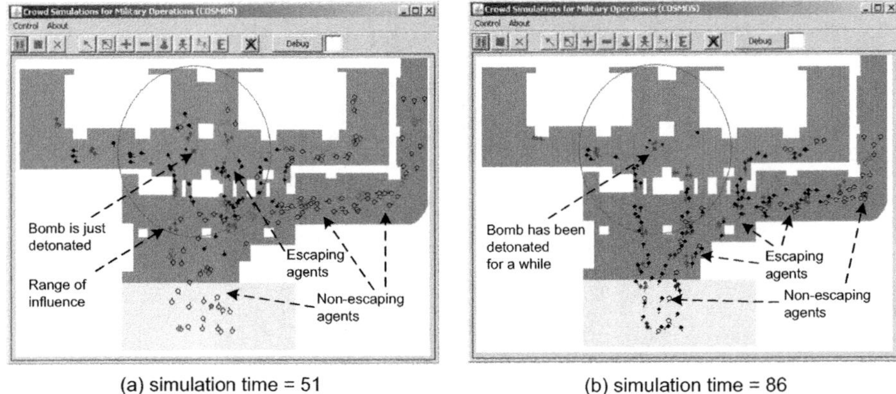

Fig. 9. 2D visualization results with bomb detonation

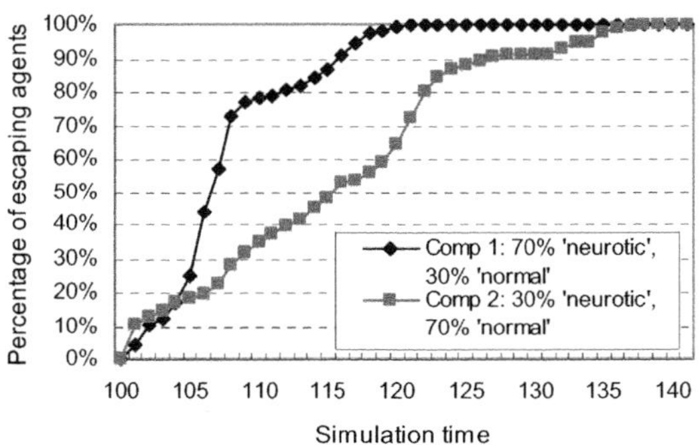

Fig. 10. Experiment 1- crowd compositions with different personalities in scenario I

neurotic people are more fearful in the face of danger and they tend to decide to escape earlier in emergency situations.

In the second experiment, the crowd population are initialized with 70% of individual agents and 30% of group agents in composition 3, and 30% of individual agents and 70% of group agents in composition 4. Fig. 11 shows the percentage of escaping agents in the station with these two crowd compositions after a bomb is detonated. It can be observed that the percentage of escaping agents increases slower when there are more group agents in the crowd composition. This is mainly because that certain kinds of the group (e.g., a family group with only adults) tend to be less susceptible to the threat and if the group members are missing, they also tend to search for the missing members first before escaping.

(b) Scenario II: food distribution scenario

Similar to the subway station scenario, the crowd simulation results for food distribution scenario are visualized in both a Java 2D display and a 3D game engine. In this scenario, for training purpose, we allow the user to directly control soldiers in the simulation through the 3D game engine. The user may control

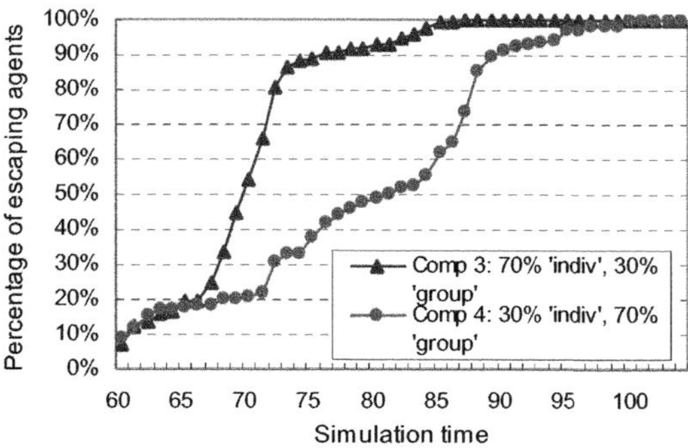

Fig. 11. Experiment 2- crowd compositions with different group and individual agents in scenario I

soldier's motion as well as issuing the commands, such as pacifying, warning, and arresting in the 3D environment. On the 2D side, the user may place the distribution points in the food distribution area, set the soldier's commands and create the instigators and criminals in real-time. Fig. 12 shows the color scheme for representing different types of agents in this scenario: the unfilled black agents - normal civilians, the filled black agents - civilians who have obtained the food packets, the red agents - angry civilians, the blue agents - soldiers, the pink agents - instigators, and the cyan agents - criminals.

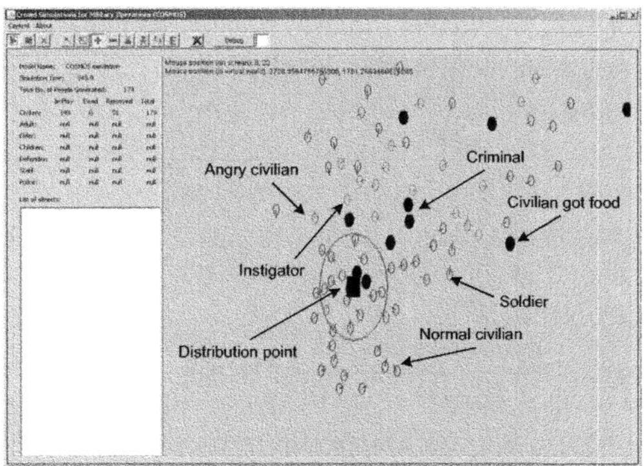

Fig. 12. Different types of agents in food distribution scenario

Fig. 13 shows the 2D and 3D visualization results in food distribution scenario. In the simulation, the civilian agents may switch from one distribution point to another one depending on the crowd density at each distribution point. The distribution point with the most agents around has more angry civilians compared to others. The instigators and criminals tend to gather around the distribution point with the most agents around. The soldiers are therefore moving around the most crowded area to control the crowd. The non-angry civilian agents may be directed by the soldiers to switch to other distribution point, whereas the angry civilian agents may grab the food from others instead of searching for alternative distribution points. The decision making process of all types of agents is modeled using HumDPM.

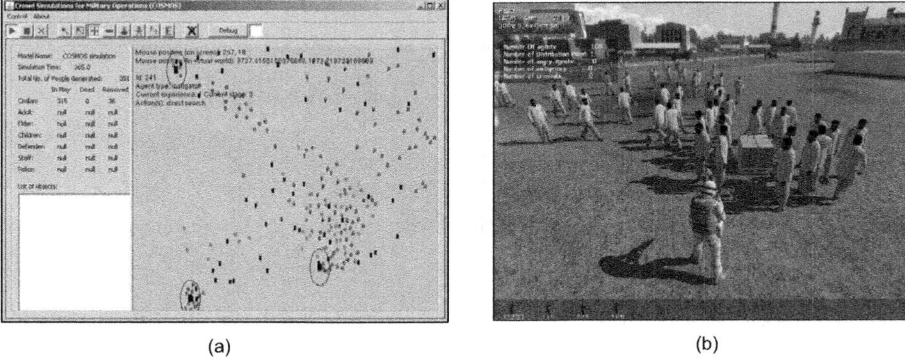

Fig. 13. Visualization results in food distribution scenario: (a) 2D result, (b) 3D result

In this scenario, the emotions of civilians are modeled using the emotion appraisal process designed in our behavior modeling framework. In the situation assessment process, a civilian agent may update the appraisal variables (e.g., goal congruence, unexpectedness, and outcome probability) based on the detection of different situational cues. In this scenario, the situation cues that may cause the changes in the appraisal variable include the incongruence to the agent's goal (e.g., the agent is not able to get the food packet due to the long waiting time or shortage of food packet), the other agent's behaviors (e.g., soldier's pacifying and warning behavior, the instigator's instigating behavior and the criminal's stealing behavior) and the emotions of other agents (e.g., group of angry agents). Due to the detection or changes of these situation cues, the appraisal variables of agents are updated. The emotion of agents (i.e., anger) is predicted from the prediction process in HumDPM.

To analyze how the different agents' behaviors and environmental factors may affect the food distribution process, we have conducted two test cases to evaluate the percentage of angry civilians under different settings in the distribution process. Fig. 14 shows the first test case, which compares the percentage of angry civilians with different number of soldiers created in the environment. The test case compares the percentage of angry civilians without any soldier's control, with one soldier, and with five soldiers. Only one distribution point is created in this test case. The position and number of food packets of the distribution is set to be the same with different settings of soldiers. It can be seen from Fig. 14 that the percentage of angry agents can reach up to 60% of the total population when there is no soldier control at all. In the case of one soldier and five soldiers' control, the maximum percentage of angry agents drops to 35% and 25% respectively.

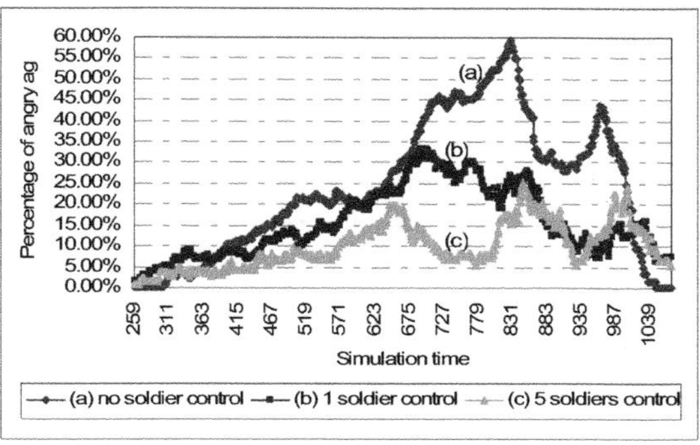

Fig. 14. Experiment 3: Setting with different number of soldiers in scenario II

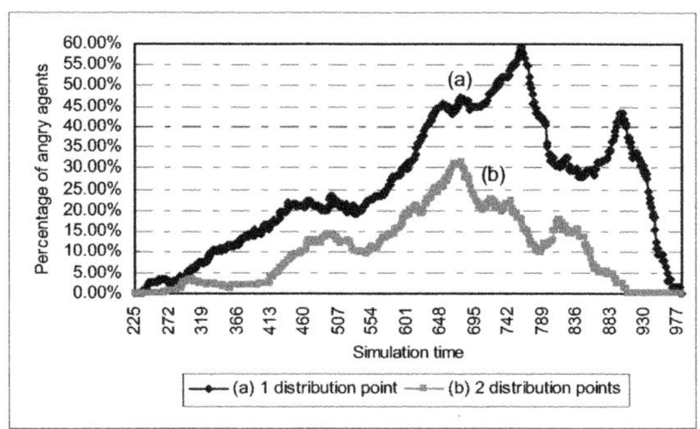

Fig. 15. Experiment 4: Setting with different number of distribution points in scenario II

In the second test case, we test how the number of distribution point may affect the distribution process. Fig. 15 shows the percentage of angry agents with one distribution point and two distribution points placed in the distribution area. In this case study, no soldier is added in the simulation. It can been seen in Fig. 15 that by creating two distribution points in the simulation the percentage of angry agents drops to almost half of the amount compared to the one distribution point case.

8 Related Work

In recent years, the decision modeling for virtual agents has been studied by the AI and computer animation communities [7], [8], [9], [19] and [21]. In our research, we focus on the study of naturalistic decision making (NDM), which studies how people actually make decisions through performing complex cognitive functions in demanding situations. One of the best known models in NDM is Klein's RPD model [11], which captures how the domain experts make decisions based on the matching of past experiences that is triggered by the recognition of the current situation. Compared to traditional theories of decision-making, the RPD model emphasizes situation recognition over the comparison of alternatives. It suggests that once the recognition is formed, corresponding course of actions can be easily enumerated and evaluated through mental evaluation. Though RPD model appears to be a valid model to simulate human-like decision making process, it only offers a high-level cognitive framework. Detailed computational models need to be developed in different application domains.

In [16], Norling has examined the integration of RPD mode into BDI framework. In [5], Fan and Yen have implemented RPD-enabled cognitive agent architecture (R-CAST) and applied it in team decision support system under Command and Control (C2) context. Ji et al. [10] have developed a fuzzy logic-based computational RPD model and applied the model to a medical application for adverse drug reaction (ADR) detection. Our work distinguishes from these works in several aspects. Firstly, our design takes a closer look on the representation of the experience case, by introducing the stages of the experience and defining the structure and content of the stage. Based on the proposed experience representation, the mechanisms on experience matching and execution are specified in HumDPM. Secondly, HumDPM explores the effects of emotions on human decision making process with the interaction of experiences. The incorporation of emotion is important, as emotion is one of the influential factors in human decision making, especially in time-critical situations. Lastly, our work is also the first attempt to apply the RPD-based decision model to crowd simulation [20].

In emotion modeling, EMA (acronym for EMotion and Adaptation [6], [14]) is one of well-established computational models based on Appraisal Theory [12]. In EMA, a plan-based representation, termed causal interpretation, is proposed to represent the person-environment relationship. The appraisal process operates over this representation to generate emotions. An agent copes with the situation by adopting a coping strategy in response to the current emotional state. In HumDPM, we focus more on how to incorporate the appraisal into RPD-based decision making process. The appraisal process relies on the assessment of the current situation and the "mental states" (e.g., goals, expectancies) of an agent, which are formed in the currently matched experience. The updates on the emotional states directly affect the agent's behaviors through changing or modifying the matched experience. As we are particularly interested in crowd simulation, how the emotional states of other agents in a crowd affect the appraisal process of the agent is modeled. The influence of others' emotions may cause an emotion propagation effect, which has been shown in our case study.

9 Conclusion

Human behaviors and decision making are complex. How an individual or group of individuals will behave in certain situations often depends on many factors for different domains, such as physical factors, psychological factors, and social factors, etc. A researcher in human behavior modeling thus faces many challenges as to which approach to use to model the human behavior in real-life, which behavior factors to consider, what level of detail to work on, which decision making mechanism to use, etc, given the objectives of the project and constraints of available resources. In our work, we aim to provide a generic decision process model to represent a human-like decision making process, which incorporates some key decision making factors, such as human's experiences and emotions.

We emphasis on the procedural realism of human behavior modeling, in the sense that model developers can follow the process defined by our model and design their own human behavior content for different applications (e.g., creating the experience cases for different behaviors of agents). Therefore, the model is intended to facilitate the work of a model developer for the development of her/his own virtual agents in different applications. In this paper, the cognitive theories related to these human factors are studied and the process model is designed accordingly. Specifically, we propose the detailed representation of experience case and investigate the mechanisms of situation assessment, experience matching and experience execution. The elicitation and effect of emotion are also incorporated. We will continue to work on the development of HumDPM by further specifying and refining the cognitive functions in each step of decision making.

References

1. Augoustinos, M., Walker, I.: Social Cognition: An Integrated Introduction. Sage Publications, Thousand Oaks (1995)
2. Clore, G.L., Gasper, K.: Feeling is believing: Some affective influences on belief. In: Frijda, N., Manstead, T., Bem, S. (eds.) Emotions and Beliefs: How Feelings Influence Thoughts, pp. 10–44 (2000)
3. Combs, W.E.: The Fuzzy Systems Handbook, 2nd edn. Academic Press, London (1999)
4. Dutta, P.K.: Strategies and games: theory and practice. MIT Press, Cambridge (1999)
5. Fan, X., Yen, J.: R-CAST: Integrating Team Intelligence for Human-Centered Teamwork. In: Proceedings of the 22nd National Conference on Artificial Intelligence, pp. 1535–1541 (2007)
6. Gratch, J., Marsella, S.: A domain-independent framework for modeling emotion. Cognitive Systems Research 5, 269–306 (2006)
7. Grimaldo, F., Lozano, M., Barber, F.: MADeM: a multi-modal decision making for social MAS. In: Proceedings of the 7th International Joint Conference on Autonomous Agents and Multiagent Systems, pp. 183–190 (2008)
8. Iglesias, A., Luengo, F.: AI Framework for Decision Modeling in Behavioral Animation of Virtual Avatars. In: Shi, Y., van Albada, G.D., Dongarra, J., Sloot, P.M.A. (eds.) ICCS 2007. LNCS, vol. 4488, pp. 89–96. Springer, Heidelberg (2007)
9. Ito, F.Y., Pynadath, D.V., Marsella, S.C.: Self-Deceptive Decision Making: Normative and Descriptive Insights. In: Proceedings of the 8th International Joint Conference on Autonomous Agents and Multiagent Systems, pp. 1113–1120 (2009)
10. Ji, Y., Massanari, R.M., Ager, J., Yen, J., Miller, R.E., Ying, H.: A fuzzy logic-based computational recognition-primed decision model. Information Sciences 177, 4338–4353 (2007)

11. Klein, G.: Sources of Power: How People Make Decisions. MIT Press, Massachusetts (1998)
12. Lazarus, R.S.: Emotion and Adaptation. Oxford University Press, USA (1991)
13. Luo, L.B., Zhou, S.P., Cai, W.T., Low, M.Y.H., Lees, M.: Toward A Generic Framework for Modeling Human-like Behaviors in Crowd Simulation. In: Proceedings of the 2009 IEEE/WIC/ACM International Conference on Intelligent Agent Technology, pp. 275–278 (2009)
14. Marsella, S., Gratch, J.: EMA: A process model of appraisal dynamics. Cognitive Systems Research 10, 70–90 (2009)
15. Neumann, J., Morgenstern, O.: Theory of Games and Economic Behavior. Princeton University Press, Princeton (1944)
16. Norling, E.: Folk Psychology for Human Modelling: Extending the BDI Paradigm. In: Proceedings of the Third International Joint Conference on Autonomous Agents and Multiagent Systems, pp. 202–209 (2004)
17. Pearl, J., Morgenstern, O.: Probabilistic Reasoning in Intelligent Systems: Networks of Plausible Inference, San Mateo, CA (1988)
18. Scherer, K.R.: Appraisal Considered as a Process of Multilevel Sequential Checking. In: Scherer, K.R., Schorr, A., Johnstone, T. (eds.) Appraisal Processes in Emotion: Theory, Methods, Research, pp. 92–120 (2001)
19. Sevin, E., Thalmann, D.: A Motivational Model of Action Selection for Virtual Humans. In: Proceedings of Computer Graphics International 2005 (2005)
20. Thalmann, D., Musse, S.R.: Crowd Simulation. Springer, Heidelberg (2007)
21. Yu, Q., Terzopoulos, D.: A Decision Network Framework for the Behavioral Animation of Virtual Humans. In: Proceedings of the 2007 ACM SIGGRAPH/Eurographics Symposium on Computer Animation, pp. 119–128 (2007)

Group-Agreement as a Reliability Measure for Witness Recommendations in Reputation-Based Trust Protocols

Sascha Hauke, Martin Pyka, and Dominik Heider

Sascha Hauke, CASED, Technical University Darmstadt, Germany
Sascha.Hauke@cased.de
Martin Pyka, Department of Psychiatry, University of Marburg, Germany
Martin.Pyka@uni-marburg.de
Dominik Heider, Center for Medical Biotechnology, University of Duisburg-Essen, Germany
Dominik.Heider@uni-due.de

Abstract. Interactions between individuals are inherently dependent upon trust, no matter if they occur in the real world or in cybercommunities. Over the past years, proposals have been made to model trust relations computationally, either to assist users or for modeling purposes in multi-agent systems. These models rely implicitly on the social networks established by participating entities (be they autonomous agents or internet users). However, state-of-the-art trust frameworks often neglect the structure of those complex networks. In this paper, we present a new approach allowing agent-based trust frameworks to leverage information from both trusted and untrusted witnesses that would otherwise be neglected. An effective and robust voting scheme based on an agreement metric is presented and its benefit is shown through simulations.

Keywords: reputation, trust, social networks, voting-derived reliability metric.

1 Introduction

This paper represents part of our work directed at developing a distributed recommendation system of (semi-) autonomous agents, aiding users in determining trustworthy service partners. We envision this system to operate on existing social structures, as, for instance, computationally represented in online social communities. Leveraging reputation-based computational trust and real-world derived social connections as a soft security mechanism, we aim at increasing overall (system) reliability in computer-mediated human interactions within cybercommunities.

Trust is an important and frequently studied concept in personal interactions and business ventures. As such, it has been examined by a multitude of scientists in diverse disciplines of study. Modern, state-of-the-art trust metrics for agent

M.L. Gavrilova et al. (Eds.): Trans. on Comput. Sci. XII, LNCS 6670, pp. 231–255, 2011.

systems [26,29,43] seek to apply the social concept of trust to computational contexts. Typically, these protocols employ a data-driven computational approach in the trust formation process, as opposed to a belief-driven cognitive methodology proposed by Castelfranchi and Falcone [7,8]. Due to the complexity of accurately representing an entity's mental state in computational adaptations, computational realizations of cognitive trust are difficult to model and thus rarely implemented.

Rather than being based on hard-to-model internal states, the computational (also: probabilistic) view of trust relies on observable data for deriving an – albeit subjective [17] – probability with which some entity will perform a particular action (this corresponds to the *level of trust* one entity puts in another to act in a particular way). By employing observed information from the past to predict behavior in the future, trust establishment becomes a data driven process. It is, thus, well-suited to computational modeling and will be the approach of choice used in this paper.

In society, *reliable* and *observable* information necessary for the derivation of a level of trust is usually procured in two different ways: Either through personal experience gathered from prior direct interactions with another entity (so-called *direct experience*) or via the reputation of an entity as reported by other members of society (so-called *witness information*). Through the provisioning of witness information in the trust formation process, knowledge about the behavior of entities is diffused through society. This represents a social security mechanism, providing *soft security* [40] to the members of a diffusing community.

2 Related Work

The communication of an entity's reputation through witness information occurs in the form of recommendations, involving three distinct parties: An agent/entity, whose reputation is to be established, that is the subject of the recommendation (the recommendee), another agent that communicates its opinion (the recommender) to a third that requires the information (the recipient).

The relationship between these three agents in the recommendation process, particularly that of recommender and recipient, is typically non-random. Expressed generally, recommender and recipient are required to be sufficiently close to each other, as per some metric.

Adjacency of recommender and recipient establishes not only the possibility of direct recommendation message exchange between these two agents, but also enables the recipient to assess the reliability of a recommender. Based on prior recommendation performance by the recommender, the recipient can estimate the expected reliability of a given recommendation, and weight it accordingly in the trust establishment process with the recommendee. This mechanism forms the foundation for the processing of witness information in state-of-the-art reputation-based trust frameworks (e.g. [26]).

As trust is an inherently social phenomenon, involving social entities and their interactions, an appropriate metric is given by the 1-neighborhoods of the social network formed by participating agents. If recommender and recipient are direct neighbors, witness information can be passed directly. Current frameworks, however, neglect the fact that the structure and dynamics of social networks are of a complex nature. Huynh [26], for instance, considers trust establishment among agents in an open multi-agent environment where vicinity of agents, and thus their ability to exchange recommendations, is determined by a 3-dimensional position and mobility system. This clearly reduces the dimensionality of inter-agent relationships in social networks, which may exhibit more complicated patterns, such as small world effects [47].

Reputation-based trust diffusion through social networks of agents has been the focus of our work in [23], while the dissemination of recommendations from non-neighboring recommenders has been discussed in [22,24]. However, one particular feature of online social networks and communities still remains an untapped source of reputation information on potential interactors: the high prevalence of weakly linked neighbors. Typically, trust frameworks rely on past experience made by the recipient with a particular recommender in order to assess the reliability of a given recommendation. This procedure requires sufficient and reasonably current prior interactions between the recommender and the recipient, so that the recipient can determine a reliability score for that recommender.

While the backbone of human relationship networks are in fact formed by those individuals with whom frequent interactions occur, the number of these *strong ties* in social networks are generally bounded. By far the larger number of links in (online) social networks are constituted by those individuals that are acquaintances, or *weak ties* [21], rather than friends. Dunbar [11] has posited, that the number of acquaintances that are actively maintained to some degree by an individual is around 150. Sociological studies [4,31,34] have shown that of these only a small number are in fact strong ties. Estimates regarding the size of strongly linked core networks go as low as 2 [34] or 3 [31] strong ties on average (for social discussion networks), although there is evidence for a methodologically induced underestimation of the core network size in these studies [4,6]. Nonetheless, social network data from *www.facebook.com* [6] supports the existence of strongly linked networks forming the core of larger social acquaintance networks, in which most direct neighbors are only weakly linked. Furthermore, Granovetter [21] argues that the main facilitators of information spread in social networks are weak ties. This phenomenon results from the highly clustered nature of the acquaintance network formed solely from strong ties, which, through its very structure, inhibits the flow of information.

Online social communities, such as Facebook, are primarily employed by their users to connect with already existing offline social contacts rather than meeting new people [12]; as such, they are a reasonable representation of human acquaintance networks. These same networks are the foundation of recommendation networks, helping to form opinions on potential future interactors. For computational reputation-based trust frameworks, online social communities thus form an adequate analog basis for real-world reputation spreading.

While several reputation-based trust protocols [18,37,41,44] have explicitly extended trust formation to take into account social networks, their work in this regard has been focused mainly on mechanisms for discovering information sources. Some fundamental work has been done by Golbeck and Kuter [20], yet a closer investigation of the impact of social network structure is relegated to future work. The very structure of human social networks and its impact on trust formation still warrants further investigation. Highly clustered network structures consisting of strong tie relationships between participants are, in fact, results of real-world causes and formation processes. As such, network structure can be considered an image of a particular (although unknown) input parameter set, implicitly containing aggregated information of actual relationships. Aside from high trust implicit in strong ties (for instance enforced by social, cultural or legal obligations), a high similarity among neighbors in human social networks can be observed [35]. Because high similarity between entities is a strong indicator for shared opinions, and thus for appropriate and high quality recommendations among these similar entities, neighbor relations and clique formation may yield information on the quality of a recommendation, without having to resort to mechanisms such as profile matching [19,45] which might not be applicable due to privacy concerns.

The impact of network structure on the proliferation of information relating an entity's reputation, and consequently on the formation of trust and extending of cooperation to a specific service partner can be gauged from figure 1. Following [23], figures 1 (a), (c) and (e) show the number of entities that have adopted a newly introduced provider, which supplies, on average, a slightly better service (+0.2 better) than the next best provider. Entities are embedded in different networks, exhibiting three different types of structure: scale free, random (respectively generated according to the methods and parameters in [27]), and highly clustered (generated according to [28]). Using only direct experience and witness trust modules, a difference in diffusive qualities between scale free and random topologies, on the one hand, and highly clustered topologies, on the other hand, is evident. Both the speed of adoption and the steady-state final number of adopters (out of 1000) is considerably lower for clustered topologies (mean ≈ 470, range $340 - 560$) than for the other two (random: mean ≈ 520, range $440 - 570$; random: mean ≈ 540, range $450 - 600$).

Contrary to the tendencies of group and community formation observed in (real-world) social networks consisting of strong ties – which may exhibit a small-world phenomenon[36] – many online social communities [14,15,51] appear to follow a – 'diffusion-friendly' – scale-free model [53]. In scale-free networks, the vast majority of nodes have but a few links to other nodes, i.e. a low node degree, while a small number of nodes have a large number of links, i.e. a very high node degree, that can be described by a node degree distribution following a power law, e.g. Zipf's law [54]. In social interactions, such a network structure arises from continuous growth combined with *preferential attachment* to popular entities [3]. At first glance, the discrepancy between highly clustered small world networks observed, for example, by [4,31,34] and scale-free

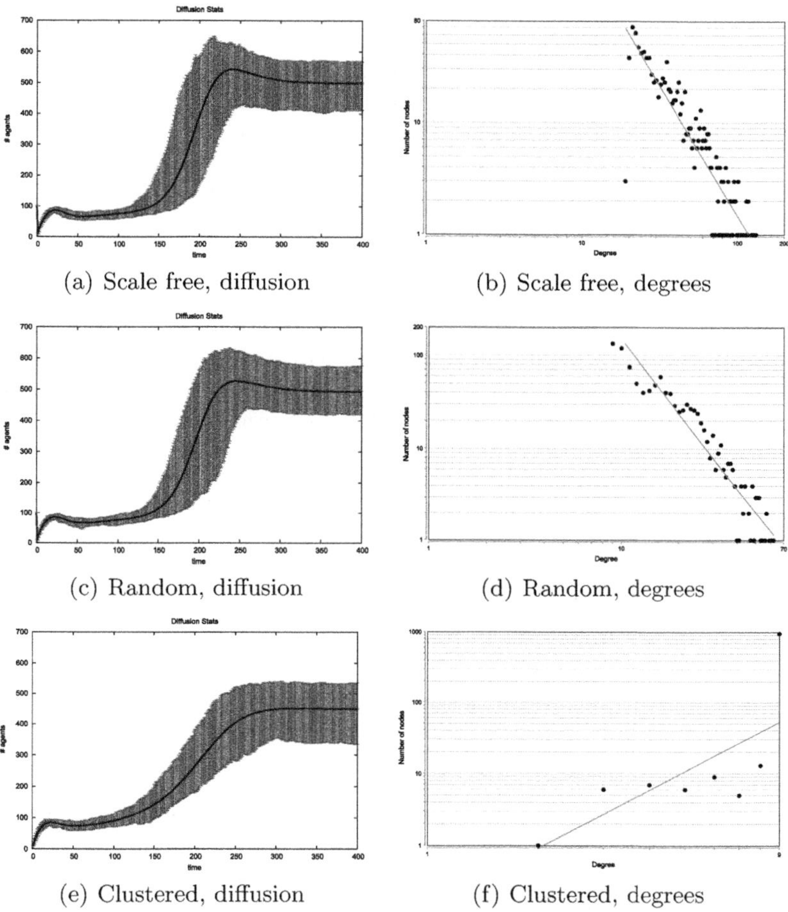

(a) Scale free, diffusion

(b) Scale free, degrees

(c) Random, diffusion

(d) Random, degrees

(e) Clustered, diffusion

(f) Clustered, degrees

Fig. 1. Diffusion characteristics (a), (c) and (e) of networks with different degree distributions (b), (d) and (f)

networks – potentially even devoid of clustering – in online cyber-communities may appear to contradict the hypothesis that online social communities might be an image of real-world acquaintance networks. However, this is easily reconciled when considering the nature of the majority of links present in online social communities. Most ties in online social networks are in fact weak ties. Consider, for example, *www.facebook.com*: Although the contacts administered are called 'Friends', a considerable majority of these are not friends but rather acquaintances ('Facebook-Friends'). As observed by [21], weak ties are particularly important in social information dissemination. It can be construed that the reason for this is twofold: On the one hand, acquaintances outnumber friends by a large margin, thus making for a bigger number of potential information sources and propagators. On the other hand, the way in which acquaintances are arrayed within the network, including central hubs serving as multiplicators of information, contributes to a more efficient dissemination process.

Given that online social networks exhibit many of the same structural features of their real-life counterparts, standard trust establishment procedures present in state-of-the-art trust frameworks for agent systems [29,43] are inadequate for leveraging reputation information supplied by those neighbors with whom only weak ties are maintained. This information is generally forfeited due to the recipient's inability to determine a meaningful reliability score for the recommender, as this score is based on prior interactions between recipient and recommender. Due to the lack of frequent interactions between weakly linked neighbors, a recommender in such a relationship is assigned a reliability score that tends to 0. While dedicated bootstrapping, i.e. applying learning rules to estimate reliability over time, can alleviate this problem, it is a time-consuming process that still discounts a significant portion of the available reputation information during the bootstrapping phase.

Furthermore, when assessing the trustworthiness of a potential interactor, e.g. for provider selection in the future internet of services [39], reputation information may be only scarcely available over an entity's social network. In such a case, consulting and integrating other sources of information, such as ratings from dedicated rating portals, can compensate that lack of information. Since the raters in such a scenario are often (pseudo-)anonymous, bootstrapping them is impractical.

3 Approach

Considering that reputation information is typically thought to be a scarce resource that justifies the application of sophisticated recommendation gathering mechanism, such as the establishment of multi-hop recommendation chains [26,38] or pro-active recommendation and recommender forwarding [22,24], leveraging weak ties for trust computation promises a useful information gain. For this, we propose augmenting reputation-based trust frameworks with a specific component designed to harness the opinions of weakly-linked direct neighbors of a particular entity, by leveraging group agreement as an approximate reliability measure in section 3.2, the scientific basis of which goes back as far as [16]. This approach can also be easily extended to leverage ratings from anonymous raters, as gathered, for instance, from dedicated rating sites or internet portals.

Furthermore, building upon the reliability measure introduced in 3.2 and taking into account the effect the network topology has on reputation diffusion (cf. figure 1), subsection 3.3 will briefly present a mechanism for directed change of the topology in highly clustered acquaintance networks used for recommending.

3.1 Implemented Notions of Trust and Reputation

In the related literature, a clear consensus on and definition of trust is yet to be reached. As the overall goal of our work is in supporting decision making, particularly in the future internet [39], an exhaustive overview of the different aspects of trust is clearly beyond the scope of this chapter. In the following, our operational definition of trust will follow [38]. Thus, trust is *a subjective expectation an agent has about another's future behavior.* While this view of trust abstracts the finer, nonetheless academically important aspects of trust, in our opinion it represents an adequate facet of trust in the context of interactor grading and selection. In particular, the calculated subjective expectation should, ideally, approximate the interactor's trustworthiness[1] in a given context/situation.

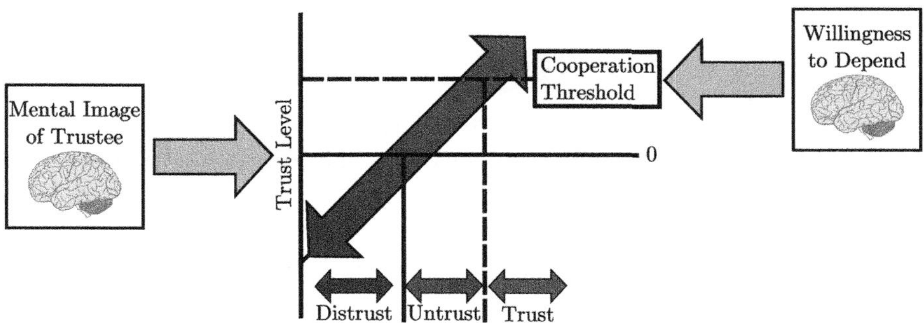

Fig. 2. Trust, Untrust and Distrust (cf. [33])

Trust, as an expectation of future behavior, will in the following be projected on the interval $[-1, 1]$, as in [26,32,33]. This mapping can be considered as the degree of some abstract *utility* [26] that an entity (the *truster*) expects to gain from an interaction with another (the *trustee*). As this expectation is only an approximation of the yet unknown future behavior of the trustee, it is determined based on a subjective image the truster has of the trustee. A trust level in $[-1, 0[$ indicates distrust, i.e. the expectation of 'losing' utility, while a trust level in $[0, 1]$ indicates the expectation of 'gaining' utility from a potential interaction. Generally, the truster requires a minimum level of trust in order to enter in an interaction with the trustee. If the trust level determined for a particular interactor exceeds this *cooperation threshold*, the truster is willing to make a positive *trust decision*[2]. For reference, cf. figure 2.

Commonly, the foundation for computing trust in another entity is the trustee's *reputation* with the truster. Reputation, according to [1], is the *expectation about an agent's behavior based on information about or observations of its past behavior.* Thus, it is based upon the recorded history of performance of the trustee in relevant past interactions that is available to the truster. These past interactions can include those

[1] While trust is a dyadic, asymmetric, only conditionally transitive relation [1], an entity's trustworthiness is an inherent property of that entity [48].

[2] In case there are multiple possible interactors, the truster will likely choose the trustee with the highest trust level. Therefore, willingness to make a positive trust decision does not necessarily lead to the truster initiating an interaction with trustees deemed suboptimal.

- that the truster itself participated in and 'remembers',
- interactions between the trustee and a third entity that the truster observed or
- interactions between the trustee and a third party that were reported to the truster by a recommender/witness.

Thus, information for reputation-based trust formation is supplied either by the truster itself or by (several) other entities. Aside from different sources of information used in its computation, trust is also both situation and context dependent [32]. That means it is both influenced by the current environment of the involved entities (their situation) and the specific planned interaction (its context). Furthermore, a context may be a composition of several sub-contexts. Therefore, establishing a reputation-based trust level for potential interactors (and selecting the best interactor) can be considered a multicriterial decision problem.

3.2 Majority-Based Reliability Assessment of Reputation

Establishing a Reliability Score. Let a be a recipient, b a recommendee and $r_1, ..., r_m \in \mathcal{R}^a$ a population of recommenders that are direct neighbors of a. $\mathcal{R}^a$ is a subset of all direct neighbors $n_1, ..., n_n \in \mathcal{N}^a$ of a, specifically those that have had prior interactions with b and can thus provide a recommendation $Rec_{r_i,b}$. We also stipulate that the recipient should not actively dis-trust any member of $\mathcal{R}^a$. Furthermore, let $Rec_{r_i,b} \in [-1, 1]$, with -1 signifying total active dis-trust of recommender r_i in b, and 1 total active trust.

The overall reliability score of the aggregate recommendation from recommenders connected via weak ties is computed based on the categorial divergence of recommendations, i.e. the agreement among the recommenders. For multi-dimensional recommendations (as would be the real-world norm, considering that trust is situation dependent [32,2]), we propose the use of Krippendorff's α-coefficient . It represents a standard reliability measure that can be used regardless of the number of observers, levels of measurement, sample sizes, and presence or absence of missing data [25]. For applying the measure to the received recommendations, each distinct trust dimension contained in the set of recommendations constitutes a unit of analysis in Krippendorff's model.

However, due to computationally constrained simulation environments, reputation gathering is often abstracted to one-dimensional opinions on a particular recommendee instead of multi-dimensional recommendation vectors. In that case, no expected disagreement between raters can be established from the single-unit empirical evidence. Thus, the reliability measure has to be computed in a different manner. If multiple units are not provided through multi-dimensional recommendations, a sufficient data base may be created by regarding multiple recommendees. In fact, this mirrors the standard procedure in determining agreement coefficients, for example in clinical studies. However, this requires either the storage of previously received series of recommendations or the concurrent querying for multiple recommendations on several recommendees.

If no such information can be procured either, the expected disagreement among recommenders is quite difficult to assess from the limited empirical data. For one-dimensional recommendations, we will therefore assume a uniform random process as the basis for determining the expected disagreement.

Each recommendation thus contains a number of trust values, one per dimension of trust, which are categorized in the following way: positive values, $\mathcal{R}^a_{pos}$, if $Rec_{r_i,b} \in [0.2, 1]$, neutral values, $\mathcal{R}^a_{neut}$, if $Rec_{r_i,b} \in]-0.2, 0.2[$ and negative values, $\mathcal{R}^a_{neg}$, if $Rec_{r_i,b} \in [-1, -0.2]$. An agreement coefficient among recommenders is established by computing:

$$agr(\mathcal{R}^a, b) = 1 - \frac{D_o}{D_e} \quad (1)$$

with D_o the observed disagreement and D_e the expected disagreement among recommenders. If sufficient multi-unit empirical data can be obtained, these values are computed according to [25], otherwise we will compute

$$D_o = |\mathcal{R}^a_{pos}| \cdot |\mathcal{R}^a_{neg}| \quad (2)$$

and

$$D_e = \left\lfloor \frac{|\mathcal{R}^a_{pos}| + |\mathcal{R}^a_{neg}|}{2} \right\rfloor \cdot \left\lceil \frac{|\mathcal{R}^a_{pos}| + |\mathcal{R}^a_{neg}|}{2} \right\rceil \quad (3)$$

If using Krippendorff's α, as this measure can be negative, we compute the total agreement of all recommendations of recommenders $r_1, ..., r_m$ on recommendee b in the following way:

$$Agr_{\mathcal{R}^a,b} = max(0, agr(\mathcal{R}^a, b)) \quad (4)$$

The resulting agreement score is independent of the total number of recommendations included in its calculation. In order to scale the final score according to the total number of informative, i.e. negative or positive, recommendations $N = |R^a_{pos} \cup R^a_{neg}|$, we multiply α by a monotonously growing scaling function $f : \mathbb{N} \mapsto [0, 1]$. This accounts for an agent's need to require a certain number of recommendations on which the reliability measure is based. While this is a somewhat ad hoc formulation of an agent's need to derive a particular certainty in dependence on the amount of evidence, it is an easily implemented heuristic. A more formal model for deriving certainty is presented in [46].

Thus, the overall reliability of recommendations included in the new trust component for weakly-linked neighbors in regard to a particular recommendee b is computed as:

$$Rel(\mathcal{R}^a, b) = f(N) \cdot Agr_{\mathcal{R}^a,b} \quad (5)$$

Computing the Expected Performance Value. While this provides an agreement-based reliability value, an actual value for the aggregate recommendation that the neighborhood of a jointly attributes to b, has yet to be determined. This aggregate recommendation serves as a representation of the expected performance of b as estimated from empirical data (provided through individual recommendations by members of $\mathcal{R}^a$). A straighforward implementation of such an estimation is used in the following – the average recommendation $Rec^{avg}_{\mathcal{R}^a,b}$ of all directly neighboring recommenders of a.

$$Rec^{avg}_{\mathcal{R}^a,b} = \frac{1}{m} \sum_{r_i=1}^{m} Rec_{r_i,b} \tag{6}$$

In order to induce a bias towards the categorial majority of recommenders, i.e. positive versus negative recommendations, the final recommendation value is adjusted by limiting it in the following way:

$$Rec_{\mathcal{R}^a,b} = \begin{cases} min(0, Rec^{avg}_{\mathcal{R}^a,b}) & \text{if } |\mathcal{R}^a_{neg}| > |\mathcal{R}^a_{pos}| \\ max(0, Rec^{avg}_{\mathcal{R}^a,b}) & \text{else} \end{cases} \tag{7}$$

3.3 Recommender Reliability Driven Topology Change

In order to make acquaintance networks more permeable to trust information, we pursue an alteration in network structure in regards to clustering, network diameter and node degree distribution. This alteration is achieved by selecting particularly reliable nodes, spreading the knowledge about these nodes through the network and allowing attachment of other members for recommendation gathering. Network evolution is driven by agent/node-level rules, addressing nomination, propagation and adoption of recommenders. Figures 3(a) and 3(b) show the impact on a highly clustered, small (50 nodes) sample network of the mechanism presented in the following. The increase in shortcut connections within the network is clearly visible.

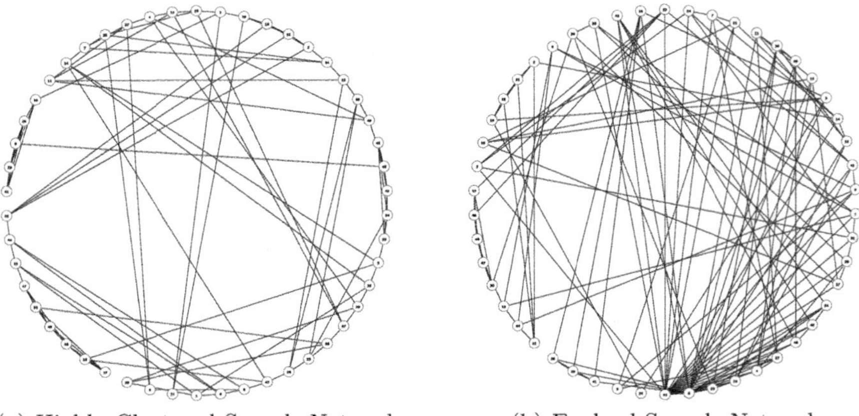

(a) Highly Clustered Sample Network (b) Evolved Sample Network

Fig. 3. Effects of the recommender propagation on a 50 node sample acquaintance network

Nomination. The initial and most critical task of the proposed mechanism lies in identifying reliable candidate agents. The nomination process involves two agents: a *nominee* and a *nominator*, which are members of each other's 0-hop neighborhood. An agent chooses whether to nominate one of its neighbors based upon a number of parameters:

- the neighbor's level of activity
- the quality of the recommendations given by the neighbor
- the number of prior recommendations given by the neighbor

Assessing a neighbor's level of activity. In a distributed system, determining another agent's level of activity, measured in interactions over a given period of time, is dependent on observations available to the assessing agent. Because the assessing agent (the potential nominator) is a direct neighbor of the potential nominee, it can heuristically derive the level of activity in the following manner: Typically, whenever an agent wishes to interact with another entity, it initiates a reputation gathering process. This process entails querying its 0-hop neighbors for recommendations, permitting these neighbors to estimate the agent's level of activity as $Act := c \cdot \frac{Q}{\Delta t}$. Q represents the number of queries for recommendations received from the agent, Δt a time interval, and c a scaling constant.

Determining recommender reliability. In order to determine the nominee's recommending reliability, the nominator, as a direct neighbor, can rely on a history of direct recommendations received from the nominee in the past. A recommender's reliability is directly dependent on the accuracy of individual recommendations it has issued. The higher the average accuracy of its recommendations, the higher the recommender's reliability. Recommendation accuracy is measured as the level of congruence between the recommendation and the interaction experience made by an agent following the recommendation. Prototypically, this can be expressed as the absolute difference between the value of the recommendation and the value resulting from the interaction. Procedures for computing recommender reliability are a staple of state-of-the-art trust models[3].

Minimum number of recommendations. Beyond the reliability value of a recommender, a nominator should also consider the absolute number of recommendations from the pertinent recommender for guaranteeing good recommending performance over several interactions. For the nomination process, we thus demand a minimum number of recommendations from which a recommender's reliability is computed.

Based on the aforementioned knowledge, an agent nominates one of its neighbors for recommender propagation if the neighbor's average activity and reliability (as computed from a minimum number of prior recommendations) exceed pre-established thresholds $T_{Activity}$ and $T_{Reliability}$. If a neighbor meets these criteria, the nominator constructs a nomination message containing the identity of the nominee and a signed token, marking the nominator's agreement. Finally,

[3] Huynh [26], for instance, proposes a simple way of computing witness reliability based on recommendation accuracy for his FIRE trust model.

the nominator forwards the nomination message to the nominee. The nominee, upon receiving a nomination message, either accepts or declines the nomination. If the nomination is declined[4], message propagation ceases. In case of acceptance, the nominator includes another signed token in the nomination message (signaling its acceptance) and concludes the nomination phase by forwarding the message to its 0-hop neighborhood.

Propagation and Adoption. For propagating nomination messages through the network formed by agent relations, we have adapted active rumor-spreading mechanisms [5,9] to reputation-based trust dissemination. The protocol for remote recommendation propagation introduced in [23], forms the foundation for message distribution, albeit with a changed subject matter and extended to permit the structural/topological evolution of the underlying network of agents.

The general message exchange between two agents, a sender s and a receiver r, regarding a nominee n is as follows: Before sending a nomination message $m_{nom}(s,r,n)$ to its neighbors, s adds a signed voting token to the message, indicating whether it agrees, disagrees or has no definite opinion on n. Upon receiving a nomination message from s, r sends a reply message $m_{nom}^{reply}(s,r,n)$ to s. This reply includes information on whether or not r has previously received nomination messages for n, and, if r is privy to such previous messages, also a composite of these previous messages.

Message evaluation by sender. The sender s compares the reply message $m_{nom}^{reply}(s,r,n)$ to the one it originally sent, $m_{nom}(s,r,n)$. This is done by first compositing the messages, i.e. creating the union of all voting tokens included in both $m_{nom}(s,r,n)$ and $m_{nom}^{reply}(s,r,n)$, computing agreement scores for $m_{nom}(s,r,n)$ and $m_{nom}^{reply}(s,r,n)$, and finally calculating the absolute difference of the agreement scores.

In order to assess agreement, we propose the same approach as outlined through equations 4 and 5, using the sets of agreeing and disagreeing voting tokens as input.

$$Agm(m_{nom}(s,r,n)) = \begin{cases} f(N)\cdot\alpha & \text{if \# agreeing} > \text{\# disagreeing} \\ -1\cdot f(N)\cdot\alpha & \text{else} \end{cases} \tag{8}$$

$$Agm(m_{nom}^{s\cup r}(s,r,n)) := Agm((m_{nom}(s,r,n) \cup m_{nom}^{reply}(s,r,n))) \tag{9}$$

$$\Delta_{s,r}Agm(m_{nom}(s,r,n)) = Agm(m_{nom}^{s\cup r}(s,r,n)) - Agm(m_{nom}(s,r,n)) \tag{10}$$

If $|\Delta_{s,r}Agm(m_{nom}(s,r,n))|$ is smaller than some threshold parameter $T_{\Delta Agm}$ (meaning that the data provided by s to r contains little to no new information on the reliability of nominee n from the point of view of receiver r) s will propagating updates of $m_{nom}(s,r,n)$ to r. This is important for stifling message propagation through the network, so as not to create an overwhelming amount of network traffic.

[4] Reasons for declining a nomination may, for instance, be bandwidth limitations or privacy concerns of the nominee.

Message evaluation by receiver. Nomination message handling by receiver r is primarily dependent on the relationship between r and nominee n. The criteria for adoption and propagation of a nominee as a future recommender differ according to whether r and n are direct neighbors or not.

0-hop neighbors. Direct neighbors of nominee n serve as the initial filtering instance for nomination messages regarding n. They have prior experiences regarding the recommendation quality of n. Thus, their decisions on propagating a nomination message are not based on the agreement score of a particular message, but on prior direct interactions with n.

A direct neighbor x of n propagates a nomination message on n under the following conditions: Either its opinion $Op(x, n)$ of n exceeds a predetermined threshold $T_{Op(x,n)}$, or the nomination contains a minimum number V of agreeing voting tokens issued by joint neighbors[5] of x and n and the opinion of x on n exceeds another, albeit lower, threshold $T^V_{Op(x,n)}$.

If one of these criteria is met, x adds a voting token signaling agreement and propagates the message to its own 0-hop neighborhood, thereby becoming a sender. If $T^V_{Op(x,n)} < Op(x, n) < T_{Op(x,n)}$ and less than V voting tokens are included, or if more than V agreeing voting tokens by neighbors are included but $0 < Op(x, n) < T^V_{Op(x,n)}$, x adds a neutral voting token and propagates the message. Otherwise, x attaches a disagreeing voting token to the nomination messages and stifles, i.e. does not propagate, the message.

Beyond the 0-hop neighborhood. Agents that are not direct neighbors of the nominee n cannot rely on prior experiences to establish the reliability of n. Instead, the level of reliability is determined from the level of agreement (cf. equation 4) computed for the nomination message $m_{nom}(s, r, n)$.

If the level of agreement exceeds threshold $T^{direct}_{Adoption}$, the agent adopts the nominee as a regular recommender, practically becoming a member of the nominee's 0-hop recommendation neighborhood. If agreement remains below $T^{direct}_{Adoption}$, but meets another, lower threshold $T^{probationary}_{Adoption}$, the agent adopts n as a recommender, but does not include its recommendations in trust computation. From this practice, the agent over time gains information regarding the recommendation reliability of n.

If, after a preset period of time, recommendation performance is satisfactory, e.g. higher than $T^{direct}_{Adoption}$, n is adopted/maintained as a regular recommender, an agreeing message token is added to $m_{nom}(s, r, n)$ and the message is propagated. Otherwise, the nominee is removed as a recommender, a disagreeing message token is added and the message is stifled.

3.4 Integration with Existing Trust Frameworks

In order to test the effectiveness of the proposed mechanism for changing network structure has to be adapted to a reputation-based trust protocol. Also, after

[5] Due to high clustering prevalent in acquaintance networks, there is a high probability of this occurring.

computing both an aggregate recommendation value and a reliability score, the information thus garnered from weakly-linked neighbors through the approach outlined in section 3.2 has yet to be integrated with the standard components of reputation-based trust frameworks. Most trust models, e.g. [26,38,42,52], already provide for compositing mechanisms when dealing with different types of reputation information, such as direct experience, witness reputation or role-based trust. Prototypically choosing the FIRE trust model [26], reputation information information from weakly-linked neighbors can be easily integrated into the trust model. FIRE relies on a generic trust formula to calculate an aggregate trust value from its components:

$$\mathcal{T}(a,b,c) = \frac{\sum_{K \in \{I,R,W,C\}} \omega_K \cdot \mathcal{T}_K(a,b,c)}{\sum_{K \in \{I,R,W,C\}} \omega_K} \tag{11}$$

According to the formula, an overall trust value is thus calculated as the sum of all available component ratings weighted by the rating relevance and normalized to $[-1, 1]$. Adaptation of weak-ties information is achieved as follows: $\mathcal{T}_K$ for the weak ties trust component is set to $Rec_{\mathcal{R}^a,b}$. ω_K for the component is the product of $Rel(\mathcal{R}^a, b)$ and a FIRE-specific component weight W.

This particular framework has been chosen for a number of reason. It is an approach to trust formation that allows for the integration of different components and is therefore conceptually modular. Including new components is easily achieved by assigning a component weight, permitting comparisons to base performance values established and verified in [26]. Furthermore, it integrates fundamental methods in both the direct experience-based trust module as well as the witness trust module, for instance the Regret reputation model [42]. It thus serves as an adequate stepping stone to build upon prior scientific findings. Additionally, it is representative of our view of trust formation as a multicriterial decision problem, the scaling of the trust value in a continuous interval (in its case in $[1-, 1]$) and its interpretation based upon an aggregation of degrees of fulfillment. Nonetheless, the results can also be transferred and represented in other trust models, based on differing approaches, e.g. CertainTrust and its accompanying representational extension HTI [41].

Once the computational mechanism is in place, the integration of information from weakly-linked recommenders into a simulation framework is rather straightforward. In the following section, we will show the benefit of augmenting a trust model with such a mechanism by presenting simulation results.

4 Simulation

In order to assess the value of adding recommendations from weakly-linked neighbors to trust models, we will evaluate the potential average gain an entity can expect when using a trust model with such a component in place (FIRE w/ Weak Ties) versus a non-augmented trust model (FIRE w/o Weak Ties). For this, we will generally follow the methodology put forth in [26], with some changes to the

Table 1. Parameter instantiations for the recommender propagation mechanism

Parameter	Value	Parameter	Value	Parameter	Value	Parameter	Value
$T_{Activity}$	0.75	$T_{Reliability}$	0.9	$T_{Op(x,n)}$	0.67	$T^{V}_{Op(x,n)}$	0.5
$T_{\Delta Agm}$	0.15	V	3	$T^{direct}_{Adoption}$	0.8	$T^{probationary}_{Adoption}$	0.25

scenario in order to better fit assumptions, for instance regarding trust dissemination [23]. In regard to simulating the effectiveness of the proposed approach for modifying network topology, an ad hoc instantiation of the required parameters is supplied in table 1. No parameter structured parameter optimization has been conducted, as the simulation results are to highlight merely the approach's power to change the network, which can still be considered work-in-progress.

An agent population consisting of consumer and provider agents is seeded in a spherical environment, as in [26]. However, differing from the FIRE standard test methodology, the placement of consumer agents is influenced by an underlying complex social network [23], either a random graph [13,27] or a highly clustered acquaintance network [28]. Long-range connections between consumer agents that could not be placed together spatially are maintained in order to simulate the small-world effects of social networks [47]. During the simulation, recommendations and published opinions are communicated harnessing the underlying social network. Provider discovery and service delivery is handled in accordance with the neighborhood-based search employed in [26]. Provider selection, from a set of discovered providers, is also handled in accordance with the proposal from the FIRE testbed, using a standard Boltzman exploration strategy [30] in order to address the exploration-vs-exploitation dilemma. Without loss of generality, we assume one-dimensional trust vectors, applying equation 3 in order to determine expected disagreement. A closer study of multi-dimensional trust vectors, although more realistic, is relegated to future work, as this would also necessitate the development of a taxonomy/ontology that adequately represents the interdependencies among different dimensions of the trust vector, as well as a robust way of aggregating sub-categories in the face of scarce and incomplete data. This would also require a closer investigation of the related work in multi-criterial decision theory, which is beyond the scope of this paper; e.g. cf. [50].

Additionally, the testing methodology for assessing the advantage of agents equipped with an augmented trust model over those without additional weak ties augmentation has largely been adopted from [26], as well. After selecting and interacting with a provider, the consumer gains or loses utility, dependent on the performance of the provider. Regarding this performance, the provider population is divided into three distinct sub-populations of consistent providers (*good*, *ordinary* and *bad*), as well as one of *intermittent* providers. Actual performance of providers is represented by a random variable, computed according to the sub-population a provider belongs to (for *bad*: $\mu \in [-10, 0], \sigma = 2$, for *ordinary*: $\mu \in [0, 5], \sigma = 2$, for *good*: $\mu \in [5, 10], \sigma = 1$, according to a normal distribution; for *intermittent*: uniformly distributed in $[-10, 10]$). A provider's expected performance μ is set at creation, its actual performance is determined per interaction. Time is measured in rounds, with events taking place during the same round occurring simultaneously. During every round, each agent chooses probabilistically, according to an individual activity level, whether or not it interacts with a provider. The utility score of every interaction is recorded by each agent, in order to assess the average utility gained or lost during each round. Default experimental variable values were retained from the standard FIRE testbed. This includes: number of simulation rounds $N = 500$, number of provider agents $N_P = 100$ (subdivided into good $N_{PG} = 10$, ordinary $N_{PO} = 40$, bad $N_{PB} = 45$ and intermittent $N_{PI} = 45$), number of consumer agents per test group $N_C = 500$. Further default parameters, such as component coefficients and reliability function parameters were also retained, with the exception of the referral length threshold n_{RL}, which was set to permit only 1-hop referrals. For simulations without the weak-ties component, the witness trust component (the FIRE component evaluating witness information from actively trusted [strong tie] neighbors) was given a weight of 0.5; simulations employing a weak-ties component, both the witness trust and weak-ties component were applied with a weight of 0.25 each. In both cases, the interaction trust component (the FIRE component harnessing direct experience) was weighed at 1. Thus, the balance between direct experience derived reputation information and witness derived reputation remains constant.

Mimicking the structure and associated dynamics of real-world acquaintance networks, the highly clustered core network, generated according to [28], exhibited a high average clustering coefficient [27] of $C = 0.42$, as well as positive assortativity. The maximum node degree, representing strong social ties (realized by setting default recommender trust for these strong ties to 0.5), was limited by the network generation procedure of the acquaintance network to 10, resulting in an average node degree of 9 across the network component consisting of 1000 nodes. For the random graph, showing a low clustering coefficient of $C = 0.02$, node degree also averaged to 9. In order to test the effectiveness of the proposed weak ties component in disseminating reputation information, additional links were inserted among the nodes comprising the core network. The number of weak ties introduced per node was probabilistically determined via a Gaussian random distribution with parameters $\mu = 45, \sigma = 20$. Each weak tie was attributed a neutral default recommender trust of 0, signifying neither positive nor negative trust.

5 Results

5.1 Majority-Based Reliability Assessment

In order to assess the benefit of augmenting a trust model with a component leveraging weak ties, we investigated the average utility received per agent and round for several groups of agents. Agents choosing providers according to direct experience derived reputation information only, without support of any witness information (either from strong or weak ties), form the baseline for comparison.

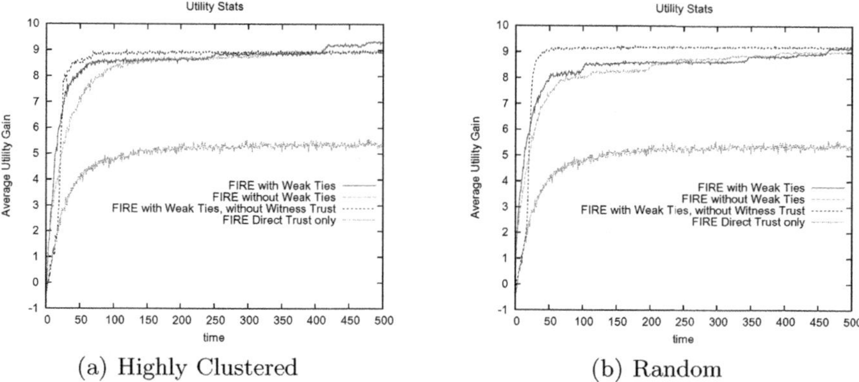

(a) Highly Clustered (b) Random

Fig. 4. Simulation Results, average utility of all interacting agents per time step

Against this baseline, two identical implementations of the FIRE trust model [26] were tested, one of these augmented with reputation communicated through weakly linked neighbors. Simulations were run under the testbed default parameters presented in [26].

In the tested scenario, consumer agents (represented as nodes in the underlying social graph) were tasked with choosing a particular provider agent. During each simulation timestep, interacting consumer agents were selected probabilistically according to an agent-specific activity threshold (specific to consumer agents, assigned at start of simulation according to a Gaussian distribution with $\mu = 0.2$, $\sigma = 0.1$). Thus, approximately 20 per cent of agents were actively engaging in interactions during each timestep – this measure was taken in order to emulate scarcity of information within the agents' immediate neighborhood. Each interaction resulted in the consumer agent receiving some level of payoff from the provider, ranging from -10 (very bad provider performance) to 10 (exceptionally good provider performance), according to the sub-population to which the chosen provider belongs (see section 4). This payoff thus represent the gain (or loss) of utility the consumer agent derives from entering into an interaction with a specific provider. Individual, agent-centric utility evaluation yields data about how well a single agent chooses his potential interactors, while a network wide averaging over all utility-scores gained by all interacting consumer agents during a single timestep permits the evaluation of both the speed of recommendation spreading and the quality of the information disseminated.

Agents equipped with a witness trust component consistently outperformed the direct experience only group by a considerable margin. The agents augmented with a weak ties component in turn outperformed the population equipped with a standard FIRE implementation by a slighter margin. Figure 4(a) shows the average utility gained by agents in a highly clustered acquaintance network. Performance advantages were significant (significance level $\alpha < 0.01$), as determined by a Wilcoxon signed-rank test [49].

While the stable steady-state average utility is higher for agents evaluating recommendations by weakly-linked neighbors versus those that do not overall, the initial growth phase in particular, i.e. the time required to reach the steady state, is markedly shorter. Especially in bootstrapping scenarios, when determining the trustworthiness of recommending neighbors, evaluating non-actively trusted recommendations is clearly beneficial over the exclusive baseline interaction experience evaluation. Remarkably, the utility gained by agents using direct experience and the weak ties component, without relying on the standard witness trust component, is inferior to the standard and the weak ties augmented standard FIRE model only in the initial phase. From around time step 20, FIRE with a weak ties but without a witness trust component consistently performs as good as or better than the competing implementations. Once a steady state utility level was achieved, the differently equipped FIRE implementations showed a trend to converge at a utility of ≈ 9. The general behavior has been observed in both highly clustered acquaintance networks (cf. figure 4(a)) and on random graphs (cf. figure 4(b)) that were evaluated for control purposes.

While this behavior highlights the power of weak ties in trust propagation, particularly by providing a broader data foundation for provider reliability evaluation, several social factors not simulated in the presented scenario still warrant the inclusion a witness trust component. As such, the simulation does not take into account a higher likelihood of strongly linked neighbors to report more accurate results. Such a higher accuracy may be attributed both to a higher motivation to report truthfully, i.e. give honest recommendations, and a higher similarity observed between strongly linked neighbors in social networks. The former is assumed to be caused by higher stakes in the recommendation process, reinforced, for instance, by pre-existing real world social/legal contracts or obligations. The latter can be considered to be due to phenomena such as homophily [35].

The effects observed in the utility simulations are congruent with findings revealed by investigating the diffusive capabilities of the different mechanisms. In the diffusion simulation [23], a new provider, performing slightly better than its competitors, was introduced to an agent population by making it, and its better expected performance, known to a fraction of the consuming agents. This simulation scenario mimics the spread of a pandemic through a population, inspired by standard epidemiological SIS (Susceptible-Infected-Susceptible) models, as adapted for information spreading [9]. While the recommendation mechanism used in this simulation is not explicitly modeled on statistical procedures discussed in [9], the underlying principles of dissemination are highly similar in their effects. Figure 5(a) shows the number of agents that choose the new provider over its competitors, clearly indicating the power of the proposed weak ties component to efficiently disseminate reputation information. Aside from displaying the increased diffusion power of the proposed mechanism, the figure also highlights how the community structure prevalent in social networks [35] – when considering only strong ties – inhibits the dissemination of information within the network. Diffusive qualities of the the acquaintance network can be seen to be generally inferior to those of the considerably lower clustered random graph

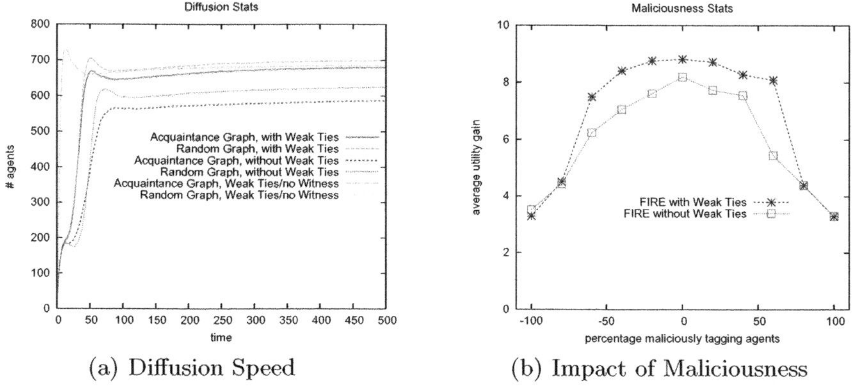

(a) Diffusion Speed (b) Impact of Maliciousness

Fig. 5. Simulation Results, (a) diffusion speed (10% seed agents, avg. activity 20%) and (b) robustness to malicious tagging (average utility of all interacting agents per time step, averaged over 200 time steps)

network generated for control purposes. This disadvantage has been largely mitigated by the proposed mechanism leveraging weak ties.

In addition to being effective in reputation information dissemination, the new proposed mechanism leveraging weak ties is also resilient to malicious tagging. This resilience can be attributed to the conservative reliability measure, based upon pairwise agreement computation, as well as its adaptability to the standard FIRE recommendation feedback mechanism, that effectively punishes unreliable recommenders and excludes them from the recommendation process. Figure 5(b) displays the response of the different implementations to malicious tagging, as per the malicious tagging procedures described in [26]. Specifically, average utility gain over 200 time steps within the recommender population is plotted against the percentage of maliciously tagging recommenders. Maliciously tagging agents either exaggerate a recommendee's reputation by adding a uniformly distributed random variable in $[0.1, 2]$ to the utility gain they expect from the recommendee, or diminish the recommendee's reputation by subtracting that random variable. A trust framework equipped with a weak ties component can be clearly seen to result in higher average utility for those agents using it, as compared to agents equipped with a non-augmented implementation of the same framework.

5.2 Recommender Reliability Driven Topology Change

The presented protocol is an effective tool for changing network topology. As shown in figures 6(a) and 6(b), the degree distribution of the base acquaintance network was significantly changed towards both a higher node degree overall and inclusion of hub-like nodes possessing a very high node degree. The average clustering coefficient [27,47] was decreased from $C = .42$ to $C = .08$ when including nodes attached through recommender nomination; for the sake of comparison, the average clustering coefficient for the random graph was $C = .02$.

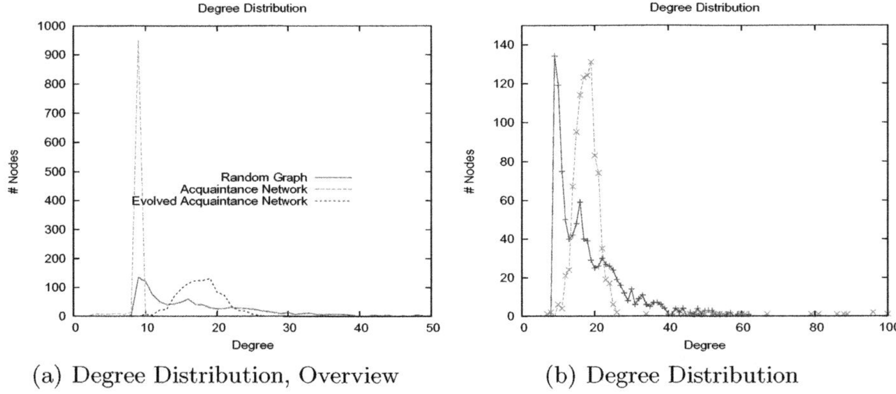

(a) Degree Distribution, Overview (b) Degree Distribution

Fig. 6. Degree distributions of the underlying networks of agents, (a) overview and (b) random graph vs. evolved acquaintance Network

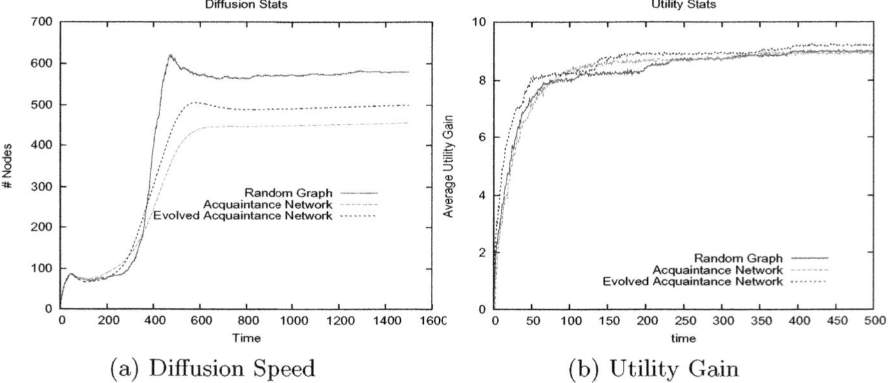

(a) Diffusion Speed (b) Utility Gain

Fig. 7. Effects of network topology on diffusion speed and utility gain

Furthermore, the network diameter (as defined by the maximum shortest path length found in the network) was reduced from 7 to 4, equivalent to that of the random graph, which also exhibited a diameter of 4. The results shown are representative and qualitatively reproducible for different initializations of the base acquaintance network.

The effect of the altered network topology in terms of the speed of information diffusion is depicted in figure 7(a). While dissemination in the random graph network is better than in the evolved acquaintance network, a significant (in the 99% confidence interval) improvement of the evolved over the base acquaintance network was observed. Diffusion performance of the evolved network is increased both in regards to the speed of information dissemination, as well as number of agents reached. A minor performance advantage of the highly clustered topology remains in the initial phase, between timesteps 100 and 300.

Results for the utility simulations, as depicted in figure 7(b), reveal performance advantages of the evolved network, ostensibly due to the strategic placements of shortcuts favoring accurate and reliable remote recommenders. While the random graph topology still shows the best diffusive qualities, average utility advantages of the evolved over the highly clustered acquaintance network are clearly observable. The Wilcoxon signed-rank test showed statistically significant effects in the same confidence interval as the diffusion tests.

6 Conclusions and Future Work

Simulation results indicate that augmenting trust models, as prototypically shown for the FIRE model [26], is beneficial to agent performance. The augmented model performed consistently better than the standard model by harnessing information that would normally be disregarded. The proposed voting mechanism, employing a standardized measure, displayed sufficient power and resilience to malicious tagging in order to guarantee the reliability of the additional reputation information.

Additionally, adding a mechanism with the power to evolve the topology of the social network recommending entities operate on, can be beneficial to agent performance. Performance of the trust model was consistently better on the evolved over the base acquaintance network. The associated communication overhead for propagating rumor-like information and network evolution can be partially mitigated by the fact that the timescale for structural network evolution is considerably bigger than for the regular network communications involved in reputation-based trust dynamics. The voting mechanism, aside from serving as a reliability measure for gauging recommendations from untrusted sources, was an adequate measure of the reliability of the propagated nominated recommenders.

With regard to our goal of developing a distributed recommendation system of (semi-) autonomous agents that harnesses the social contacts users establish and maintain in cybercommunities, it was shown that the high number and prevalence of only weakly linked neighbors does not represent a hurdle, but rather a large pool of information that can be successfully tapped for the recommendation process. While motivated by structural features of social networks, the proposed weak tie mechanism can also be applied to judge the reliability of information gathered through recommendation referral methods [38], which suffers from long-distance effects when calculating reliability in the traditional manner, i.e. by multiplying all reliability scores along the referral chain.

In conjunction with other methods we have previously proposed for making highly clustered acquaintance networks more permeable to reputation information [23,22,24], the presented results promise to be a solid foundation for developing an effective and efficient distributed application for cybercommunities. By leveraging the social filtering capabilities of social networks, these methods not only increase the speed with which information is distributed among agents, but also guarantee acceptable recommendation quality necessary to make a correct trust decision by minimizing uncertainty about potential interactors.

The simulation results, while highlighting the effectiveness of the proposed approach, do not yet take several of the semantics of human social network formation into account. Shared tastes and experiences among members of social cliques, i.e. those that maintain strong ties among each other and are part of the same cluster, both arise from and give rise to high similarity between these members. Also, particular shared norms, cultural backgrounds or differing kinds of contracts among entities, such as legal or familial obligations, provide incentives and a basis of familiarity between the parties involved, again having a direct impact on recommendation quality. Modeling this accurately in the simulation process represents a challenge requiring further study.

Furthermore, both the exact structure of human acquaintance and trust networks in computational contexts, as well as the way that reputation information is communicated over them, still leave considerable room for future investigations. Modeling these networks and applying computational procedures to them can not only serve to better understand human action, but also to assist online users, for instance by offering an automated, distributed (p2p) recommendation network. With the persistent popularity of social networking sites, the integration of socially augmented information systems is a logical next step in assuring reliable service in cybercommunities and internet commerce. Simulations, however, can only serve to elucidate and explore a subset of the dangers and potentials that come with employing formalized recommendation mechanisms in cybercommunities. Model assumptions, while derived from sociological research and prior work on computational social trust, have yet to be confirmed in existing social networks. This requires the acceptance and adoption of trust evaluation and establishment applications by real users in considerable numbers, close monitoring and study of their behavior and the – possibly emergent – effects caused by their use, as well as continual iterative improvement of both assumptions and applications.

Beyond the scope of trust models, the simulation of human behavior by agents operating on complex networks, modeled after human social networks, has considerable power to study other social or epidemiological spreading phenomena. Recently, increased attention has, for example, been paid to the spread of new influenza epidemics, such as H1N1, and its computational modeling [10]. This approach to predict the course of potential global pandemics may, for instance, help policy makers to implement effective countermeasures.

References

1. Abdul-Rahman, A., Hailes, S.: A distributed trust model. New Security Paradigms 97 (1997)
2. Abdul-Rahman, A., Hailes, S.: Supporting Trust in Virtual Communities. In: Proceedings of the Hawaii's International Conference on Systems Sciences, Maui Hawaii (2000)
3. Barabasi, A.-L., Albert, R.: Emergence of scaling in random networks. Science 286, 509–512 (1999)

4. Bearman, P., Parigi, P.: Cloning Headless Frogs and Other Important Matters: Conversation Topics and Network Structure. Social Forces 83(2), 535–557 (2004)
5. Bettencourt, L., Cintron-Arias, A., Kaiser, D., Castillo-Chavez, C.: The power of a good idea. Physica A 364, 512–536 (2006)
6. Byron, L., Lento, T., Marlow, C., Rosenn, I.: Maintained Relationships on Facebook. Tech. rep., Facebook Data Group (2009)
7. Castelfranchi, C., Falcone, R.: Principles of trust for MAS: cognitive anatomy, social importance, and quantification. In: Proceedings of the Third International Conference on Multi-Agent Systems (1998)
8. Castelfranchi, C., Falcone, R.: Social Tru. In: Trust and Deception in Virtual Societies, pp. 55–90. Kluwer Academic Publishers, Dordrecht (2001)
9. Castellano, C., Fortunato, S., Loreto, V.: Statistical Physics of Social Dynamics. Reviews of Modern Physics 81, 591–646 (2009)
10. Chao, D.L., Halloran, M.E., Obenchain, V.J., Longini Jr., I.M.: FluTE, a Publicly Available Stochastic Influenza Epidemic Simulation Model. PLoS Computational Biology 6(1), e1000,656 (2010)
11. Dunbar, R.I.M.: Neocortex Size as a Constraint on Group Size in Primates. Journal of Human Evolution, 469–493 (1992)
12. Ellison, N.B., Steinfeld, C., Lampe, C.: The Benefits of Facebook "Friends": Exploring the relationship between college students' use of online social networks and social capital. Journal of Computer Mediated Communciation 12(3) (2007)
13. Erdős, P., Rényi, A.: On the Evolution of Random Graphs. Publications of the Mathematical Institut of the Hungarian Academy of Sciences 5, 17–61 (1960)
14. Faraj, S., Wasko, M., Johnson, S.L.: Electronic Knowledge Networks: Processes and Structure. In: Knowldge Management: An Evolutionary View of the Fiield, pp. 270–291 (2008)
15. Finin, T., Ding, L., Zhou, L., Joshi, A.: Social networking on the semantic web. The Learning Organization 12(5), 418–435 (2005), http://www.emeraldinsight.com/10.1108/09696470510611384, doi:10.1108/09696470510611384
16. Galton, F.: Vox populi. Nature 75, 450–451 (1907)
17. Gambetta, D.: Can We Trust Trust?. In: Gambetta, D. (ed.) Trust: Making and Breaking Cooperative Relations, pp. 213–237. Basil Blackwell, Oxford (1988)
18. Golbeck, J.: Computing and applying trust in web-based social networks. Doctoral thesis, University of Maryland (2005)
19. Golbeck, J.: Trust and nuanced profile similarity in online social networks. ACM Transactions on the Web (TWEB), 1–30 (2009)
20. Golbeck, J., Kuter, U.: Computing with Social Trust. Human-Computer Interaction, 169–181 (2009)
21. Granovetter, M.S.: The Strength of Weak Ties. American Journal of Sociology 78(6), 1360–1380 (1973)
22. Hauke, S., Pyka, M., Borschbach, M., Heider, D.: Augmenting Reputation-Based Trust Metrics with Rumor-Like Dissemination of Reputation Information. IFIP Advances in Information and Communication Technology 330, 136–147 (2010)
23. Hauke, S., Pyka, M., Borschbach, M., Heider, D.: Reputation-based Trust Diffusion in Complex Socio-Economic Networks. In: Soro, A., Eloisa, Vargiu, A.G., Paddeu, G. (eds.) Information Retrieval and Mining in Distributed Environments, pp. 21–40. Springer, Heidelberg (2010)
24. Hauke, S., Pyka, M., Heider, D.: Towards Improved Trust Diffusion Through Active Recommender Propagation. SIWN 8 (2010)

25. Hayes, A.F., Krippendorff, K.: Answering the Call for a Standard Reliability Measure for Coding Data. Communication Methods and Measures 1(1), 77–89 (2007)
26. Huynh, T.D.: Trust and Reputation in Open Multi-Agent Systems. Ph.D. thesis, University of Southampton (2006)
27. Jackson, M.O., Rogers, B.W.: Search in the Formation of Large Networks: How Random are Socially Generated Networks? Game Theory and Information 0503005, EconWPA (2005)
28. Jin, E.M., Girvan, M., Newman, M.E.J.: Structure of growing social networks. Phys. Rev. E 64(4), 46,132 (2001), doi:10.1103/PhysRevE.64.046132
29. Jøsang, A., Ismail, R., Boyd, C.: A survey of Trust and Reputation Systems for Online Service Provision. Decision Support Systems 43(2), 618–644 (2007)
30. Kaelbling, L.P., Littman, M.L., Moore, A.W.: Reinforcement Learning: A survey. Journal of Artificial Intelligence Research 4, 237–285 (1996)
31. Marsden, P.V.: Core Discussion Networks of Americans. American Sociological Review 52(1), 122–131 (1987)
32. Marsh, S.: Formalising Trust as a Computational Concept. Ph.D. thesis, Department of Computing Science and Mathematics, University of Stirling (1994)
33. Marsh, S., Dibben, M.R.: Trust, Untrust, Distrust and Mistrust – An Exploration of the Dark(er) Side. In: Herrmann, P., Issarny, V., Shiu, S.C.K. (eds.) iTrust 2005. LNCS, vol. 3477, pp. 17–33. Springer, Heidelberg (2005)
34. McPherson, M., Smith-Lovin, L., Brashears, M.E.: Social Isolation in America: Changes in Core Discussion Networks over Two Decades. American Sociological Review 71, 352–375 (2006)
35. McPherson, M., Smith-Lovin, L., Cook, J.M.: Birds of a Feather: Homophily in Social Networks. Annual Review of Sociology 27, 415–444 (2001)
36. Milgram, S.: The small world problem. Psychology Today 2, 60–67 (1967)
37. Mui, L.: Computational Models of Trust and Reputation: Agents, Evolutionary Games and Social Networks. Doctoral dissertation, Massachusetts Institute of Technology (2003)
38. Mui, L., Mohtashemi, M., Halberstadt, A.: A computational model of trust and reputation. In: Proceedings of the 35th Hawaii International Conference on System Science, pp. 280–287 (2002)
39. Papadimitriou, D.: Future Internet – The Cross-ETP Vision Document. Tech. rep., European Future Internet Portal (2009)
40. Rasmusson, L.: Socially Controlled Global Agent Systems. Master's thesis, Royal Institute of Technology, Dept. of Computer and Systems Science, Stockholm (1996)
41. Ries, S.: Trust in Ubiquitous Computing. Doctoral thesis, TU Darmstadt (2009)
42. Sabater, J.: Trust and Reputation for Agent Societies. Ph.D. thesis, Universitat Autonoma de Barcelona (2003)
43. Sabater, J., Sierra, C.: Review on Computational Trust and Reputation Models. Artificial Intelligence Review 24, 33–60 (2005)
44. Walter, F.E., Battiston, S., Schweitzer, F.: Coping with Information Overload through Trust-Based Networks. In: Helbing, D. (ed.) Managing Complexity: Insights, Concepts, Applications, pp. 273–300. Springer, Heidelberg (2008)
45. Walter, F.E., Battiston, S., Schweitzer, F.: Personalised and Dynamic Trust in Social Networks. In: RecSys 2009 – Proceedings of the Third ACM Conference on Recommender Systems (2009)
46. Wang, Y., Singh, M.P.: Formal Trust Model for Multiagent Systems. In: Proceedings of the 20th International Joint Conference on Artifical Intelligence, pp. 1551–1556 (2007)

47. Watts, D.J.: Small Worlds. Princeton University Press, Princeton (1999)
48. Wierzbicki, A.: Trust and Fairness in Open, Distributed Systems. Springer, Berlin (2010), doi:10.1007/978-3-642-13451-7
49. Wilcoxon, F.: Individual comparisons by ranking methods. Biometrics 1, 80–83 (1945)
50. Xu, D., Yang, J., Wang, Y.: The evidential reasoning approach for multi-attribute decision analysis under interval uncertainty. European Journal of Operational Research 174(3), 1914–1943 (2006)
51. Xu, J.: Mining Communities of Bloggers: A Case Study on Cyber-Hate. In: ICIS 2006 Proceedings, pp. 135–144 (2006)
52. Yu, B., Singh, M.: Towards a Probabilistic Model of Distributed Reputation. In: Proceedings of the Fourth Workshop on Deception, Fraud and Trust in Agent Societies, Montreal (2001)
53. Zhou, L., Ding, L., Finin, T.: How is the Semantic Web evolving? A dynamic social network perspective. Computers in Human Behavior (2010)
54. Zipf, G.K.: Human behavior and the principle of least effort. Addison Wesley, Cambridge (1949)

Real-Time EEG-Based Emotion Recognition and Its Applications

Yisi Liu, Olga Sourina, and Minh Khoa Nguyen

Nanyang Technological University
Singapore
{LIUY0053,EOSourina,RaymondKhoa}@ntu.edu.sg

Abstract. Since emotions play an important role in the daily life of human beings, the need and importance of automatic emotion recognition has grown with increasing role of human computer interface applications. Emotion recognition could be done from the text, speech, facial expression or gesture. In this paper, we concentrate on recognition of "inner" emotions from electroencephalogram (EEG) signals. We propose real-time fractal dimension based algorithm of quantification of basic emotions using Arousal-Valence emotion model. Two emotion induction experiments with music stimuli and sound stimuli from International Affective Digitized Sounds (IADS) database were proposed and implemented. Finally, the real-time algorithm was proposed, implemented and tested to recognize six emotions such as fear, frustrated, sad, happy, pleasant and satisfied. Real-time applications were proposed and implemented in 3D virtual environments. The user emotions are recognized and visualized in real time on his/her avatar adding one more so-called "emotion dimension" to human computer interfaces. An EEG-enabled music therapy site was proposed and implemented. The music played to the patients helps them deal with problems such as pain and depression. An EEG-based web-enable music player which can display the music according to the user's current emotion states was designed and implemented.

Keywords: emotion recognition, EEG, emotion visualization, fractal dimension, HCI, BCI.

1 Introduction

Nowadays, new forms of human-centric and human-driven interaction with digital media have the potential of revolutionising entertainment, learning, and many other areas of life. Since emotions play an important role in the daily life of human beings, the need and importance of automatic emotion recognition has grown with an increasing role of human computer interface applications. Emotion recognition could be done from the text, speech, facial expression or gesture. Recently, more researches were done on emotion recognition from EEG [6,18,27,28,33,43]. Traditionally, EEG-based technology has been used in medical applications. Currently, new wireless headsets that meet consumer criteria

M.L. Gavrilova et al. (Eds.): Trans. on Comput. Sci. XII, LNCS 6670, pp. 256–277, 2011.

for wearability, price, portability and ease-of-use are coming to the market. It makes possible to spread the technology to the areas such as entertainment, e-learning, virtual worlds, cyberworlds, etc. Automatic emotion recognition from EEG signals is receiving more attention with the development of new forms of human-centric and human-driven interaction with digital media. In this paper, we concentrate on recognition of the "inner" emotions from EEG signals as humans could control their facial expressions or vocal intonation.

There are different emotion classifications proposed by researchers. We follow two-dimensional Arousal-Valence model [38]. This model allows mapping of the discrete emotion labels to the Arousal-Valence coordinate system. One of emotion definitions is as follows: "The bodily changes follow directly the perception of the exciting fact, and that our feeling of the changes as they occur is the emotion" [20]. Our hypothesis is that the feeling of changes can be noticed from EEG as fractal dimension changes. We focused on study of fractal dimension model and algorithms, and proposed a fractal based approach to emotion recognition.

To evoke emotions, different stimuli could be used: visual, auditory, and combined. They activate different areas of the brain. Our hypothesis is that emotions have spatio-temporal location. There is no easily available benchmark databases of EEG labeled with emotions. But there are labeled databases of audio stimuli for emotion induction - International Affective Digitized Sounds (IADS) [8] and visual stimuli - International Affective Picture System (IAPS) [26]. Thus, we proposed and carried out one experiment on emotion induction using IADS database of labeled audio stimuli. We also proposed and implemented an experiment with music stimuli to induce emotions by playing music pieces. Both experiments were carried out with prepared questionnaires for the participants to label the recorded EEG with the corresponding emotions.

There are a number of algorithms for recognizing emotions. The main problem of such algorithms is a lack of accuracy. Research is needed to be carried out to evaluate different algorithms and propose algorithms with the improved accuracy. As emotion recognition is a new area, a benchmark database of EEG signals for different emotions is needed to be set up, which could be used for further research on EEG-based emotion recognition. Until now, only limited types of emotions could be recognized. Research could be done on more types of emotions recognition. Additionally, most of the emotion recognition algorithms were developed for off-line data processing. In our paper, we target on real-time emotion recognition and its applications. The user emotions are recognized and visualized in real time on his/her avatar. We add one more so-called "emotion dimension" to human computer interfaces. Also an EEG-based music therapy and a music player are implemented with our real-time emotion recognition algorithm. Although in this paper, we describe standalone implementations of emotion recognition and its applications, it could be easily extended for further use in collaborative environments/cyberworlds.

In Section 2.1, review on emotion classification is given. In Section 2.2, emotion induction experiments are introduced. In Section 2.3, emotion recognition algorithms from EEG are reviewed. In Section 2.4, a fractal dimension algorithm proposed by Higuchi is described. Our approach, emotion induction experiments, a novel fractal-based emotion recognition algorithm, data analysis and results are given in Section 3. Real-time emotion recognition and visualization of human emotions on 3D avatar using Haptek system [2], the EEG-based music therapy and the EEG-based music player are described in Section 4. In Section 5, conclusion and future work are given.

2 Related Works

2.1 Emotion Classification

There are different emotion classification systems. The taxonomy can be seen from two perspectives: dimensional and discrete one [32]. Plutchik defines eight basic emotion states: anger, fear, sadness, disgust, surprise, anticipation, acceptance and joy. All other emotions can be formed by these basic ones, for example, disappointment is composed of surprise and sadness [36]. Another approach towards emotion classification is advocated by Paul Ekman. He revealed the relationship between facial expressions and emotions. In his theory, there are six emotions associated with facial expressions: anger, disgust, fear, happiness, sadness, and surprise. Later he expanded the basic emotion by adding: amusement, contempt, contentment, embarrassment, excitement, guilt, pride in achievement, relief, satisfaction, sensory pleasure, and shame [13].

From the dimensional perspective, the most widely used one is the bipolar model where arousal and valence dimensions are considered. This emotion classification approach is advocated by Russell [38]. Here, the arousal dimension ranges from not aroused to excited, and the valence dimension ranges from negative to positive. Another fundamental dimension is an approach-withdrew dimension which is based on the motivating aspects of the emotion [32]. For example, in this theory, anger is an approach motivated emotion in some cases, as it could encourage the person to make effort to change the situation.

The dimensional model is preferable in emotion recognition experiments due to the following advantage: dimensional model can locate discrete emotions in its space, even when no particular label can be used to define a certain feeling [10,32].

2.2 Emotion Induction Experiments

In order to obtain affective EEG data, experiments are carried out with different kinds of stimuli such as audio, visual, and combined ones to induce emotions.

Among the EEG-based emotion recognition works which implemented experiments using audio stimuli to collect EEG data, there are some works where subjects' emotions were elicited by pre-labeled music with emotions. For example, in [28], it was reported that emotions were induced in 26 subjects by pre-assessed music pieces to collect EEG data. A 90% classification accuracy rate was received to distinguish four kinds of emotions: joy, anger, sadness and pleasure. 32 channels were used, and a Multiclass Support Vector Machine was applied for the classification.

Another types of audio stimuli used in works on emotion recognition from EEG are retrieved from the labeled databases of audio stimuli, IADS database. For example, in [6], four kinds of emotion states including positive/aroused, positive/calm, negative/calm and negative/aroused were induced by sounds clips from IADS. The Binary Linear Fisher's Discriminant Analysis was employed to do the classification. They achieved 97.4% maximum rate for arousal levels recognition and 94.9% maximum rate for valence levels with Fpz and F3/F4 channels.

For experiments using visual stimuli, IAPS is a preferred choice. [6] also selected pictures from IAPS, however, it was reported that the EEG data collected with visual stimuli experiments are more difficult to classify. eNTERFACE project described in [42] established an EEG database using pictures selected from IAPS as stimuli, and 3 emotional states were elicited as follows: exciting positive, exciting negative, and calm state. Though this project did not target emotion recognition from EEG signals, they published EEG data labeled with emotions that were cited by other works on EEG-based emotion recognition as a benchmark. For example, [22] combined correlation dimension with statistical features which improved the results from 66.6% to 76.66%. [43] also used IAPS and it was reported that valence level was recognized with 62.07% accuracy. Another form of visual stimuli that was used in [35] employed a Mirror Neuron System. Pictures of affective facial expressions were used.

Combined stimuli were used in [51]. Films were selected to be the stimuli in that work.

2.3 Emotion Recognition Algorithms

There are an increasing number of researches done on EEG-based emotion recognition algorithms. In [28], short-time Fourier Transform was used to calculte the power difference between 12 symmetric electrodes pairs with 6 different EEG waves for feature extraction and Support Vector Machine (SVM) approach was employed to classify the data into different emotion modes. The result was 90.72% accuracy to distinguish the feelings of joy, sadness, anger and pleasure. A performance rate of 92.3% was obtained in [6] using Binary Linear Fisher's Discriminant Analysis and emotion states among positive/arousal, positive/calm, negative/calm and negative/arousal were differentiated. SVM was applied in [18] for emotion classification with the accuracy for valence and arousal identification as 32% and 37% respectively. By applying lifting based wavelet transforms to extract features and Fuzzy C-Means clustering to do classification, sadness, happiness, disgust, and fear were recognized in [33]. In [43], optimization such as different window sizes, band-pass filters, normalization approaches and dimensionality reduction methods were investigated, and it achieved an increase in accuracy from 36.3% to 62.07% by SVM after applying these optimizations. Three emotion states: pleasant, neutral, and unpleasant were distinguished. By using Relevant Vector Machine, differentiation between happy and relaxed, relaxed and sad, happy and sad with a performance rate around 90% was obtained in [27].

Although the number of researches done on EEG-based emotion recognition algorithms in recent years has been increasing, EEG-based emotion recognition is still a new area of research. The effectiveness and the efficiency of these algorithms, however, are somewhat limited. Some unsolved issues in current algorithms and approaches are listed as follows:

1. Time constrains. The performance time consists from the time of feature extraction and time of classification. The number of electrodes used in the emotion recognition puts another time constrain on the algorithms. As a result, to our best knowledge, most of the algorithms were proposed for off-line emotion recognition.
2. Accuracy. The accuracy of the EEG-based emotion recognition still needs to be improved. The accuracy decreases when more emotions are needed to be recognized.
3. Number of electrodes. From the perspectives of the time to set up the EEG device, the comfort level of the users who wear the device and the amount of features to process, the number of electrodes should be reduced. However, most of the current works still require relatively big number of electrodes. For example, 16 channels were used in [43], and more than 32 channels were used in [11,12,28].
4. Number of the recognized emotions. Although there are varieties of emotional states to describe the human's feelings, until now only limited types of emotions can be recognized using EEG. The best performance obtained was reported in [35] where 3 channels were used and 83.33% maximum accuracy was achieved for differentiating 6 emotions.
5. Benchmark EEG affective databases. So far, a very few benchmark EEG databases with labeled emotions are available. EEG affective databases with different stimuli such as visual and audio are needed to be set up for future researches.

Additionally, as the brain is a complicated system, the EEG signal is nonlinear and chaotic [23,24]. However, little has been done to investigate chaos of brain for emotion recognition. [6,18,27,28,33,43] were based on linear analysis, however, linear analysis such as Fourier Transform only preserves the power spectrum in the signal, but destroys the spike-wave structure [49].

A fractal dimension analysis is suitable for analyzing nonlinear systems and could be used in real-time EEG signal processing [4,29]. Early work such as [37] showed that fractal dimension could reflect the change of EEG signal; [30] showed that fractal dimension varied for different mental tasks; a more recent work like [24] revealed that when brain processed tasks which were of emotional difference only, fractal dimension can be used to differentiate these tasks. [44,46] used music as stimuli to elicit emotions, and applied fractal dimension for the analysis of the EEG signal. [47,52] applied fractal dimension to detect the concentration level of the subjects and developed EEG-based "serious" games. All these works show that fractal dimension is a potentially promising approach to investigate EEG-based emotion recognition. In our research, fractal dimension model is explored to provide better accuracy and performance in EEG-based emotion recognition.

2.4 Fractal Dimension Model

For calculation of fractal dimension value, we implemented and analyzed two well-known algorithms: box-counting [5] and Higuchi [17]. Both of them were evaluated using Brownian and Weierstrass functions where "true value" is known [31]. Higuchi algorithm was chosen to process the data since it gave a better accuracy as it was closer to the theoretical FD values [53] and it outperformed in the processing of EEG data [45].

Let us describe the Higuchi algorithm as we apply it for FD calculation in our proposed fractal-based emotion recognition algorithm shown in Section 3.

Let $X(1), X(2), \ldots, X(N)$ be a finite set of time series samples, the new time series is constructed as follows:

$$X_k^m : X(m), X(m+k), X(m+2k), \ldots, X(m+[\frac{N-m}{k}]\cdot k)\ . \tag{1}$$

where $m = 1, 2, \ldots, k$, m is the initial time and k is the interval time.

Then, k sets of $L_m(k)$ are calculated as follows:

$$L_m(k) = \frac{\left\{ (\sum_{i=1}^{[\frac{N-m}{k}]} |X(m+ik) - X(m+(i-1)\cdot k)|) \frac{N-1}{[\frac{N-m}{k}]\cdot k} \right\}}{k}\ . \tag{2}$$

$\langle L(k)\rangle$ denotes the average value over k sets of $L_m(k)$ and the relationship exists as follows:

$$\langle L(k)\rangle \propto k^{-D}\ . \tag{3}$$

Finally, the fractal dimension can be obtained by logarithmic plotting between different k and its associated $\langle L(k)\rangle$ [17].

3 Fractal Dimension Based Approach to Emotion Recognition

In this paper, we proposed a fractal dimension based approach to EEG-based emotion recognition. First, we designed and implemented emotion induction experiments using two-dimensional model to describe emotions. Then, we analyzed EEG recordings with Higuchi algorithm and our proposed algorithm for online emotion recognition.

3.1 Emotion Induction Experiments

As we mentioned in Introduction, there is no easily available benchmark database of EEG recordings with labeled emotions. We designed two experiments to elicit emotions with audio stimuli. Five truncated songs which lasted for 1 minute each were used in Experiment 1. The following music was chosen for the targeted emotions: "Wish to Wish" (S.E.N.S) for negative/low aroused (sad), "Cloud Smile" (Nobue Uematsu) for positive/low aroused (pleasant), "A Ghost in the Machine" (Angelo Badalamenti) for negative/high aroused (fear), "William Tell Overture: Finale" (Gioachino Rossini) for positive/high aroused (happy) and "Disposable Hero" (Metallica) for negative/high aroused (angry). 10 participants, 2 female and 8 male students whose ages ranged from 23 to 35, participated in the music experiment.

Sound clips selected from the International Affective Digitized Sounds (IADS) were used in Experiment 2. All the sounds in the IADS database are labeled with their arousal and valence values. IADS provides a set of standardized sound stimuli to evoke emotions that could be described by Arousal-Valence model. For example, positive valence and high arousal values define happy emotion. 27 clips were chosen to induce five emotional states: 3 clips for neutral with mean arousal ratings ranging between 4.79 and 5.15, mean valence rating ranging from 4.83 to 5.09; 6 clips for each of positive/low aroused, positive/high aroused, negative/low aroused and negative/high aroused emotions with mean arousal rating ranging between 3.36 and 4.47; 6.62 and 7.15; 3.51 and 4.75; 7.94 and 8.35 respectively, and mean valence rating ranging between 6.62 and 7.51; 7.17 and 7.90; 4.01 and 4.72; 1.36 and 1.76 respectively. 12 subjects, 3 female and 9 male students whose ages ranged from 22 to 35, participated in the sound experiment. None of the subjects participated in both experiments had history of mental illness.

The procedures for both experiments are described as follows. After a participant was invited to the project room, he/she was briefly introduced to the experiment protocol and the usage of self-assessment questionnaire. Then, the participant was seated in front of the computer which played the audio stimuli. He/she was required to keep still and eyes closed during experiment sessions to avoid muscle movement and eye blinking artifacts.

As shown in Fig. 1, Experiment 1 was consisted from five sessions. Each session targeted one type of emotions. There was a 60 seconds silent period at the beginning of each session for the subject to calm down and get ready for the session. After that, one piece of music truncated to 1 minute duration was played to the subject.

For Experiment 2 using stimuli from IADS, 5 sessions, namely: session 1 - neutral, session 2 - positive/low aroused, session 3 - positive/high aroused, session 4 - negative/low aroused, session 5 - negative/high aroused were prepared as shown in Fig. 2. In each session, there was a 6 seconds' silent break, then 6 clips of IADS stimuli aiming at one particular emotion were played to the subjects. For neutral state, since only three sounds clips were available, each clip was played twice in order to keep the same interval of each session.

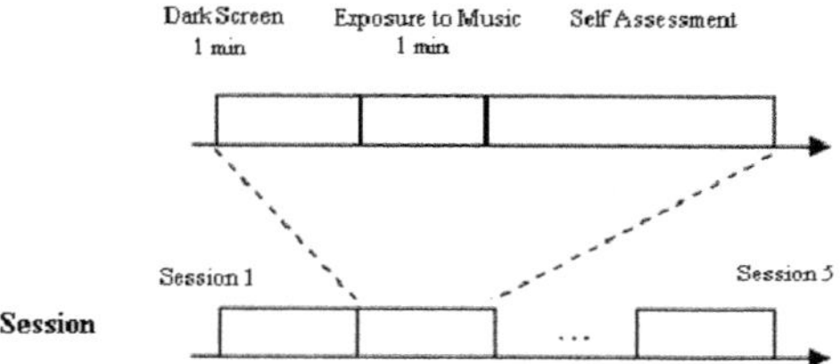

Fig. 1. The procedure of music experiment

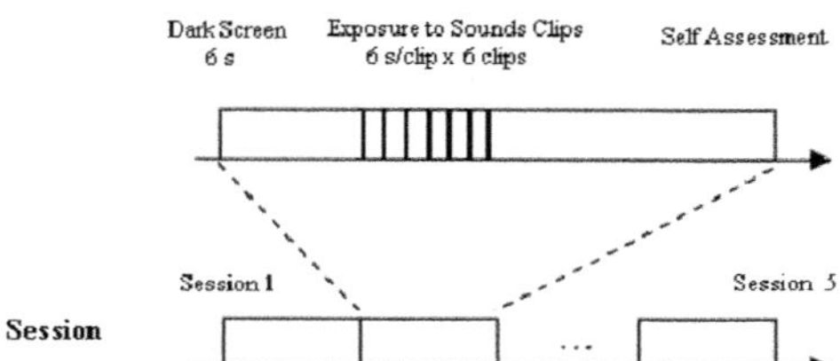

Fig. 2. The procedure of IADS experiment

For both experiments, the subjects needed to complete the questionnaire after listening to music/sounds. In the questionnaire, the Self-Assessment Manikin (SAM) technique [7] with two dimensions: arousal and valence and five levels indicating the intensity of both dimensions, was employed for emotion state measurement. Additionally, the subjects were asked to write down their feelings in a few words such as happy, sad, etc.

Emotiv wireless headset [1] was used for carrying out experiments. Emotiv device has 14 electrodes locating at AF3, F7, F3, FC5, T7, P7, O1, O2, P8, T8, FC6, F4, F8, AF4 (CMS/DRL as references) following the American Electroencephalographic Society Standard [3]. The sampling rate of the Emotiv headset is 128Hz. The bandwidth of the device is 0.2-45Hz, and digital notch filters are at 50Hz and 60Hz. The A/D converter is with 16 bits resolution.

3.2 Data Analysis and Results

The data collected in our two experiments using the Emotiv headset were analyzed to find spatio-temporal emotion patterns of high and low arousal level with positive and negative valence level.

In the following analysis, we only focus on the analysis of data with negative/high aroused, positive/high aroused, negative/low aroused, and positive/low aroused labels. Since we have the questionnaires which give us the true reaction of the subjects to the stimuli, we ignore the cases in our processing when the subjects' feelings were not compatible with the targeted emotions.

A 2 to 42 Hz band-pass filter was applied to the raw data as the major EEG waves (alpha, theta, beta, delta, and gamma) lie in this band [40,41]. Then, Higuchi fractal dimension algorithm described in section 2.4. was applied for FD values calculations. We implemented the algorithm with a sliding window where the window size was 1024 samples and 99% overlapping was applied to calculate FD values of the filtered data.

In the first experiment using music stimuli, the data from 13th to 24th seconds of recording were processed. In the second experiment using IADS clips, the data from 2nd to the 13th seconds of recording were processed.

The arousal level could be identified from different electrode locations. FC6 was selected for the arousal level recognition as the FD values computed from it gave better arousal difference compared to other channels. The mean of FD values computed from FC6 aiming at recognizing the arousal level for all subjects in music and IADS experiments are shown in Table 1 and Table 2. Two FD values for high arousal level with negative and positive valence, and two FD values for low arousal level with negative and positive valence are presented in the tables as follows: negative high aroused (N/HA), positive high aroused (P/HA), negative low aroused (N/LA), and positive low aroused (P/LA). In Table 1, it is shown that 10 subjects participated in the Music experiment. In Table 2, it is shown that 12 subjects participated in IADS experiment. Based on the self-assessment questionnaires, 46 pairs of comparisons from different subjects between high aroused and low aroused states regardless of the valence level were used. 39/46 (84.9%) showed that the higher arousal was associated with the larger FD values. This phenomenon is illustrated in Table 1 and 2 as the mean of FD values for the high aroused states (N/HA and P/HA) is larger than the low aroused states (N/LA and P/LA). For example, for the subject #1 N/HA value 1.9028 is larger than N/LA value 1.7647 and P/LA value 1.8592, and P/HA value 1.9015 is larger than N/LA value 1.7647 and P/LA value 1.8592. In Table 1 and 2, we denoted the FD value as X if the subject's feeling was different from the targeted emotion by self-assessment questionnaire report. Thus, we eliminated such cases from our analysis.

The asymmetrical frontal EEG activity may reflect the valence level of emotion experienced. Generally, right hemisphere is more active during the experience of negative emotions while left hemisphere is more active during positive emotions. It was found that when one is watching a pleasant movie scene, a greater EEG activity is appeared in the left frontal lobe, and with unpleasant scene, right frontal lobe shows relatively higher EEG activity [21]. Another set of evidence supporting this hypothesis is described in work done by Canli et al. [9]. They used fMRI to investigate the human brain's response to visual stimuli, and got the results that greater left hemisphere activity is shown during the positive picture exposure but greater right hemisphere activity for negative pictures. However, there are also studies such as [25,34] that oppose this hypothesis.

In our study, the difference between FD values from electrode pair AF3 (left hemisphere) and F4 (right hemisphere) were used to identify valence level and test the lateralization theory. It was found that more stable pattern can be

Table 1. Mean FD values for arousal level analysis of music experiment

Music	Emotion State FD Value			
	N/HA	P/HA	N/LA	P/LA
Subject #1	1.9028	1.9015	1.7647	1.8592
Subject #2	X	1.9274	X	1.9268
Subject #3	1.9104	1.9274	1.7579	1.8426
Subject #4	1.9842	1.956	1.8755	1.9361
Subject #5	1.7909	1.8351	1.8242	X
Subject #6	1.9111	X	X	1.9127
Subject #7	1.9352	1.9231	1.9106	1.9204
Subject #8	X	X	X	X
Subject #9	1.9253	1.939	X	1.9203
Subject #10	1.8507	1.8842	X	1.8798

Table 2. Mean FD values for arousal level analysis of IADS experiment

IADS	Emotion State FD Value			
	N/HA	P/HA	N/LA	P/LA
Subject #1	1.8878	1.8478	X	1.8042
Subject #2	1.9599	1.922	1.938	1.8237
Subject #3	1.9418	1.9507	1.9201	X
Subject #4	1.9215	1.917	1.9265	1.9118
Subject #5	1.7215	1.9271	X	1.8659
Subject #6	1.8898	1.8902	1.888	X
Subject #7	X	X	X	X
Subject #8	X	X	X	X
Subject #9	1.9261	1.9241	1.7437	X
Subject #10	1.9223	1.9333	X	1.9026
Subject #11	1.8796	1.8543	1.7183	X
Subject #12	X	X	X	X

achieved by differentiating valence level within the same arousal level, either high aroused or low aroused. In Fig. 3, a hierarchical scheme we used to distinguish emotions is shown.

The results of the valence level recognition also revealed partial support for the asymmetric frontal hypothesis. Although not all the subjects' dominant hemisphere for positive or negative emotions was the same as expected in the asymmetric hypothesis, the pattern of lateralization for a particular subject was consistent among different experiments with similar arousal level. 10 subjects' data were available for comparison of positive and negative emotion states with similar arousal level. 9/10 (90%) has shown the consistent pattern as follows. For example, one subject's EEG data showed the larger difference between AF3 and F4 FD values for negative emotions than for positive emotions in all experiment trails with different valence levels but similar arousal levels. Five subjects had the larger difference of FD values between left hemisphere and right hemisphere

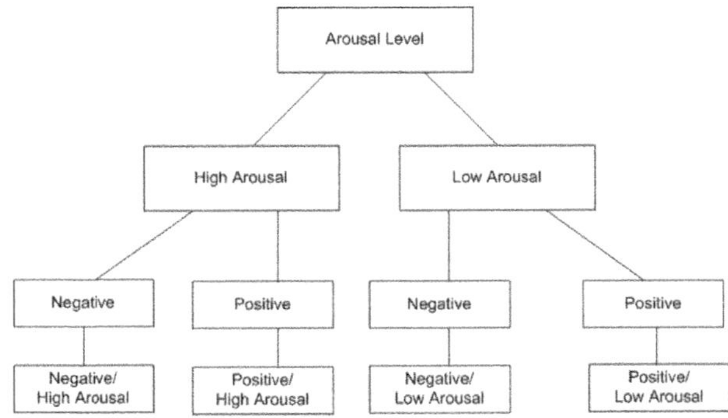

Fig. 3. The heriachical scheme of emotion recognition

(AF3-F4) during the experiencing of negative emotion, while 4 subjects had the larger difference of FD values when they experienced positive emotions. This phenomenon may indicate that the frontal lateralization exists with individual differences, and it may not be applicable for everyone that the left hemisphere is more active for positive and right hemisphere is more active for negative emotions. It could be opposite for some individuals, and this outcome complies with the conclusion made in [16] that individual difference may affect the processing of emotion by brain.

Based on the result of our analysis, we developed the following real-time emotion recognition algorithm described in the next section.

3.3 Real-Time Emotion Recognition Algorithm

As it was mentioned in Introduction, we follow two-dimensional Arousal-Valence model described in section 2.1. This model allows the mapping of the discrete emotion labels in the Arousal-Valence coordinate system as shown in Fig. 4.

The advantage of using this model is that we can define arousal and valence levels of emotions with the calculated FD values. For example, the increase in arousal level corresponds to the increase of FD values. Then, by using ranges of arousal and valence level, we could obtain discrete emotions from the model. Finally, any emotion that can be represented in the Arousal-Valence model can be recognized by our emotion recognition algorithm.

The emotion recognition algorithm for real time is illustrated in Fig. 5. The raw EEG data gathered from AF3, F4 and FC6 are the input to the 2 to 42 Hz band-pass filter. Then, Higuchi fractal dimension algorithm with a sliding window of window size 1024 and 99% overlapping is applied to the filtered data. The benefit of the usage of the sliding window is that it enables real-time processing.

The FD value calculated from FC6 is used to distinguish the arousal level independently by comparing with a default threshold extracted from our experiments' results described in section 3.2. As shown in Fig. 4, the change of FD could be mapped along the arousal axis since our experiments revealed that higher arousal level was associated with larger FD values. Based on this

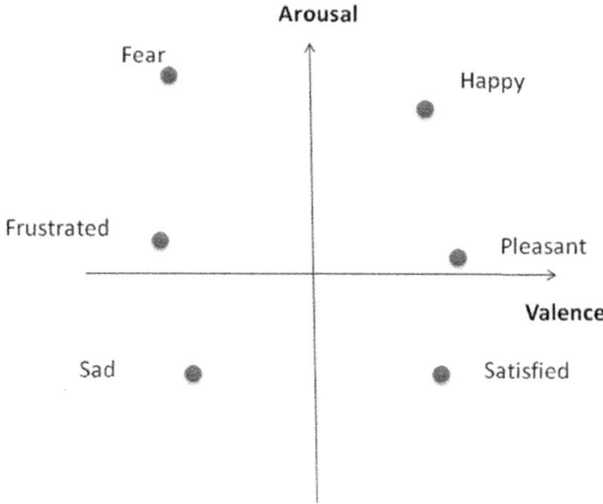

Fig. 4. Emotion labels in arousal-valence dimension(Adapted from Russell's circumplex model [39])

observation, continuous recognition of changing emotion from low arousal to high arousal is enabled. For example, satisfied, pleasant, and happy are all positive emotions but with different arousal levels - ranging from low arousal to high arousal level, and their corresponding FD values also ranges from small one to large one.

The difference of FD values between left hemisphere and right hemisphere (AF3-F4) is computed simultaneously. After the arousal level has been identified, the valence level of emotions is recognized within the similar arousal level by comparing the difference of FD with another threshold which is set for valence level recognition.

Finally based on the arousal level and valence level, the emotions are mapped into 2D model.

In the algorithm, we set default thresholds for real-time emotion recognition based on the experiments' results. However, because of the existence of individual difference which means the pattern of emotion for one particular subject is consistent but FD values may vary among different subjects, a training session is needed to be introduced in order to improve the accuracy.

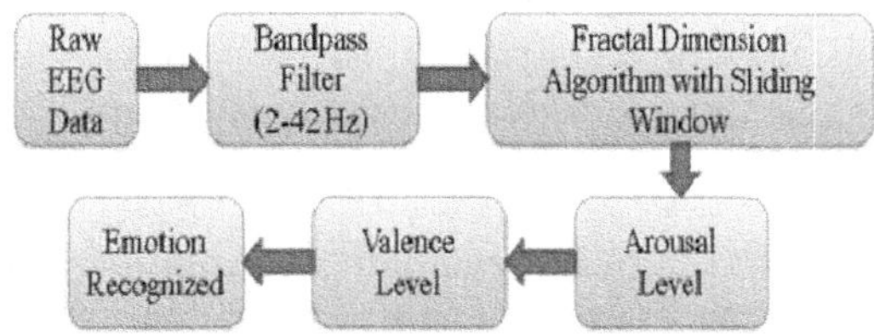

Fig. 5. The emotion recognition algorithm overview

The training session scheme is illustrated in Fig. 6. The procedure is similar to the real-time scheme, except the input is EEG data of the labeled emotions of the particular subject. Then, thresholds are calculated and the lateralization pattern is found based on the data collected from the training session for each subject. When the subject wants to use this system after training, the procedure is illustrated as the lower part below the dash line in Fig. 6. The pattern of newly collected EEG data is recognized according to the comparisons with the calculated thresholds obtained from the training session.

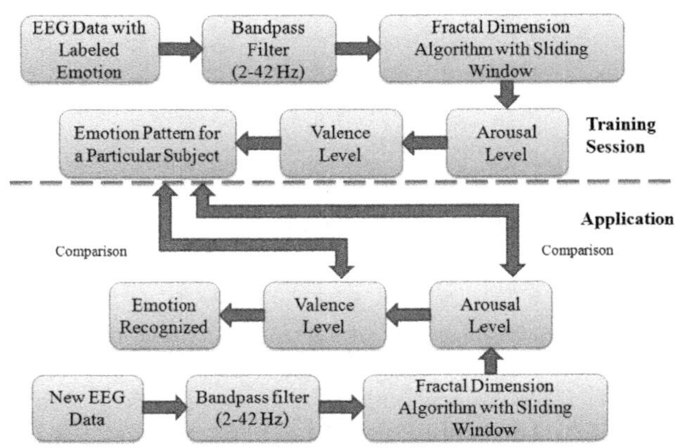

Fig. 6. An emotion recognition scheme with training session

4 Real-Time Applications

The real-time EEG-based emotion recognition can be applied to many fields such as entertainment, education, medicine, etc. In our work, we implemented three applications: an emotional avatar, EEG-based music therapy, and EEG-based music player.

4.1 Emotional Avatar

In order to visualize the recognized emotions, we implemented our algorithm with Haptek activex control system [2]. Microsoft Visual C++ 2008 was used in this project. Haptek software is a 3D model with predefined parameters for controlling facial muscles visualization, thereby enables users to create customized emotions and expressions. Haptek supports stand-alone and web-based application.

Data Acquisition. EEG data are acquired using Emotiv headset at 128 Hz. We used Emotiv Software Development Kit for acquiring raw data from the device. Three out of fourteen Emotiv's channels at locations AF3, F4 and FC6 are fed into the algorithm for the emotion recognition process.

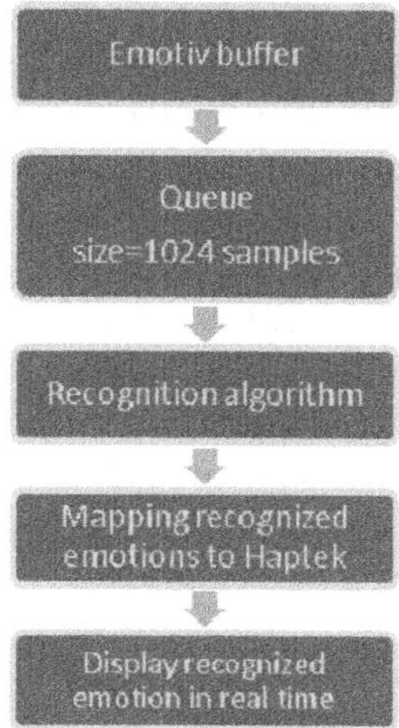

Fig. 7. Implementation process of the emotional avatar application

Data Processing. A data stream from the Emotiv device is stored in a buffer. Every time a read command is triggered, all the samples in the buffer are taken out and the buffer is cleared. Therefore, the number of data obtainable at a time depends on how long the samples have accumulated in the buffer.

The fractal algorithm requires data to be fed in a bunch of 1024 samples at a time for one channel. Therefore, we use a queue to buffer the data from Emotiv's buffer to the algorithm. The queue is refreshed by the current number of samples in Emotiv's buffer every time the read command is triggered as shown in Fig. 7. In the algorithm, those obsolete values in the queue are replaced by latest values in the Emotiv buffer at the time.

Emotion Mapping to Haptex. Haptek Activex control provides functions and commands to change facial expressions of 3D avatars. We used an avatar

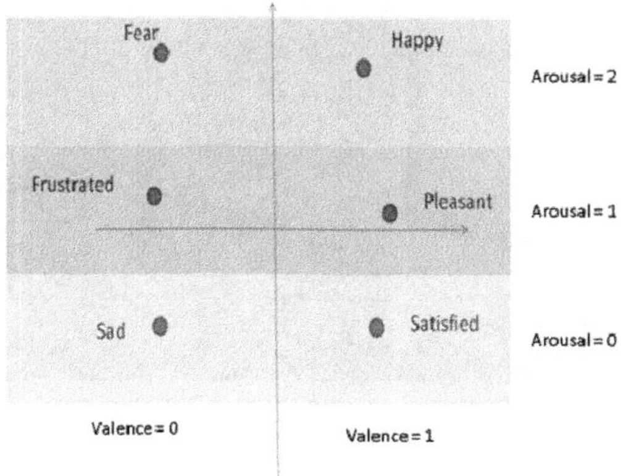

Fig. 8. Illustration of discrete arousal and valence levels

available with version of Haptek development package [2]. The face of the avatar can be changed according to the photo image of the user's face. We defined six emotions by changing the parameters controlling the facial muscles of the Haptek emotional avatar. Those emotions are: fear, frustrated, sad, happy, pleasant and satisfied. The above emotions can be recognized by the proposed emotion recognition algorithm described in the section 3.

For the mapping, arousal and valence levels are transformed into discrete values using thresholds. After this step, arousal level can only take one of the following values 0, 1 or 2 and valence 0 or 1 as shown in Fig. 8. Combination of discrete values of arousal and valence level gives six types of emotions.

Mapping of discrete values of (Arousal level, Valence level) into 6 emotions is shown in Table 3.

Table 3. Mapping of combinations of (Valence, Arousal) and corresponding emotions

(Valence, Arousal)	Emotion
(0,0)	Sad
(0,1)	Frustrated
(0,2)	Fear
(1,0)	Satisfied
(1,1)	Pleasant
(1,2)	Happy

Picture of the user with the Emotiv headset and emotional avatar and pictures of six emotions created using Haptek are shown in Fig. 9 and Fig. 10 respectively.

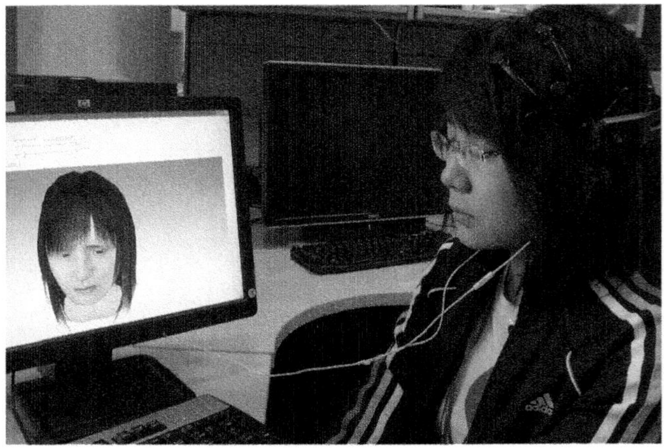

Fig. 9. User with Emotiv headset and emotional avatar

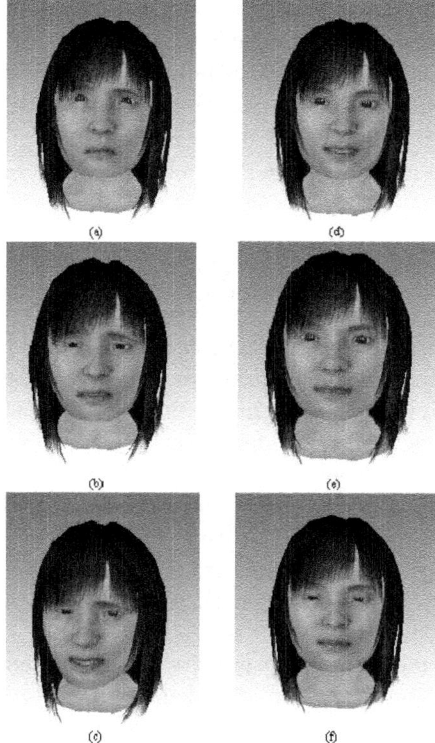

Fig. 10. Six visualized emotions with Haptek (a) Fear (b) Frustrated (c) Sad (d) Happy (e) Pleasant (f) Satisfied

4.2 EEG-Based Music Therapy

Music therapy is considered as a nonpharmacological intervention to help the patients deal with the stress, anxiety and depression problems. In [50], the patients reported that their anxiety was released by listening to music during their surgery. [15] also gave a positive support to the effectiveness of the music therapy in the treatment of the patients who suffered from Alzeheimer's disease. Their anxiety level dropped sharply after the music therapy session.

Since music therapy is proved to be a helpful approach in the medical area, we combined it with our real-time EEG-based human emotion recognition algorithm. By this, we can identify the patient's current emotional state and adjust the music therapy based on the patient brain feedback in real time.

The general music therapy can be described as follows. First, the patient needs to choose the type of music therapy such as pain management, depression, etc. Then the corresponding music is selected and played. The emotion state of the patient is continuously checked by his/her EEG in real time, and if the currently playing music does not effectively evoke the targeted emotion of the therapy, the music is changed and another song is played. If the present emotion is in the accordance to the targeted emotion state, the music is played to maintain the target emotion. When a piece of music is over, another piece of music is played.

The length of the music therapy usually lasts from 25 to 40 minutes [14]. In our application, we denote the duration of maintaining the patient in the targeted emotion for one particular music therapy as $t1$. Then, $t1$ is compared with t which is accumulated by the efficient time slices te_i of each music piece played and evoked the targeted emotion in a whole therapy treatment as follows:

$$t = \sum_{i=1}^{n} te_i \ . \tag{4}$$

where n is the number of the music pieces played in one music therapy, and te_i defines the efficient time slices of all music pieces played during the music therapy. It can be a whole piece of music duration, or only part of a song. For example, music 1 keeps the patient feel positive for 2 minutes, and then, it fails to induce the positive emotion, so it is replaced by music 2. Suppose music 2 is displayed for 4 minutes as it can induce the positive state through that duration. Then, these 2 minutes and 4 minutes time-intervals compose two components in te_i as $te_1 = 2$, $te_2 = 4$, and they are reckoned in t. When the constantly accumulating summation of t is larger than $t1$, the music will be stopped, and the end of one music therapy session is reached.

Fig. 11 shows the music therapy website we implemented. For implementation, emotion recognition algorithm is packaged as an ActiveX Component so it can be used in the Internet Explorer environment. Visual C++ was used to integrate the ActiveX Component. The user's "inner" emotion is recognized from the EEG signals in real time. For music therapy on pain management, happy (positive/high aroused) songs are played to the user to distract his/her attention

Fig. 11. Subject is accessing the music therapy website

from the pain he/she is suffering. This strategy is compatible with [48] which implemented EEG-based games to switch patient attention from the pain feeling. The user's emotion state is checked in real time by his/her EEG data. If the happy emotion is not evoked by the current song, the player automatically switches to another one. For music therapy dealing with depression, pleasant (positive/low aroused) songs are played to the user to make him/her feel relaxed. The song is changed according to the EEG feedback.

4.3 EEG-Based Music Player

Another application of real-time EEG-based emotion recognition is an EEG-based music player website. In this application, the user's current emotion state is recognized, and then, the corresponding music is played according to the identified emotion. The user's emotion is detected by the algorithm running behind the scene. Songs are categorized into six emotion types: fear, sad, frustrated, happy, satisfied and pleasant.

The design of the player is shown in Fig. 12. Information about the current emotional state of the user and the music being played is given on the display of the player. For example, as shown in Fig. 12, the emotion state is recognized as pleasant, and the music which is categorized as pleasant music is played to the user.

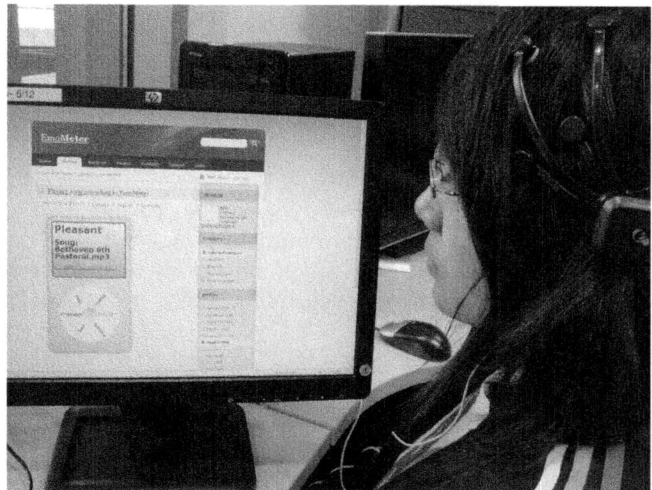

Fig. 12. Subject with EEG-based music player

5 Conclusion and Future Work

In this paper, emotion classifications, emotion evoking experiments and emotion recognition algorithms were reviewed. We proposed and implemented a novel fractal dimension based algorithm for recognition of emotions from EEG in real time. We implemented our algorithm with Haptek system. The system allows visualization of emotions as facial expressions of personalized avatars in 3D collaborative environments in real time. We also developed a prototype for an EEG-based music therapy and one EEG-based music player. Compared with other works, our algorithm uses fewer electrodes. We recognized emotions with AF3, F4 and FC6 electrodes, however, for example, in [33] and [43], 63 and 16 electrodes were used respectively. Until now, to our best knowledge there is no real-time EEG-based emotion recognition algorithms reported. We implemented a novel real-time emotion recognition algorithm based on fractal dimension calculation. In this paper, we implemented recognition of six emotions: fear, frustrated, sad, happy, pleasant and satisfied. However, our approach based on FD calculation allows recognize even more emotions that can be defined in 2-dimensional Arousal-Valence model.

Currently, the real-time emotion recognition and its applications are standalone implementations. The next step of the project is an integration of our tools in Co-Spaces on the Web targeting entertainment industry.

Short videos about the emotion recognition algorithm implemented in real time with the Haptex system and the music player, and more information about our project EmoDEx are presented in [19].

Acknowledgments. This project is supported by grant NRF2008IDM-IDM004-020 "Emotion-based personalized digital media experience in Co-Spaces" of National Research Fund of Singapore.

References

1. Emotiv, `http://www.emotiv.com`
2. Haptek, `http://www.haptek.com`
3. American electroencephalographic society guidelines for standard electrode position nomenclature. Journal of Clinical Neurophysiology 8(2), 200–202 (1991)
4. Accardo, A., Affinito, M., Carrozzi, M., Bouquet, F.: Use of the fractal dimension for the analysis of electroencephalographic time series. Biological Cybernetics 77(5), 339–350 (1997)
5. Block, A., Von Bloh, W., Schellnhuber, H.J.: Efficient box-counting determination of generalized fractal dimensions. Physical Review A 42(4), 1869–1874 (1990)
6. Bos, D.O.: EEG-based emotion recognition (2006), `http://hmi.ewi.utwente.nl/verslagen/capita-selecta/CS-Oude_Bos-Danny.pdf`
7. Bradley, M.M.: Measuring emotion: The self-assessment manikin and the semantic differential. Journal of Behavior Therapy and Experimental Psychiatry 25(1), 49–59 (1994)
8. Bradley, M.M., Lang, P.J.: The international affective digitized sounds (2nd edition; IADS-2): Affective ratings of sounds and instruction manual. Tech. rep., University of Florida, Gainesville (2007)
9. Canli, T., Desmond, J.E., Zhao, Z., Glover, G., Gabrieli, J.D.E.: Hemispheric asymmetry for emotional stimuli detected with fMRI. NeuroReport 9(14), 3233–3239 (1998)

10. Chanel, G.: Emotion assessment for affective-computing based on brain and peripheral signals. Ph.D. thesis, University of Geneva, Geneva (2009)
11. Chanel, G., Kierkels, J.J.M., Soleymani, M., Pun, T.: Short-term emotion assessment in a recall paradigm. International Journal of Human Computer Studies 67(8), 607–627 (2009)
12. Chanel, G., Kronegg, J., Grandjean, D., Pun, T.: Emotion assessment: Arousal evaluation using EEG's and peripheral physiological signals (2006)
13. Ekman, P.: Basic emotions. In: Dalgleish, T., Power, M. (eds.) Handbook of Cognition and Emotion. Wiley, New York (1999)
14. Grocke, D.E., Wigram, T.: Receptive Methods in Music Therapy: Techniques and Clinical Applications for Music Therapy Clinicians, Educators and Students, 1st edn. Jessica Kingsley Publishers (2007)
15. Guetin, S., Portet, F., Picot, M.C., Defez, C., Pose, C., Blayac, J.P., Touchon, J.: Impact of music therapy on anxiety and depression for patients with alzheimer's disease and on the burden felt by the main caregiver (feasibility study). Interets de la musicotherapie sur l'anxiete, la depression des patients atteints de la maladie d'Alzheimer et sur la charge ressentie par l'accompagnant principal 35(1), 57–65 (2009)
16. Hamann, S., Canli, T.: Individual differences in emotion processing. Current Opinion in Neurobiology 14(2), 233–238 (2004)
17. Higuchi, T.: Approach to an irregular time series on the basis of the fractal theory. Physica D: Nonlinear Phenomena 31(2), 277–283 (1988)
18. Horlings, R.: Emotion recognition using brain activity. Ph.D. thesis, Delft University of Technology (2008)
19. IDM-Project: Emotion-based personalized digital media experience in co-spaces (2008), http://www3.ntu.edu.sg/home/eosourina/CHCILab/projects.html
20. James, W.: What is an emotion. Mind 9(34), 188–205 (1984)
21. Jones, N.A., Fox, N.A.: Electroencephalogram asymmetry during emotionally evocative films and its relation to positive and negative affectivity. Brain and Cognition 20(2), 280–299 (1992)
22. Khalili, Z., Moradi, M.H.: Emotion recognition system using brain and peripheral signals: Using correlation dimension to improve the results of eeg. In: Proceedings of the International Joint Conference on Neural Networks, pp. 1571–1575 (2009)
23. Kulish, V., Sourin, A., Sourina, O.: Analysis and visualization of human electroencephalograms seen as fractal time series. Journal of Mechanics in Medicine and Biology 26(2), 175–188 (2006)
24. Kulish, V., Sourin, A., Sourina, O.: Human electroencephalograms seen as fractal time series: Mathematical analysis and visualization. Computers in Biology and Medicine 36(3), 291–302 (2006)
25. Lane, R.D., Reiman, E.M., Bradley, M.M., Lang, P.J., Ahern, G.L., Davidson, R.J., Schwartz, G.E.: Neuroanatomical correlates of pleasant and unpleasant emotion. Neuropsychologia 35(11), 1437–1444 (1997)
26. Lang, P., Bradley, M., Cuthbert, B.: International affective picture system (IAPS): Affective ratings of pictures and instruction manual. Tech. rep., University of Florida, Gainesville, FL (2008)
27. Li, M., Chai, Q., Kaixiang, T., Wahab, A.: EEG emotion recognition system. In: In-Vehicle Corpus and Signal Processing for Driver Behavior, pp. 125–135. Springer, US (2009)

28. Lin, Y.P., Wang, C.H., Wu, T.L., Jeng, S.K., Chen, J.H.: EEG-based emotion recognition in music listening: A comparison of schemes for multiclass support vector machine. In: Proceedings of the IEEE International Conference on Acoustics, Speech and Signal Processing, ICASSP, Taipei, pp. 489–492 (2009)
29. Liu, Y., Sourina, O., Nguyen, M.K.: Real-time EEG-based human emotion recognition and visualization. In: Proc. 2010 Int. Conf. on Cyberworlds, Singapore, pp. 262–269 (2010)
30. Lutzenberger, W., Elbert, T., Birbaumer, N., Ray, W.J., Schupp, H.: The scalp distribution of the fractal dimension of the EEG and its variation with mental tasks. Brain Topography 5(1), 27–34 (1992)
31. Maragos, P., Sun, F.-K.: Measuring the fractal dimension of signals: morphological covers and iterative optimization. IEEE Transactions on Signal Processing 41(1), 108–121 (1993)
32. Mauss, I.B., Robinson, M.D.: Measures of emotion: A review. Cognition and Emotion 23(2), 209–237 (2009)
33. Murugappan, M., Rizon, M., Nagarajan, R., Yaacob, S., Zunaidi, I., Hazry, D.: Lifting scheme for human emotion recognition using EEG. In: International Symposium on Information Technology, ITSim 2008, vol. 2 (2008)
34. Pardo, J.V., Pardo, P.J., Raichle, M.E.: Neural correlates of self-induced dysphoria. American Journal of Psychiatry 150(5), 713–719 (1993)
35. Petrantonakis, P.C., Hadjileontiadis, L.J.: Emotion recognition from EEG using higher order crossings. IEEE Transactions on Information Technology in Biomedicine 14(2), 186–197 (2010)
36. Plutchik, R.: Emotions and life: perspectives from psychology, biology, and evolution, 1st edn. American Psychological Association, Washington, DC (2003)
37. Pradhan, N., Narayana Dutt, D.: Use of running fractal dimension for the analysis of changing patterns in electroencephalograms. Computers in Biology and Medicine 23(5), 381–388 (1993)
38. Russell, J.A.: Affective space is bipolar. Journal of Personality and Social Psychology 37(3), 345–356 (1979)
39. Russell, J.A.: A circumplex model of affect. Journal of Personality and Social Psychology 39(6), 1161–1178 (1980)
40. Sammler, D., Grigutsch, M., Fritz, T., Koelsch, S.: Music and emotion: Electrophysiological correlates of the processing of pleasant and unpleasant music. Psychophysiology 44(2), 293–304 (2007)
41. Sanei, S., Chambers, J.: EEG signal processing. John Wiley & Sons, Chichester (2007)
42. Savran, A., Ciftci, K., Chanel, G., Mota, J., Viet, L., Sankur, B., Akarun, L., Caplier, A., Rombaut, M.: Emotion detection in the loop from brain signals and facial images (2006), `http://www.enterface.net/results/`
43. Schaaff, K.: EEG-based emotion recognition. Ph.D. thesis, Universitat Karlsruhe (TH) (2008)
44. Sourina, O., Kulish, V.V., Sourin, A.: Novel tools for quantification of brain responses to music stimuli. In: Proc. of 13th International Conference on Biomedical Engineering, ICBME 2008, pp. 411–414 (2008)
45. Sourina, O., Liu, Y.: A fractal-based algorithm of emotion recognition from EEG using arousal-valence model. In: Biosignals 2011, Rome, Italy (accepted, 2011)
46. Sourina, O., Sourin, A., Kulish, V.: EEG data driven animation and its application. In: Gagalowicz, A., Philips, W. (eds.) MIRAGE 2009. LNCS, vol. 5496, pp. 380–388. Springer, Heidelberg (2009)

47. Sourina, O., Wang, Q., Liu, Y., Nguyen, M.K.: A real-time fracal-based brain state recognition from EEG and its application. In: Biosignals 2011, Rome, Italy (accepted, 2011)
48. Sourina, O., Wang, Q., Nguyen, M.K.: EEG-based "serious" games and monitoring tools for pain management. In: Proc. MMVR18, Newport Beach, California (accepted, 2011)
49. Stam, C.J.: Nonlinear dynamical analysis of EEG and MEG: Review of an emerging field. Clinical Neurophysiology 116(10), 2266–2301 (2005)
50. Stevens, K.: Patients' perceptions of music during surgery. Journal of Advanced Nursing 15(9), 1045–1051 (1990)
51. Takahashi, K.: Remarks on emotion recognition from multi-modal bio-potential signals. In: IEEE International Conference on Industrial Technology, IEEE ICIT 2004, vol. 3, pp. 1138–1143 (2004)
52. Wang, Q., Sourina, O., Nguyen, M.K.: EEG-based "serious" games design for medical applications. In: Proc. 2010 Int. Conf. on Cyberworlds, Singapore, pp. 270–276 (2010)
53. Wang, Q., Sourina, O., Nguyen, M.K.: Fractal dimension based algorithm for neurofeedback games. In: Proc. CGI 2010, Singapore, p. SP25 (2010)

Author Index

Batch number: 09490872

Printed by Printforce, the Netherlands